Photonics, Plasmonics and Information Optics

Photonics, Plasmonics and Information Optics

Research and Technological Advances

Edited by
Arpan Deyasi
Pampa Debnath
Asit Kumar Datta
Siddhartha Bhattacharyya

CRC Press
Taylor & Francis Group
Boca Raton London New York

CRC Press is an imprint of the
Taylor & Francis Group, an **informa** business

First edition published 2021
by CRC Press
6000 Broken Sound Parkway NW, Suite 300, Boca Raton, FL 33487-2742

and by CRC Press
2 Park Square, Milton Park, Abingdon, Oxon, OX14 4RN

CRC Press is an imprint of Taylor & Francis Group, LLC

ISBN: 978-0-367-49734-7 (hbk)
ISBN: 978-0-367-49952-5 (pbk)
ISBN: 978-1-003-04719-3 (ebk)

Typeset in Times
by SPi Global, India

Arpan Deyasi would like to dedicate this book to his wife Munmun Deyasi and son Saranyo Deyasi

Pampa Debnath would like to dedicate this book to her respected father, Mr. Parimal Debnath; his mother, Mrs Krishna Debnath; her beloved husband, Mr. Snehasis Roy; and kid, Master Deeptanshu Roy

Siddhartha Bhattacharyya would like to dedicate this book to Dr Fr Paul Achandy, CMI, the Honourable Chancellor of CHRIST (Deemed to be University), Bangalore

Contents

Preface

Of late, there has been an upsurge in the technological developments and current research trends in the field of photonics and EBG structures. These are promising areas in the twenty-first century where electronic counterparts are replaced by all-optical devices and components, and henceforth, the latest novel innovations and proposals in the various fields of photonic engineering are assuming greater importance. Photonic crystals are considered as building blocks in optical circuits and high-frequency communication systems due to their novel ability of restricting electromagnetic wave propagation in some precise spectra and allowing other spectra. This feature can be used to design various optical devices and passive components like optical filters. Metamaterials are now used to improve SNR in optical 2D devices and antennas which are very important to filter out the signal from a noisy spectrum. Optical logic gates are rapidly replacing the conventional digital logic gates owing to superior speed of operation, and this is one of the prime requirements of making supercomputers. Silicon photonics nowadays are very popular in the low-dimensional device fabrication due to lower cost and existing fabrication technologies, and thus plays important role in today's world of photonics. Ultra-fast photonic components such as optical modulators and detectors are finding widespread use in establishing high-speed interconnects.

This volume is an attempt to report the latest research domains in this direction provided by the experts/researchers in the respective fields with a major focus to highlight the basic physics as well as progress of the works. Since detailed theoretical/analytical works are not generally contained in similar type of projects, so that area will be focused and tinted with major care instead of only providing experimental outputs for some particular applications.

The volume comprises twelve well versed contributory chapters on the recent developments in these fields apart from the introductory and concluding chapters.

In the age of quantum computing, where photonic independence is the key for making superposition states, the property of the photons leads to the ability of quantum particles to inhibit two contradictory states at the same instant. Therefore, a major emphasis is given to utilize this property of photon in realizing information technology at extremely high-frequency through practical implementation of logic gates, and the corresponding branch is referred to as information optics. Chapter 1 provides a bird's eye view on the latest status of research in photonics and information optics with reference to an in-depth analysis of optical logic gates.

Chapter 2 discusses the optical characterizations of TiO_2 thin film as-deposited and annealed samples on glass substrate. TiO_2 solution is prepared by sol–gel technique using titanium (IV) iso-propoxide (TTIP), acetic acid and ethanol. Thin films are deposited by spin coating of the solution on clear glass substrates and annealed at 150°C, 300°C and 450°C for 30 minutes. Ellipsometry measurement reveals thickness of TiO_2 films to be 87.34 nm for as-deposited samples but goes on decreasing as the annealing temperature is increased. X-Ray diffraction (XRD) exhibits formation of amorphous TiO_2 thin films for all the samples on the clear glass substrate. It is also established that the extinction coefficient at a wavelength of 633 nm is minimum for as-deposited sample but increases with the annealing temperature.

Electromagnetic metamaterials, also popularly known as metamaterials, are structural material exhibiting artificial electromagnetic properties, which are not readily present in nature under electromagnetic excitation. The electromagnetic properties associated with the metamaterials can be considered to possess effective medium parameters instead of the bulk parameters. These effective parameters can vary over an extremely wide range; ranging from extremely high value to a very low value even with a negative sign which has not been feasible in any naturally available materials. Chapter 3 throws light on the variations in the values of effective permittivity and permeability which are characteristics of the metamaterials.

Chapter 4 deals with the dispersion diagram of two-dimensional photonic crystal for effective design of superlens in the UV as well as visible ranges. The authors consider both rectangular and triangular structures for simulation, where the plane wave expansion method is adopted for evaluation purpose. Both TE and TM modes are considered, and normal incidence is generally excluded. Fundamentally, rectangular geometry is taken into account, and all the simulated data are compared with a triangular structure with identical formation. Results show that the virtual formation of superlens is possible mainly in the IR frequency spectrum, and some extensions can be made in visible zone. Obtained findings will be important for designing of all photonic-integrated circuit which may be considered as a building block of next-generation communication technology.

Kerr and Pockels materials are so much important in the application domain of a super-fast switching mechanism. Several optical/opto-electronic devices have already been developed using those materials. Basically, the second- and first-order non-linearities of those materials, respectively, are responsible for the above switching. Numerous all-optical schemes are proposed in the last some decades, where those electro-optical materials are massively used. In Chapter 5, the authors propose some new schemes where the Kerr and Pockels material-based switches are used for implementing new modulation schemes/techniques. The authors also show the strong advantages of the above schemes over the conventional methods.

Slotted photonic crystal waveguide (SPCW) is a unique waveguide geometry in integrated photonics platform that simultaneously provides tight spatial as well as temporal mode-field and, thereby, offers ultra-high optical confinement. This facilitates the technologists to deliver compact solutions to realize light-matter interactions, for applications ranging from sensing; electro-optic modulation; nonlinear optics to quantum optics, with an unprecedented miniaturization of underlying physical footprint and operating power. Chapter 6 highlights some of the recent theoretical developments of active all-optical devices based on enhanced light-matter interactions inside SPCWs, especially focusing to the Stimulated Raman scattering (SRS) interaction. The SRS has been adopted here as the key nonlinear optical phenomenon as it has shown potentialities to alleviate some important active optical functionalities in photonics. However, the potentialities of SRS in silicon photonics have actually been hindered over the years by the requirement of larger effective length of waveguide and threshold power, as compared to the need in photonic integrated circuits (PICs). The chapter also shows that the hindrance can effectively be exterminated drastically by enhancing the high SRS gain of materials like silicon nanocrystals through utilizing the ultra-high optical confinement in SPCWs. The developments are amenable to PICs and may serve as building blocks of all-optical signal processors.

Chapter 7 presents comparative studies of RAMAN amplifier embedded optical fibre communication system designed at 1310 nm and 1550 nm. Simulation studies reveal that bit detection threshold is higher for 1310 nm, which articulates for lower sensitivity, when both the architectures are fed with simultaneous co- and counter-propagating pumping lasers. However, both gain and Q-factor are higher when system is designed at 1310 nm, which speaks in favour of the system for delivering output at larger distance or achieving better gain at lower input level, when calculated at predefined length and bit rate. But for both the design, peak value of Q is achieved at 0.8 sec of bit period. For equal input power, better output power level is observed at 1310 nm owing to flatter gain profile. For a particular operating wavelength, output signal level can be hold for larger distance if bit rate is enhanced. Results speak for pros and cons of both the operating wavelength, which is required for specific communication applications.

The optical comb filters have shown tremendous application in the field of optical signal processing, wavelength division multiplexing, multi-wavelength laser, microwave photonics, etc. The sampled fibre Bragg grating is an essential device for generation of comb spectrum. In Chapter 8, a detailed study on the design of comb filter based on sampled-chirped fibre Bragg grating has been provided. The multiple phase shift technique and spectral Talbot effect have been explained and demonstrated for achieving narrowband comb spectrum. A special design of ultra-narrowband comb filter using Gaussian-sampled chirped fibre Bragg grating has also been shown.

Aerospace structures experience severe amount of stress over their life cycle. For an example, structures located on the exterior surface of a space shuttle often undergo severe stress due to the presence of variable pressure and high gravity. Additionally, with a huge influx of small and micro-sized satellites, the probability of an impact in the outer-space orbits is now a concerning factor. There is also a presence of similar disturbing events for commercial flights. Only within the USA, the incidents of planes striking birds are more than 40 times on average per day, as per a report published by the Federal Aviation Administration (FAA) in 2019. Now, the skin of the planes and the rockets are usually made of various composite light-weight materials and multiple layers of specially engineered tempered glasses to withstand excessive stress due to the variation of pressure and temperature and unwanted impacts of a smallscale. However, repetitive occurrences of impacts on the skin of the aerospace structures, even in the small-scale, may lead to the generation of severe mechanical stress which can lead to a catastrophic incident. Hence, it is imperative to identify, locate, analyse, and monitor any accidental impact on aerospace structures in realtime. Chapter 9 presents a realtime and wireless structural health-monitoring method for analysing impact responses of aerospace structures using FBG optical sensors. A theoretical background is also followed up by a discussion on a laboratory scale experimental measurement.

Optical spatial solitons in photorefractive media have been very attractive for research because of their relative ease of formation at low laser powers. Photonic lattices embedded in photorefractivecrystals can mimic structures and properties of photonic crystals.In Chapter 10, the authors have discussed the stability and existence of gap solitons in different types of photorefractive optical lattices, i.e., optical lattices in noncentrosymmetric photorefractives, centrosymmetric photorefractives and pyroelectric photorefractives. A theoretical foundation using the Helmholtz equation has been laid which serves as a general framework for photorefractive crystals having different nonlinearities and configurations.

For investigating phenomena like wave guiding, radiation and scattering, Maxwell's equations are required to be solved numerically. The finite-difference in time domain (FDTD) method for solving Maxwell's equation wasfirst proposed by Kane S. Yee. As the computational capabilities of computers are increased with time, the FDTD method becomes more and more pertinent for studying a variety of electromagnetic problems. In Chapter 11, the FDTD method is discussed briefly and then the spectral analysis of Photonic Bandgap Structure (PBGS) is carried out using of FDTD method. PBGS plays an important role in the modern optical communication system for manipulating light energy. In the last section of the chapter, it is shown how the PBGS with predetermined materials widths is used to design a narrowband opticalfilter at 1550 nm.

An ultra-light lens that can change the whole imaging systems leaves many designing opportunities for researchers. The old concept of diffraction relying on the very known phenomenon rainbow is formulated in such a manner that ultra-thin glass or even a plastic material (ten times less than a human hair) is capable of capturing image pattern and video activities. Multi-level flat lenses are kept in a sequence that helps to bend all the seven rays (diffraction) in a single focal point so that an image can be formed and accuracy are also verified by the computational methodologies as well. Chapter 12 discusses such types of lenses that are referred to as multi-level diffractive lenses (MDL).MDLs are often called an achromatic lens, because of their special capability of nullifying chromatic aberration, a distortion caused when glass splits white light into multiple colour wavelengths in the spectrum. In comparison to non-corrected singlet lenses, flat lenses (Super Achromatic Lens) can produce much clearer images, which gives a more accurate perception and improved quality pictures.

Chapter 13 reports various application-based case studies of peristaltic pump with optical measuring unit. The chapter also presents a design of an efficient controller to deliver a constantflow rate of peristaltic pump for precise fluidic applications considering all the components. An optical sensor-based measurement system is proposed for flow sensing of the pump. A pump model based on input–output data set along with modelling uncertainty has been considered. Internal model principle-based repetitive control (RC) is featured to deal with the periodical load disturbances encountered by the pump and which is to be attenuated on a desired level for precise flow rate control of the same. The repetitive control law is modified with an augmented adaptive control law which corrects thevariations in the flow rate based on the parameters updated by that control law by compensating a varying frequency disturbance signal.

Chapter 14 draws a line of conclusion while focusing on future trends and research directions in these rapidly emerging fields.

The volume would certainly come in use to the graduate researchers and doctoral fellows in the field of photonics. Sine theoretical understanding with detailed background physics in most of the contents would certainly help experimental workers also to get the significance of their practical outcome.

Arpan Deyasi
Pampa Debnath
Asit Kumar Datta
Siddhartha Bhattacharyya
September 2020
Kolkata, Bangalore

Editors

Arpan Deyasi is an Assistant Professor in the Department of Electronics and Communication Engineering in RCC Institute of Information Technology, Kolkata, India. He has 13 years of professional experience in academics and industry. He earned a BSc (Hons), a BTech, an MTech at the University of Calcutta. He is working in the area of semiconductor nanostructure and semiconductor photonics. He has published more than 150 peer-reviewed research papers and a few edited volumes under the banner of CRC Press, IGI Global, etc. His major teaching subjects are solid state device, electromagnetics, photonics. He has organized international and national conferences, faculty development programmes, workshops, laboratory and industrial visits, seminars and technical events for students under the banner of IE(I) Kolkata section. He is also associated with a few reputed conferences. He is reviewer of a few journals of repute and some prestigious conferences in India and abroad. He has delivered a few talks and conducted hands-on sessions on nanoelectronics, photonics and electromagnetics in various FDPs, workshops, seminars. He is the editor of various conference proceedings and edited volumes. He is a member of IEEE Electron Device Society, IE(I), Optical Society of India, IETE, ISTE, ACM, etc. He is working as SPOC of RCCIIT Local Chapter (NPTEL course), Nodal Coordinator of e-outreach programme and Internshala, Faculty Advisor of the student chapter of Institution of Engineers (India) in ECE Department.

Pampa Debnath is an Assistant Professor in the Department of Electronics and Communication Engineering in RCC Institute of Information Technology, Kolkata, India. She has more than 12 years of professional teaching experience in academics. She earned a BTech and an MTech at the University of Burdwan. Her research interest covers the area of microwave devices, microstrip patch antennas, SIW based circuit and antenna. She has published several research papers in IEEE Xpore, Microsystem Technologies-Springer, CRC Press and in some national and international conferences, and a few edited volumes under the banner of CRC Press, IGI Global, etc. Her major teaching subjects are electromagnetics, RF andmicrowave and antenna. She has already served as editor as well as technical chair of One International Conference (ICCSE 2016), which is published by CRC Press, coordinated a few faculty development programmes, workshops, laboratory and industrial visits, seminars and technical events under the banner of the Institution of Engineers (India) Kolkata section. She is also associated with a few national and international conferences. She is a reviewer of few journals of repute and some national and international conferences. She is the editor of various conference proceedings and edited volumes. She has conducted hands on session on photonics, electromagnetics and microwaves in various FDPs, workshops and seminars. She is a member of the Institution of Electronics and Telecommunication Engineers (IETE), Indian Society for Technical Education (ISTE), International Association for Engineers (IAENG).

Asit Kumar Datta is a former Professor and former Head of the Department of Applied Physics and later of the Department of Applied Optics and Photonics, University of Calcutta. He earned a PhD in power electronics at the University of Calcutta. After his retirement from Calcutta University, he acted as Dean and mentor of three engineering colleges in West Bengal. He was teaching and conducting research for more than 40 years in the Department of Applied Physics and the Applied Optics and Photonics. Professor Datta has contributed significantly in the area of power electronics, control instrumentation, photonic image processing and photonics computing. He executed 12 funded and sponsored research projects and has published more than 120 papers in national and international journals and conference proceedings and also guided 14 research students toward their PhD degrees. He has also published two books through CRC Press in the area of face recognition and information photonics. He represented India at the International Commission for Optics (ICO), Paris and was the Indian representative in two divisions of the International Commission of Illumination (CIE), Geneva. He was an Advisor of UPSC, Member of Photonics programme of DRDO, expert of NBA in electronics and communication.

Siddhartha Bhattacharyya earned his Bachelors in Physics, Bachelors in Optics and Optoelectronics and Masters in Optics and Optoelectronics from University of Calcutta, India in 1995, 1998 and 2000 respectively. He completed PhD in Computer Science and Engineering from Jadavpur University, India in 2008. He is the recipient of the University Gold Medal from the University of Calcutta for his Masters. He is the recipient of several coveted awards including the Distinguished HoD Award and Distinguished Professor Award conferred by Computer Society of India, Mumbai Chapter, India in 2017, the Honorary Doctorate Award (D. Litt.) from The University of South America and the South East Asian Regional Computing Confederation (SEARCC) International Digital Award ICT Educator of the Year in 2017. He has been appointed as the ACM Distinguished Speaker for the tenure 2018-2020. He has been inducted into the People of ACM hall of fame by ACM, USA in 2020. He has been appointed as the IEEE Computer Society Distinguished Visitor for the tenure 2021-2023.

He is currently serving as a Professor in the Department of Computer Science and Engineering of Christ University, Bangalore. He served as the Principal of RCC Institute of Information Technology, Kolkata, India during 2017-2019. He has also served as a Senior Research Scientist in the Faculty of Electrical Engineering and Computer Science of VSB Technical University of Ostrava, Czech Republic (2018-2019). Prior to this, he was the Professor of Information Technology of RCC Institute of Information Technology, Kolkata, India. He served as the Head of the Department from March, 2014 to December, 2016. Prior to this, he was an Associate Professor of Information Technology of RCC Institute of Information Technology, Kolkata, India from 2011-2014. Before that, he served as an Assistant Professor in Computer Science and Information Technology of University Institute of Technology, The University of Burdwan, India from 2005-2011. He was a Lecturer in Information Technology of Kalyani Government Engineering College, India during 2001-2005.

He is a co-author of 5 books and the co-editor of 72 books and has more than 300 research publications in international journals and conference proceedings to his credit. He has got two PCTs to his credit. He has been the member of the organizing and technical program committees of several national and international conferences. He is the founding Chair of ICCICN 2014, ICRCICN (2015, 2016, 2017, 2018), ISSIP (2017, 2018) (Kolkata, India). He was the General Chair of several international conferences like WCNSSP 2016 (Chiang Mai, Thailand), ICACCP (2017, 2019) (Sikkim, India) and (ICICC 2018 (New Delhi, India) and ICICC 2019 (Ostrava, Czech Republic).

He is the Associate Editor of several reputed journals including Applied Soft Computing, IEEE Access, Evolutionary Intelligence and IET Quantum Communications. He is the editor of International Journal of Pattern Recognition Research and the founding Editor in Chief of International Journal of Hybrid Intelligence, Inderscience. He has guest edited several issues with several international journals. He is serving as the Series Editor of IGI Global Book Series Advances in Information Quality and Management (AIQM), De Gruyter Book Series Frontiers in Computational Intelligence (FCI), CRC Press Book Series(s) Computational Intelligence and Applications & Quantum Machine Intelligence, Wiley Book Series Intelligent Signal and Data Processing, Elsevier Book Series Hybrid Computational Intelligence for Pattern Analysis and Understanding and Springer Tracts on Human Centered Computing.

His research interests include hybrid intelligence, pattern recognition, multimedia data processing, social networks and quantum computing.

Dr. Bhattacharyya is a life fellow of Optical Society of India (OSI), India, life fellow of International Society of Research and Development (ISRD), UK, a fellow of Institution of Engineering and Technology (IET), UK, a fellow of Institute of Electronics and Telecommunication Engineers (IETE), India and a fellow of Institution of Engineers (IEI), India. He is also a senior member of Institute of Electrical and Electronics Engineers (IEEE), USA, International Institute of Engineering and Technology (IETI), Hong Kong and Association for Computing Machinery (ACM), USA. He is a life member of Cryptology Research Society of India (CRSI), Computer Society of India (CSI), Indian Society for Technical Education (ISTE), Indian Unit for Pattern Recognition and Artificial Intelligence (IUPRAI), Center for Education Growth and Research (CEGR), Integrated Chambers of Commerce and Industry (ICCI), and Association of Leaders and Industries (ALI). He is a member of Institution of Engineering and Technology (IET), UK, International Rough Set Society, International Association for Engineers (IAENG), Hong Kong, Computer Science Teachers Association (CSTA), USA, International Association of Academicians, Scholars, Scientists and Engineers (IAASSE), USA, Institute of Doctors Engineers and Scientists (IDES), India, The International Society of Service Innovation Professionals (ISSIP) and The Society of Digital Information and Wireless Communications (SDIWC). He is also a certified Chartered Engineer of Institution of Engineers (IEI), India. He is on the Board of Directors of International Institute of Engineering and Technology (IETI), Hong Kong.

Contributors

Ahsan Aqueeb
South Dakota School of Mines and
  Technology
Rapid City, South Dakota, United States

Rajorshi Bandyopadhyay
Department of Applied Optics and
  Photonics
University of Calcutta
Calcutta, India

Moumita Banerjee
Department of Electronics and
  Communications Engineering
RCC Institute of Information
  Technology
Kolkata, India

Siddhartha Bhattacharyya
Department of Computer Science and
  Engineering
Christ University
Bangalore, India

Somak Bhattacharyya
Indian Institute of Technology
Banaras Hindu University
Varanasi, India

Benjamin Braaten
Department of Electrical and Computer
  Engineering
North Dakota State University
Fargo, North Dakota, United States

Ellie Burczek
South Dakota School of Mines and
  Technology
Rapid City, South Dakota, United States

Rajib Chakraborty
Department of Applied Optics and
  Photonics
University of Calcutta
Calcutta, India

Agnijita Chatterjee
Department of Physics
University of Burdwan
Burdwan, India

Tanmoy Datta
Indian Institute of Technology (Indian
  School of Mines) Dhanbad
Dhanbad, India

Pampa Debnath
Department of Electronics and
  Communication Engineering
RCC Institute of Information
  Technology
Kolkata, India

Naiwrita Dey
Department of Applied Electronics and
  Instrumentation Engineering
RCC Institute of Information
  Technology
Kolkata, India

Arpan Deyasi
Department of Electronics and
  Communication Engineering
RCC Institute of Information
  Technology
Kolkata, India

Rajarshi Dhar
Indian Institute of Engineering Science
  and Technology
Shibpur, India

Subhashish Dolai
Department of Electrical and Computer
    Engineering
University of Utah
Salt Lake City, Utah, United States

Aavishkar Katti
Department of Physics
BanasthaliVidyapith
Rajasthan, India

Suranjan Lakshan
Department of Physics
University of Burdwan
Burdwan, India

Soham Lodh
Department of Electronics and
    Communication Engineering
MCKV Institute of Engineering
Kolkata, India

Minakshi Mandal
Department of Physics
University of Burdwan
Burdwan, India

Ekata Mitra
Department of Physics
Techno India
Kolkata, India

Ujjwal Mondal
Department of Applied Physics
Calcutta University
Kolkata, India

Sourangshu Mukhopadhyay
Department of Physics
University of Burdwan
Burdwan, India

Sayan Roy
Department of Electrical Engineering
South Dakota School of Mines and
    Technology
Rapid City, South Dakota,
    United States

Debashri Saha
University of Burdwan
Burdwan, India

Mrinal Sen
Department of Electronics Engineering
Indian Institute of Technology (Indian
    School of Mines) Dhanbad
Dhanbad, India

Anindita Sengupta
Department of Electrical Engineering
Indian Institute of Engineering Science
    and Technology
Shibpur, India

Somnath Sengupta
Department of Electronics and
    Communication Engineering
Birla Institute of Technology Mesra
Ranchi, India

Priya Singh
BanasthaliVidyapith
Vanasthali, India

Mijia Yang
Department of Civil and Environmental
    University
North Dakota State University
Fargo, North Dakota, United States

1 Foundation, Progress and Future of Photonics, Plasmonics and Information Optics
Researchers Perspective

Pampa Debnath, Arpan Deyasi and Siddhartha Bhattacharyya

CONTENTS

1.1 INTRODUCTION

Electron–electron interaction is the backbone of semiconductor and quantum-scale devices, which is already not suitable for high-speed reliable information processing [1] owing to severe inherent scattering mechanisms and leakage current flow. Though several new architectures are proposed for nanoscale MOSFETs to get rid of the scattering and leakage problems, the other alternative technology looks brighter where photons are considered as the medium of information transmission instead of electron [2]. Design of photonic integrated circuit is a revolution today and is already accepted by several research communities [3,4] as the future novel technology for chip design. Twenty-first century can be considered as the age of photonics, and this becomes physically realizable with the modernization of fabrication technology and

1

successful implementation of several all-optical components afterwards. Photonic crystal is now considered as the building block of this revolution, after the novel proposal of Loudon [5]; and with the further investigation of Yablonovitch[6], it has now come to the forefront. The present chapter deals with the successful journey of the photonic crystal so far, after the brief introduction of electromagnetic (EM)bandgap structure. Both the unbounded and bounded medium transmission is henceforth discussed in the first two sections, followed by the world of plasmonics and its applications. Computational electromagnetic (CEM) here plays a crucial role in designing the theoretical prediction of performance and experimental model verification. The chapter ends with a glimpse of information optics and quantum information processing, which will be considered as the future of robust telecommunication.

The demand of market is to get room temperature interaction between photons, as well as making this possible in known materials. Fortunately, these incredible things are reported in silicon, which opened a new door of progress in photonics research. Also, oscillation of surface electrons according to the external radiofrequency/optical input leads to another promising arena called plasmonics. Twenty-first century will be governed by progress of these emerging technologies, and therefore a close bird-eye view of their research progress so far and possible future directions needs a systematic understanding. Here lies the significance of this proposal.

1.1.1 ELECTROMAGNETIC BANDGAP STRUCTURE

Advancement of high-frequency communication systems claims more developed and exceptional features of EM materials for highly efficient applications which developed a new era in the field of EM [7]. The progress of metamaterial perception announced a revolution and achieved a prodigious consideration from the investigators. An enormously huge scientific area has been covered by metamaterials, which varies from material science engineering to antenna design and optical communication to nanotechnology [8–11]. A unique feature has been observed in this composite material. A composite structure which is periodic in nature reveals left hand (LH) feature in one band of frequency and the property of bandgap in other frequency band [12–14]. Thus, researchers devote their attention to the unique properties of these materials and put them in various EM fields and antenna applications [15]. These exceptional features make EBG structure to consider as a distinct form of metamaterial. Interaction of EM waves with EBG stirring miracles observe and result in a remarkable feature. Some special features such as different frequency bands (pass band, stop band as well as bandgaps) might be recognized. Implementation of EBG in the design of antenna array is familiarized as a radical expansion with their adaptability malleability.

1.1.2 PHOTONIC CRYSTAL

One special form of the physical manifestation of EM bandgap structure is the photonic crystal, which is nothing but the intermittent arrangement of dielectric–dielectric or metal-dielectric materials [16] with periodic variations of lower and higher refractive indices (alternatively be looked in terms of dielectric constants) dielectric constants in one, two or three dimensions. Photonic bandgap is formed due to this assembly [17] which allows only a few selected spectra for propagation inside, and restricts all others. Among different variations, one-dimensional structure is preferred for analysis and implementation for applications owing to its structural simplicity (analogous to grating structure) and easier fabrication technologies. Among various applications, it is already considered as an established candidate for developing fibre, conventionally known as photonic crystal fibre[18]. This has proved its candidature due to very low loss over a longer distance, lower attenuation and dispersion; compared to the optical fibre.

Among other devices, photonic crystal is already utilized to develop laser [19], photodiode [20], sensor [21], waveguide [22], filter [23], etc. Tunnelling devices are fabricated using photonic crystal heterostructure[24]. Vasiliev[25] investigated high-frequency (optical) properties of one-dimensional MPC when EM wave is incident in any of the polarized form and presented the speculative investigation of the transmittance, reflectance, and Faraday rotation spectra. Ternary photonic crystals are already suggested [26] as remarkable refractometric sensing elements, and filters with multiple windows are also experimentally appreciated for image processing applications [27].

However, works based on the photonic crystal are applicable mostly at higher terahertz region, precisely beyond 100 THz. Therefore, a void is generated between microwave frequency spectrum and high-frequency optical communication, which is eventually filled up by plasmonic devices. In the next section, plasmonics is therefore briefly covered up with focus on diverse applications.

1.1.3 PLASMONICS

Plasmonics is the new branch of research which primarily deals with generation, modulation and detection of polaritons, precisely surface plasmonpolaritons. Considering the spectrum of operation, it works as the bridge between electronic devices and photonic devices, i.e., its working region lies between 100 GHz and 10 THz. The physics of trapping the EM wave, alternatively called waveguiding, can be accomplished at the interface between two metals. Surface plasmon resonance can be controlled by the nature of the constituent materials as well as their size and shape [28]. This waveguiding nature leads to the physical realization of biosensors [29], photovoltaic devices [30], and optical detectors with higher sensitivity and accuracy compared to electronic or photonic counterparts [31]. Plasmonic nanoparticles are already become the exclusive choice of researchers for nanosensor design [32] and also in the field of quantum optics.

Generation of localized surface plasmon resonance modes is one of the interesting aspects already gifted by this very young domain of research, which is the result of oscillations of quantized free electrons of a few selected materials. This exclusive property gives birth to wider optical applications in higher frequency region, precisely beyond 100 GHz. Not only the traditional metals exhibit this property, but several novel materials like lanthanum hexaboride[33] also exhibit the effect with better absorbance property. One of the major achievements that can be obtained using surface plasmon is the third harmonic generation [34], which is extremely essential for nonlinear imaging [35]. Efficiency of the perovskite solar cells can greatly be enhanced using plasmonicnanomaterials[36], where surface plasmon can be generated by natural source, i.e., sunlight at a specific frequency. It is already reported that photocurrent can be increased as per the demand of the industry in otherwise ordinary solar cell, when suitable nanoparticles are deposited in the top surface [37] or at the internal layers.

Localized surface plasmon can produce sub-wavelength confinement which is critically important for optical nanoantenna design [38]. Extraordinary optical transmission (EOT) can be obtained [39] with proper coupling of surface plasmons. Better uniform field distribution is reported with nanoaperture antenna [40] where surface-enhanced fluorescence property is utilized. Sub-wavelength confinement for this device is investigated [41] with proper metal-dielectric interface. Plasmonicnanomaterials have already exhibited several applications over sum-frequency generation [42], and therefore can be considered as one of the leading research topic waiting to be explored in the twenty-first century.

1.1.4 Computational Electromagnetics

All devices include electrical and electronics are directed by the principle of EM. Therefore, better understanding of EM is very crucial for any field-based engineer who can properly evaluate problems concerning several phenomena such as satellite communication, wireless LAN, fiber-optic network. In digital circuit's solutions, emission (radiation) of wires often ignored and analysis has been limited to the level of two voltages. It helps to diminish complexity. But to achieve more precise solution, simplifying the hypothesis is no longer convincing, and more complete analysis is required. It is difficult to implement if the knowledge of physics behind the experimental model is not cleared. The most significant motivation for learning EM is that mathematical confidence can be achieved and one can apply this knowledge to real and practical problems [43–46].

An accurate and precise depiction of a specific problem has been offered by Maxwell's equations in addition to particular boundary conditions. Most of the practical problems related to EM are partly resolved by considering simplified assumptions providing explication with few applications before the arrival of intelligent computers. Exact analysis of a system was only feasible by constructing and experimenting many prototypes, which was extortionate and time-consuming. Numerical analysis of Maxwell equations became feasible after the invention of intelligent computers in the year of 70s. The electromagnetic problems of different types can be analyzed and solved using a computer, therefore gives rise to a new branch termed as 'computational electromagnetics' [47–51]. Since structures have different shapes and materials, there was no best technique to solve all issues with the highest perfection and shortest computational time. Therefore, several techniques have to be taken into consideration where each one has distinct features in terms of precision, accurateness, effectiveness and complexity to implement [51–57]. A bulky and complicated assembling has been investigated using nowadays with CAD (computer-assisted design) tools. The time for developing the prototype has been reduced by this robust package and it is also cost-effective.

Several techniques such as fundamental numerical method, method of moments, finite-difference method, finite-element method andfinite-difference time domain (FTTD)[58–62] are the basic components of CEM. These techniques have discrete ways of implementation. The basic numerical method includes integration, Monte Carlo analysisand generation of random numbers. Method of Moment (MoM refers to the integration depictions of EM issues [62–67]. Many problems have been solved by a general method known as finite-element method, which is a very powerful technique. The fundamental theory behind the technique is that to partition the domain into finite elements and based on the equations for each element, the full set of equations for the composite systems can be extracted. Maxwell's equations have been directly referred to FDTD. Several archetypal EM complications can be recognized, if one observes at the developing history of CEM. The revolutions of advancement in CEM are accomplished by tracking improved resolutions for these problems. Moreover, in some fields innovative models have been building up by amalgamation of both technologies. The major prototype swing recently observed in the field of microwave photonic signal processing.

1.1.5 Information Optics

From the childhood class where 'light' is considered as the fundamental necessary component for living, the concept has been reshaped in the framework of modern science where 'light' works as the source of energy for information. Engineering the present and next-generation civilization is greatly influenced and controlled by information optics [68] which is the combination of sensing, measurement and conceptual visualization of information using photonic devices and components. Role of optics in current information technology is critically important for next-generation data processing, and creates a new sub-domain of research termed as 'optical information processing'. With the advent of technology, this is manifested into the technology of 'quantum information processing'[69]. The introduction of information optics also explored a new horizon in biophotonics research [70] with real-time optical manipulation in a three-dimensional reference frame.

Information optics is nothing but the application of photonics in the vast domain of information technology. This is a big revolution for the efficient design of future telecommunication network. It opens the gateway of making almost infinite bandwidth network, with very little usage of power, and also travelling of information at a much longer distance. Though the latest research clearly acknowledges that photonic neuromorphic computing is still unparallel for solving big-data classification problems [71], this approach can be equivalently designed in hardware level using photonic integrated circuits. The same objective is also earlier implemented in a different approach [72] at 1310 nm without using repeater, and the result is supported by higher throughput. Different soft computing methodologies are associated with the present-day information transfer through all-optical mode [73,74]. It enables to incorporate the parallelism approach in the optical frequency domain from space domain, where nonlinear phenomena can be involved for better and efficient data transmission [75]. Photonic crystal is the most unit level block for this physical realization process, and effective hardware implementation of photonic chip gives birth to next-generation computing technologies, be termed as quantum information processing.

1.1.6　QUANTUM INFORMATION PROCESSING

Quantum optics basically deal with the interaction between radiation and matter, and due to the advent of single-photon devices both at transmitter and detector levels, this concept is now broadly extended for physical transfer of information [76]. Concept of entanglement and superposition principle leads to the realization of quantum information processing, which is a potentially ground-breaking concept, leading to the path of more secured communication [77]. Laws and postulates of quantum mechanics are used for the sole purpose of computation [78], which speaks in favour of nonlinear photonic components for circuit-level implementation. Owing to the inherent property of negligible noise generated at the time of photon movement as well as free of decoherence[79], photonic integrated circuit may work as the future teleportation network. Design of deterministic logic gates is still difficult using photons due to their non-interactive nature [80], but several models are proposed to overcome this burden, and in future, realization of universal quantum computer may not be a distant dream.

Since all-optical architectures are yet to be capable of designing two-photon logic gates,the concept of plasmonicsis utilized by a group of researchers [81] which can successfully able to measure the photon–polariton interaction [82], and the limitations imposed by the Quantum Zeno effect will not be applicable. Here optical nonlinearity playsa crucial role in designing photonic circuits at sub-wavelength level. Design of waveguide with field confinement property plays an integral part of the real-world implementation, and very recently grapheme is proposed as the appropriate candidate [83] for that purpose. With suitable material, quantum information processing will change the present model of information transfer, and ultimately lead towards a secured communication.

1.2　CONCLUSION

The present decade is the exploration of photonics and plasmonics devices for the search of robust telecommunication network design for accurate and efficient information transfer. Owing to the present development of nanophotonic devices and their experimental realization, researchers may able to achieve the target of quantum information processing, and may pursue towards the photonic neuromorphic computing. Solution of big data-based classification problems is the need of the hour, and the present researches may be considered as an inspiring force for achieving the desired goal. Henceforth, the present chapter can be culminated with the aspiration of making a perfect choice of high-frequency devices that will lead to the path of a secured communication network in near future.

REFERENCES

1. Kim, Y.-B.(2010) Challenges for nanoscaleMOSFETs and emerging nanoelectronics.*Transactions on Electrical and Electronic Materials*,11(3), 93–105
2. Helkey, R., Saleh, A. A. M., Buckwalter, J., Bowers, J. E. (2019) High-Performance photonic integrated circuits on silicon, *IEEE Journal of Selected Topics in Quantum Electronics*, 25(5), 1–15
3. Chovan, J., Uherek, F. (2018) Photonic integrated circuits for communication systems, *Radioengineering*, 27(2), 357–363
4. Smit, M. K., Williams, K. A. (2020) Indium Phosphide Photonic Integrated Circuits, Optical Fiber Communication Conference, W3F.4
5. Loudon, R. (1970) Thepropagation of electromagnetic energy through an absorbing dielectric, *Journal of Physics A*, 3, 233–245
6. Yablonovitch, E. (1987) Inhibited spontaneous emission in solid-state physics and electronics, *Physical Review Letters*, 58, 2059–2061
7. Faruque, M. R. I., Islam, M. T., Misran, N. (2012) Design analysis of new metamaterial for EM absorption reduction, *Progress in Electromagnetics Research*, 124, 119–135
8. Elsheakh, D. M. N., Elsadek, H. A., Abdullah, E. A. (2012) Antenna Designs with Electromagnetic Bandgap Structures, *Metamaterial*, Ed. Jiang, X. Y., InTech, Rijeka, Croatia
9. Faruque, M. R. I., Islam, M. T., Misran, N. (2010) Evaluation of em absorption in human head with metamaterial attachment, *Applied Computational Electromagnetics Society Journal*, 25(12), 1097–1107
10. Yang, F., Rahmat-Samii, Y. (2009) *Electromagnetic Bandgap Structures in Antenna Engineering*, Cambridge University Press, Cambridge, UK
11. Ziolkowski, R. W., Engheta, N. (2006) *Introduction, History and Selected Topics in Fundamental Theories of Metamaterials, Metamaterials: Physics and Engineering Explorations*, Eds. Engheta, N., Ziolkowski, R., 1, John Wiley & Sons, New York, NY, USA
12. Caloz, C., Itoh, T. (2006) *Electromagnetic Metamaterials: Transmission Line Theory and Microwave Applications*, John Wiley & Sons, Toronto, Canada
13. de Maagt, P., Gonzalo, R., Vardaxoglou, Y. C., Baracco, J. M. (2003) Electromagnetic bandgap antennas and components for microwave and (sub)millimeter wave applications, *IEEE Transactions on Antennas and Propagation*, 51(10I), 2667–2677
14. Yang, F., Rahmat-Samii, Y. (2002) Applications of electromagnetic bandgap (EBG) structures in microwave antenna designs, Proceedings of the 3rd International Conference on Microwave and Millimeter Wave*Technology*, 528–531
15. Alam, M. S., Islam, M. T., Misran, N. (2011) Design analysis of an electromagnetic bandgapmicrostrip antenna, *The American Journal of Applied Sciences*, 8(12), 1374–1377
16. Gao, Y., Chen, H., Qiu, H., Lu, Q., Huang, C. (2011) Transmission spectra characteristics of 1D photonic crystals with complex dielectric constant, *Rare Metals*, 30, 150–154
17. Aghajamali, A., Akbarimoosavi, M., Barati, M. (2014) Properties of the bandgaps in 1D ternary lossyphotonic crystal containing double-negative materials, *Advances in Optical Technologies*, 2014, 1–7
18. Belhadj, W., Malek, A. F., Bouchriha, H. (2006) Characterization and study of photonic crystal fibres with bends, *Material Science and Engineering: C*, 26, 578–579
19. D'Orazio, A., De Palo, V., De Sario, M., Petruzzelli, V., Prudenzano, F. (2003) Finite difference time domain modeling of light amplification in active photonic bandgapstructures, *Progress in Electromagnetics Research*, 39, 299–339

20. Nozaki, K., Matsuo, S., Fujii, T., Takeda, K., Ono, M., Shakoor, A., Kuramochi, E., Notomi, M. (2016) Photonic-crystal nano-photodetector with ultrasmall capacitance for on-chip light-to-voltage conversion without an amplifier, *Optica*, 3, 483–492

21. Shanthi, K. V., Robinson, S. (2014) Two-dimensional photonic crystal based sensor for pressure sensing, *Photonic Sensors*, 4, 248–253

22. Zhang, Z., Qiu, M. (2004) Small-volume waveguide-section high Q microcavities in 2D photonic crystal slabs, *Optics Express*, 12, 3988–3995

23. Maity, A., Chottopadhyay, B., Banerjee, U., Deyasi, A. (2013) Novel band-pass filter design using photonic multiple quantum well structure with p-polarized incident wave at 1550 nm, *Journal of Electron Devices*, 17, 1400–1405

24. Istrate, E., Sargent, E. H. (2002) Photonic crystal heterostructures - resonant tunnelling, waveguides and filters, *Journal of Optics A: Pure and Applied Optics*, 4, S242–S246

25. Vasiliev, M., Belotelov, V. I., Kalish, A. N., Kotov, V. A., Zvezdin, A. K., Alameh, K. (2006) Effect of oblique light incidence on magnetooptical properties of one-dimensional photonic crystals, *IEEE Transactions on Magnetics*, 42, 382–388

26. Banerjee, A. (2009) Enhanced temperature sensing by using one-dimensional ternary photonic bandgap structures, *Progress in Electromagnetics Research*, 11, 129–137

27. Deyasi, A., Sarkar, A. (2019) Computing optical bandwidth of bandpass filter using metamaterial-based defected 1D PhC, AIP Conference Proceedings, 2072, 020003

28. Bryche, J.-F.; Tsigara, A.; Bélier, B.; Lamy de la Chapelle, M.; Canva, M.; Bartenlian, B.; Barbillon, G. (2016) Surface enhanced Raman scattering improvement of gold triangular nanoprisms by a gold reflective underlayer for chemical sensing, *Sensor and Actuator B*, 228, 31–35

29. Dolci, M.; Bryche, J.-F.; Leuvrey, C.; Zafeiratos, S.; Gree, S.; Begin-Colin, S.; Barbillon, G.; Pichon, B. P. (2018) Robust clicked assembly based on iron oxide nanoparticles for a new type of SPR biosensor, *Journal of Material Chemistry C*, 6, 9102–9110

30. Vangelidis, I., Theodosi, A., Beliatis, M. J., Gandhi, K. K., Laskarakis, A., Patsalas, P., Logothetidis, S., Silva, S. R. P., Lidorikis, E. (2018) Plasmonicorganic photovoltaics: Unraveling plasmonicenhancement for realistic cell geometries, *ACS Photonics*, 5, 1440–1452

31. Salamin, Y., Ma, P., Baeuerle, B., Emboras, A., Fedoryshyn, Y., Heni, W., Cheng, B., Josten, A., Leuthold, J. (2018) 100 GHz Plasmonicphotodetector, *ACS Photonics*, 5, 3291–3297

32. Lee, Y., Lee, J., Lee, T.K., Park, J., Ha, M., Kwak, S.K., Ko, H. (2015) Particle-on-film gap plasmons on antireflective ZnOnanoconearrays for molecular-level surface-enhanced ramanscattering sensors, *ACS Applied Materials and Interfaces*, 7, 26421–26429

33. Mattox, T. M., Urban, J. J. (2018) Tuning the surface plasmonresonance of lanthanum hexaboride to absorb solar heat: A review, *Materials*, 11, 2473

34. Monat, C., Grillet, C., Collins, M., Clark, A., Schroeder, J., Xiong, C., Li, J., O'faolain, L., Krauss, T. F., Eggleton, B. J. (2014) Integrated optical auto-correlator based on third-harmonic generation in a silicon photonic crystal waveguide, *Nature Communication*, 5, 3246

35. Olivier, N., Aptel, F., Plamann, K., Schanne-Klein, M. C., Beaurepaire, E. (2010) Harmonic microscopy of isotropic and anisotropic microstructure of the human cornea, *Optical Express*, 18, 5028–5040

36. Hajjiah, A., Kandas, I., Shehata, N. (2018) Efficiency enhancement of Perovskitesolar cells with plasmonicnanoparticles: A simulation study, *Materials*, 11, 1626

37. Atwater, H., Polman, A. (2010) Plasmonics for improved photovoltaic devices, *Nature Materials*, 9, 865

38. Ignatov, A. I., Merzlikin, A. M., Baryshev, A.V. (2017) Wood anomalies for s-polarized light incident on a one-dimensional metal gratingand their coupling with channel plasmons, *Physical Review A*, 95, 053843

39. Baburin, A. S., Gritchenko, A. S., Orlikovsky, N. A., Dobronosova, A. A., Rodionov, I. A., Balykin, V. I., Melentiev, P. N. (2019) State-of-the-art plasmonic crystals for molecules fluorescence detection, *Optical Material Express*, 9, 1173–1179

40. Shen, H. M., Lu, G. W., Zhang, T. Y., Liu, J., Gong, Q. H. (2013) Enhanced single-molecule spontaneous emission in an optimized nanoantenna with plasmonicgratings, *Plasmonics*, 8, 869–875

41. Rigneault, H., Capoulade, J., Dintinger, J., Wenger, J., Bonod, N., Popov, E., Ebbesen, T. W., Lenne, P. F. (2005) Enhancement of single-molecule fluorescence detection in subwavelength apertures, *Physical Review Letters*, 95, 17401

42. Han, X., Liu, K., Sun, C.-S. (2019) Plasmonics for Biosensing, *Materials*, 12, 1411

43. Taflove, A. (1995) *Computational Electrodynamics: The Finite Difference Time Domain Approach*, Artech House, Norwood, MA.

44. Berenger, J. (1994) A perfectly matched layer for the absorption of electromagnetic waves.*Journal of Computational Physics*, 114(10), 185–200.

45. Peterson,A. F.,Ray, S. L.,Mittra, R. (1998) *Computational Methods for Electromagnetics*, IEEE Press, New York.

46. Stratton, J. A. (1941) *Electromagnetic Theory*, McGraw-Hill, New York.

47. Harrington, R.F. (1961) *Time-Harmonic Electromagnetic Fields*, McGraw-Hill, New York.

48. Tai, C. T. (1971) *Dyadic Green's Functions in Electromagnetic Theory*, International Textbook Company, Scratton, PA.

49. Sheng, X.Q., Yung, E. K. N. (2002) Implementation and experiments of a hybrid algorithm of the MLFMA-enhanced FE-BI method for open-region inhomogeneous electromagnetic problems, *IEEE Transactions on Antennas and Propagation*, 50(2), 163–167

50. Umashankar, K., Taflove, A. (1982) A novel method to analyze electromagnetic scattering of complex objects, *IEEE Transactions Electromagnetic Compatibility*, 24, 397–405

51. Wilcox, C.H. (1956) An expansion theorem for electromagnetic fields. Communications on Pure and Applied Mathematics, M, 115–134

52. Joseph, R. M., Hagness, S. C., Taflove, A. (1991) Direct time integration of Maxwell's equations in linear dispersive media with absorption for scattering and propagation of femtosecond electromagnetic pulses, *Optics Letters*, 16, 1412–1414

53. Chang, D. C., Zheng, J. X. (1992) Electromagnetic modeling of passive circuit elements in MMIC, *IEEE Transactions on Microwave Theory and Techniques*, 40(9), 1741–1747

54. Chew, W. C., Weedon, W. (1994) A 3D perfectly matched medium from modified Maxwell's equations with stretched coordinates, *Microwave and Optics Technology Letters*, 13, 599–604

55. Silvester, P. P., Ferrari, R. L. (1983) *Finite Elements for Electrical Engineering*, Cambridge University Press, Cambridge.

56. Lee, J. F., Sun, D. K., Cendes, Z. J. (1991) Full-waves analysis of dielectric waveguides using tangential vector finite elements, *IEEE Transactions on Microwave Theory and Techniques*, 39(7), 1262–1271

57. Ise, K., Inoue, K., Koshiba, M. (1990) Three-dimensional finite-element solution of dielectric scattering obstacles in a rectangular waveguide, *IEEE Transactions on Antennas and Propagation*, 38(9), 1352–1359

58. Thiele, G. A., Newhouse, T. H. (1975) A hybrid technique for combining moment methods with the geometrical theory of diffraction, *IEEE Transactions on Antennas and Propagation*, 23(1), 62–69

59. Gedney, S. D. (1996) An anisotropic perfectly matched layer-absorbing medium for the truncation of FDTD lattices. *IEEE Transactions on Antennas and Propagation*, 44(12), 1630–1639

60. Kashiwa, T.Fukai, I. (1990) A treatment by FDTD method of dispersive characteristics associated with electronic polarization. *Microwave and Optics Technology Letters*, 3, 203–205

61. Sui, W., Christensen, D. A., Durney, C. H. (1992) Extending the two dimensional FDTD method to hybrid electromagnetic systems with active and passive lumped elements, *IEEE Transactions on Microwave Theory and Techniques*, 40(4), 724–730

62. Naishadham, K.Lin, X. P. (1994) Application of spectral domain Prony's method to the FDTD analysis of planar microstrip circuits, *IEEE Transactions on Microwave Theory and Techniques*, 42(12), 2391–2398

63. Khayat, M. A.Wilton, D. R. (2008) An improved transformation and optimized sampling scheme for the numerical evaluation of singular and near-singular potentials, *IEEE Antennas and Wireless Propagation Letters*, 7, 377–380

64. Donepudi, K.C., Jin, J.M., Velamparambil, S.et al.(2001) A higher order parallelized multilevel fast multipole algorithm for 3D scattering, *IEEE Transactions on Antennas and Propagation*, 49(7), 1069–1078

65. Hu, F. G., Wang, C. F., Gan, Y. B. (2007) Efficient calculation of interior scattering from large three-dimensional PEC cavities, *IEEE Transactions on Antennas and Propagation*, 55(1), 167–177

66. Velamparambil, S., Chew, W. C., Song, J. (2003) 10 million unknowns: Is it that big?*IEEE Antennas and Propagation Magazine*, 45, 43–58

67. Pan, X. M.Sheng, X. Q. (2008) A sophisticated parallel MLFMA for scattering by extremely large targets, *IEEE Antennas and Propagation Magazine*, 50(3), 129–138

68. Leith, E. N. (2000) The evolution of information optics, *IEEE Journal of Selected Topics in Quantum Electronics*, 6(6), 1297–1304

69. Pan, J. (2009) *Experimental manipulation of photons and cold atoms: Towards scalable quantum information processing, European Conference on Lasers and Electro-Optics and the European Quantum Electronics Conference*, IEEE, 14–19 June, Munich, Germany

70. Glückstad, J. (2009) *Information optics in biophotonics. 9th Euro-American Workshop on Information Optics*, IEEE, 12-16 July, Helsinki, Finland

71. Lugnan, A., Katumba, A., Laporte, F., Freiberger, M., Sackesyn, S., Ma, C., Gooskens, E., Dambre, J., Bienstman, P. (2020) Photonic neuromorphic information processing and reservoir computing. *APL Photonics*, 5, 020901

72. Sheehan, R., Gallet, A., Ghorbel, I., Eason, C., Carroll, L., O'Brien, P., Shen, A., Duan, G-H., Gunning, F. C. G. (2018) Repeaterless data transmission at 1310 nm using silicon photonic integrated circuit. Silicon Photonics: From Fundamental Research to Manufacturing, 10686, 106860J, Strasbourg, France

73. O'Brien, J. L., Furusawa, A., Vučković, J. (2009) Photonic quantum technologies. *Nature Photonics*, 3, 687–695

74. Chen, C.-Y., Murmann, B., Seo, J.-S., Yoo, H.-J. (2019) Custom sub-systems and circuits for deep learning: Guest editorial overview, *IEEE Journal of Emerging Selected Topics in Circuits and Systems*, 9, 247–252

75. Shen, Y., Harris, N. C., Skirlo, S., Prabhu, M., Baehr-Jones, T., Hochberg, M., Sun, X., Zhao, S., Larochelle, H., Englund, D., Soljačić, M. (2017) Deep learning with coherent nanophotonic circuits. *Nature Photonics*, 11, 441–446

76. Simon, C. (2011) *Photonic quantum information processing: From quantum memories to photon-photon gates. International Conference on Information Photonics*, IEEE, 18–20 May, Ottawa, ON, Canada

77. Furusawa, A. (2017) *Hybrid quantum information processing, a way for large-scale optical quantum information processing. Conference on Lasers and Electro-Optics Europe & European Quantum Electronics Conference*, IEEE, 25–29 June, Munich, Germany

78. Slussarenko, S., Pryde, G. J. (2019) Photonic quantum information processing: A concise review. *Applied Physics Reviews*, 6, 041303

79. Peters, N., Altepeter, J., Jeffrey, E., Branning, D., Kwiat, P. (2003) Precise creation, characterization, and manipulation of single optical qubits. *Quantum Information and Computation*, 3, 503–517

80. Gyongyosi, L., Imre, S. (2019) Quantum circuit design for objective function maximization in gate-model quantum computers. *Quantum Information Processing*, 18, 225

81. Calafell, I. A., Cox, J. D., Radonjić, M., Saavedra, J. R. M., de Abajo, F. J. G., Rozema, L. A., Walther, P. (2019) Quantum computing with grapheneplasmons. *Quantum Information*, 5, 37

82. Fang, Y., Sun, M. (2015) Nanoplasmonic waveguides: Towards applications in integrated nanophotonic circuits, *Light: Science & Applications*, 4, e294

83. de Abajo, F. J. G. (2014) Grapheneplasmonics: Challenges and opportunities. *ACS Photonics*, 1, 135–152

2 Bandgap Engineering of Sol–Gel Spin-Coated TiO$_2$ Thin Film on Glass Substrate

Soham Lodh and Rajib Chakraborty

CONTENTS

2.1 INTRODUCTION

In recent years, TiO$_2$ has come out as an attractive research material in the field of photonics and electronic devices for its attractive characteristics and easy availability in rocks and minerals [1]. TiO$_2$ exists in amorphous and crystalline (anatase, rutile and brookite) forms. Dielectric constant of TiO$_2$ ranges in between 10 and 80, with large resistivity of 10^{19}–10^{24} Ω.m making it very suitable for MOS device applications [2,3]. TiO_2 thin films are widely used for different optical purposes, as it has appealing physical characteristics. TiO_2 thin films provide reliable optical coating with high refractive index, good transmittance in the visible as well as near-infrared region, large bandgap along with an excellent thermal, physical and chemical stability [1,4–7], which makes it suitable for photonics applications, such as gas sensors [8], antireflective coating [9], photocatalyst[10], solar cells [11], multilayer transmissive and reflective optical applications [12].

Anatase TiO_2 film used as a CO gas sensor was studied by *Al-Homoud et al.*[8]. TiO_2 films with two different thicknesses were grown on glass, Si and sapphire substrate by employing sputtering. Presence of CO was measured by the change in the film resistance in N_2 environment. At 300°C, all the TiO_2 films (250 nm) exhibited square root dependency of resistance in the presence of CO gas flow. The decrease in resistance of 250nm film was found to be higher for the sapphire substrate in comparison to the Si and glass substrate. Thick films on the other hand show faster response time over the thin films. *Vicente et al.* [9] suggested sol–gel dip-coated TiO_2 antireflective coating on monocrystalline textured Si solar cells. Decrease in hemispherical reflectance was observed with the introduction of TiO_2 coating on textured Si solar cells. Moreover, vacuum sintering of TiO_2 film reduces the average reflectance significantly. Solar cell made by employing transparent TiO_2 film of 10μm thickness coated with charge transfer dye was reported by *O'Regan et al.* [11]. More than 80% of incident photon conversion to electrical current was observed.

Optical applications of TiO_2 require a uniform, dense, high refractive index film along with very low extinction coefficient and optical anisotropy, so in several applications amorphous TiO_2 is used for its anisotropic nature [13]. Some reports suggest that thin-film deposition not only lowers the cost by reducing the material usage, but also provides excellent performance [14,15]. So, preparation of good quality thin film can be very useful for photonics applications.

Quality and structure of TiO_2 thin film strongly depends on the substrate quality and process of deposition. Several techniques like sputtering [16], chemical bath deposition (CBD) [17], sol–gel dip/spin coating [18], vacuum evaporation [19], molecular beam epitaxy (MBE)[20], electrodeposition[21], liquid-phase deposition [22], atomic layer deposition (ALD) [23], etc., are employed to develop TiO_2 thin film on different substrates.

Sputtering is a very effective vacuum deposition technique for thin films. In this process, the target (source) is positioned in the line of sight of the substrate. The target is bombarded with inert gas plasma (argon/nitrogen). The substrate is biased in such a way that the positive ions of the inert gas plasma hit the target, which then erupts and gets deposited on the substrate positioned in front of it. CBD is another efficient way for the thin-film deposition. In this process, the solution of the material to be deposited is prepared with the help of precursor and other required chemicals. The substrate is submerged in the solution with the help of a holder under continuous stirring of the solution. Deposition speed and the thickness can be controlled by tuning the solution composition, time and other parameters. Another useful line-of-sight deposition method is a vacuum evaporation process. In this method, the coating material is melted by heating (using electrical signal/electron beam). Then the coating material evaporates and gets deposited on the substrate. The vapour deposited on the substrate then cools down slowly and forms solid thin film. The thickness can be controlled by controlling the deposition time. In the sol–gel process, the coating solution is prepared with precursor and other required materials. For dip coating process, the substrate on which film is to be deposited is immersed inside the solution and withdrawal with a fixed rate to get desired thickness of coated film. On the other hand, spin coating of solution on the substrate is done with a pipette and a spin coater. Small amount of solution is spin-coated to a fixed rotation per minute (rpm). Thickness of the film can be varied by varying the rpm of the spin coater. Loss of chemicals is less for the sol–gel spin coating process in comparison to the CBD and sol–gel dip-coating techniques. MBE is a very efficient ultra-high vacuum (UHV) method to deposit good quality epitaxial layers by employing physical vapour deposition or vacuum evaporation technique. The ALD method is used to deposit ultra-thin high-performance films on a substrate. ALD process is employed by using a chemical vapour deposition technique. In ALD method, film thickness can be controlled in angstrom level with variable film composition. Both MBE and ALD processes require very costly equipment, whereas sol–gel and CBD require inexpensive equipment. Among all the thin-film deposition techniques, sol–gel is the most cost-effective technique where low-cost TiO_2 thin films are developed by spin/dip coating of solution made from precursor.

Bandgap plays an important role in photonic device and solar cell applications. Bandgap of TiO_2 depends on its phases (amorphous, anatase, ruitile and brookite). It can also be changed for a particular phase which basically depends upon the growth procedure, annealing temperature, substrate material and precursor [24–29]. *Shi et al.* [24] reported change in TiO_2 optical bandgap from 3.718 eV to 3.417 eV owing to the change in film thickness from 2.55 to 20.61 nm by varying the ALD cycle from 50 to 600. It is also reported that the optical bandgap depends on annealing temperature. *Yoo et al.* [26] reported changes in optical bandgap with different annealing methods. Bandgap of as-deposited sample (3.29 eV) annealed at 700°C and 800°C changes to 3.39 and 3.43 eV, respectively, for conventional thermal annealing (CTA) as well as 3.38 and 3.32 eV, respectively, for rapid thermal annealing (RTA). Thus, TiO_2 thin-film bandgap can be varied with different techniques, which can be very useful for photonic device applications.

In this chapter, TiO_2 thin-film preparation using sol–gel spin coating process is discussed in detail for the benefit of future researchers. Titanium(IV) iso-propoxide (TTIP) is mixed with ethanol and acetic acid to get precursor solution of TiO_2. Prepared solution is spin-coated on clear glass substrate for 15 sec at 5000 rpm (revolutions per minute) to form as-deposited TiO_2 thin films, which are further annealed at different temperatures. Optical and material characteristics are also studied for as-deposited and annealed samples.

2.2 PREPARATION OF TIO$_2$ SOLUTION AND THIN FILM

TTIP is the most common precursor used for the preparation of TiO_2 solution. Here, TiO_2 solution is prepared by TTIP, acetic acid and ethanol (C_2H_5OH). 0.8573 mol of ethanol and 0.0874 mol of acetic acid are continuously stirred in a confined glass beaker for 15 minutes, then 0.0213 mol of TTIP is added to the solution drop wise under continuous magnetic stirring. Then, the solution is stirred for another 3 hours, which results in a transparent yellowish solution. All the chemicals are manufactured by Sigma-Aldrich and used as supplied. The glass substrate is made by Borosil glass works.

During the TiO_2 solution preparation process, care should be taken so that the beaker and measuring cylinders are cleaned and dried very well. Solution preparation is done in an air-tight container to prevent evaporation of alcohol as well as reaction of TTIP with atmospheric moisture, which results in a contaminated TiO_2 solution. Prepared TiO_2 solution must be taken from the beaker very fast using a pipette, as long exposure to air may damage the solution. After taking the solution it should be immediately poured on the previously cleaned glass substrate followed by spin coating, otherwise uneven film with large white particles may be seen on the substrate. Tip of pipette should also be properly cleaned and dried. A tip can be used only once. Reuse of tip may produce uneven film with small particles on the glass substrate. The solution will also be contaminated if a tip is inserted to it more than once. Presence of very small amount of water in any apparatus will damage the thin film preparation process badly.

Glass substrates are cleaned inside a sonicator with the standard cleaning procedure. First, they are cleaned with soap water and rinsed in **deionized**(DI) water. Next, acetone in an air-tight container at 50°C is used to wash substrates for 10 minutes and methanol is used to clean them for another 5 minutes. After that, substrates are rinsed in DI water and blow dried with nitrogen gas. Afterwards, concentrated H_2SO_4 wash is done for 5 minutes. To eliminate surface contaminants, methanol and HCL mixture (1:1) cleaning is performed for another 30 minutes [30]. Glass substrates are then again rinsed in DI water followed by drying under Nitrogen gas flow. Finally, piranha solution is used to remove organic residues from the substrate surface followed by rinsing in DI water and drying with nitrogen gas.

Samples are made with the help of spin coating of TiO_2 solution (30μL) at 5000 rpm for 15 sec on clear glass substrate to form TiO_2 thin film of ~ 86 nm thickness. After spin coating, the solution is dried for 20 minutes with nitrogen gas, then the samples are annealed at 150°C, 300°C and 450°C for 30 minutes. The flowchart of preparation of TiO_2 solution and thin film is given in Figure 2.1.

Material characterization is done by employing X-ray diffraction (XRD) technique, surface quality is studied by scanning electron microscope (SEM), and thickness and refractive index of the samples is measured with the help of spectroscopic ellipsometry (SE). Optical transmission is measured with spectrophotometer. Optical bandgap is calculated using Tauc plot [31].

2.3 RESULTS AND DISCUSSION

2.3.1 SEM Image

SEM images are shown in Figure 2.2 for all four samples. Figure 2.2 reveals that all the four samples are having good quality TiO_2 thin film deposited on the glass substrate. It is seen that the grain size of TiO_2 film increases as the annealing temperature is increased. Imperfect TiO_2 film formation on the glass substrate by spin coating is shown in Figure 2.3. It can be noticed from Figure 2.3(a) that TiO_2 film shows a big

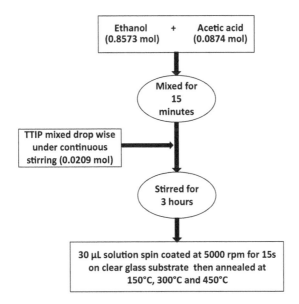

FIGURE 2.1 Flow chart of TiO$_2$ solution preparation and deposition.

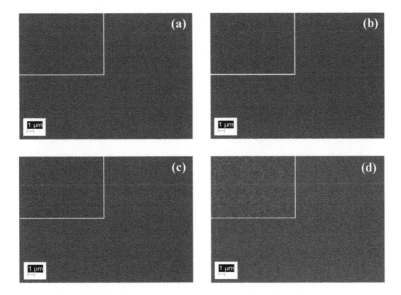

FIGURE 2.2 SEM images of (a) as deposited, (b) annealed at 150°C, (c) annealed at 300°C and (d) annealed at 450°C samples.

crack with several black spots after annealing at 350°C, which may be due to the inhomogeneous TiO$_2$film formation during spin coating. Figure 2.3(b) shows as-deposited TiO$_2$ thin film with a white particle of ~ 1 μm size (pointed with orange arrow), although rest part of the film is with good continuity. If the white particle was present on the glass substrate prior to spin coating, then formation of TiO$_2$ film with good

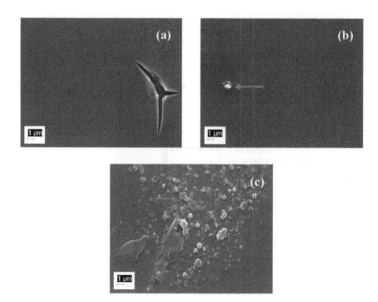

FIGURE 2.3   SEM images of imperfect TiO$_2$ film.

uniformity around the particle was not possible. Thus, it can be concluded that the particle might have dropped on the as-deposited TiO$_2$ film immediately after or just prior to completion of spin coating. Improper solution preparation and/or substrate cleaning may produce imperfect TiO$_2$ thin film as shown in Figure 2.3(c). Thus, proper measures should be taken for the solution preparation and thin-film deposition.

2.3.2   X-RAY DIFFRACTION (XRD)

TiO$_2$ is available in amorphous and three crystalline forms: rutile, anatase and brookite[25,32–37]. XRD pattern is studied to know the nature of TiO$_2$ thin film deposited on glass substrate. The study of XRD is also useful to observe changes in the XRD pattern owing to annealing of TiO$_2$ film [32–34]. The substrate plays an important role for the film type and quality. TiO$_2$ solution prepared in the same manner but coated on different substrates like glass, ITO/FTO-coated glass, semiconductors, etc., results in different type of XRD patterns [36]. *Zhang et al.* [28] reported amorphous as-deposited TiO$_2$ film on glass substrate, which transformed to anatase after 1 hour annealing at 600°C. *Hosseini et al.* [36] studied TiO$_2$ film on glass as well as ITO-coated glass and reported that TiO$_2$ films on glass substrate exhibit amorphous nature, whereas the same show crystalline nature on ITO-coated glass substrate. *Liu et al.*[32] reported rutile TiO$_2$ formed on GaAs substrate at an annealing temperature of 700°C.

It has been also reported that the nature of XRD pattern of TiO$_2$ film also depends on the number of coating cycles [33,36]. Another factor that is responsible for different types of XRD patterns of TiO$_2$ film is the sample preparation technique. XRD pattern for four samples are shown in Figure 2.4, where Figure 2.4(a) corresponds to the as-deposited sample and Figure 2.4(b), 2.4(c) and 2.4(d) correspond to the annealed (150°C, 300°C and 450°C) samples. All the four samples show amorphous nature. Among them, the as-deposited sample reveals the lowest intensity and the intensity increases with increase in annealing temperature. This increase in intensity with increase in annealing temperature could be attributed to the improved packing density of amorphous TiO$_2$.

2.3.3 ELLIPSOMETRY AND SPECTROPHOTOMETRY RESULTS

SE is a useful non-destructive tool to know the precise sample thickness along with some optical properties of the thin film formed on glass or any other substrate. SE measurement can reveal the change in film thickness and optical properties due to annealing and/or other factors. The model structure considered is air/TiO$_2$/glass to determine the thickness as well as the refractive index. Transmission within the wavelength range of 300–800 nm is measured with spectrophotometer.

2.3.3.1 Thickness

All the samples are prepared on glass substrate of dimension 26mm × 25mm × 1 mm using spin coating technique. The measurement of thickness along with refractive index is made within 3 mm radius around the centre of each sample with the help of SE. SE *is a non-destructive technique that can be used to measure thin film thickness accurately.* In non-destructive measurement technique, the sample under measurement can be further used for other characterization. Multiple measurements of

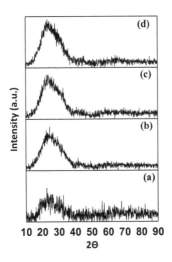

FIGURE 2.4 X-ray diffraction patterns of (a) as deposited, (b) annealed at 150°C, (c) annealed at 300°C and (d) annealed at 450°C samples.

TABLE 2.1
TiO₂ Film Thickness

Sample type	Thickness (nm)
As deposited	86.24
Annealed at 150°C	58.69
Annealed at 300°C	50.10
Annealed at 450°C	43.12

thickness at different locations for each sample can be very useful to know the uniformity of film thickness. In this study, six different points are taken within 3 mm radius around the centre to find any changes in thickness. The film uniformity for all the samples is within the range of ± 0.5 nm. Thus, thickness uniformity is within the acceptable range. Thickness of as-deposited and annealed samples are listed in Table 2.1.

Significant film thickness shrinkage for annealed samples can be seen from Table 2.1. Film thickness shrinkage of 450°C annealed sample is found to be ~ 50% in comparison to the as-deposited sample. Densification of annealed TiO₂ thin films is due to the condensation reaction of acetic acid and elimination of residual water [38].

2.3.3.2   Optical Transmission

Spectrophotometry measurement is done in the UV–VIS spectra to study the transmittance and optical bandgap of the as-deposited and annealed TiO₂ thin films. Transmittance spectra of TiO₂ thin films on glass substrate are shown in Figure 2.5. It is evident from Figure 2.5 that in the visible region all the four films are transparent and sharp decrease in transmission is observed in the ultraviolet (UV) region owing to fundamental light absorption [29]. Figure 2.5 reveals that the transmission is roughly around 85% for as-deposited film throughout the visible range. It can also be seen that the transmission value decreases with the rise in the annealing temperature. The possible reason for the decrease in transmission with the rise in annealing

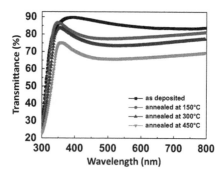

FIGURE 2.5   Transmittance spectra for four different samples.

temperature is due to the fact that the grain size of TiO$_2$ film increases, which might have caused greater scattering from the surface. Another reason for the decrease in transmission with the increase in annealing temperature may be due to the increase in refractive index, which is discussed in the next section.

2.3.3.3 Refractive Tndex (n)

Refractive index (n_r) is a combination of real n(λ) and imaginary k(λ) part and is represented as follows:

$$n_r = n(\lambda) + k(\lambda) \tag{2.1}$$

The real part of refractive indices (n) at 2.255 eV (550 nm wavelength) for as-deposited and annealed samples is shown in Figure 2.6. Its value for the range of 1.550–4.133 eV is shown in inset of Figure 2.6. It can be seen from Figure 2.6 that the refractive index is lowest for the as-deposited sample, which further increases with the rise in annealing temperatures. The higher refractive index is attributed to the higher packing density of TiO$_2$ film [38]. So, it can be concluded that the packing density of TiO$_2$ film gradually increases with rise in the annealing temperatures.

Extinction coefficient is calculated from the transmittance data received from spectrophotometer by employing the following equations [39]:

$$\alpha = \frac{1}{d} \ln\left(\frac{1}{T}\right), \tag{2.2}$$

where α is absorption coefficient, d is the thickness of TiO$_2$ thin film and T represents the transmittance. Absorption coefficient is related to the extinction coefficient (k) by the following equation [39]:

$$k = \frac{\alpha\lambda}{4\pi} \tag{2.3}$$

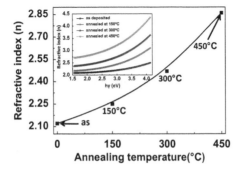

FIGURE 2.6 Real part of refractive index (n) for as-deposited and annealed samples at 2.255 eV (550 nm wavelength). Inset: Its value over entire measurement range.

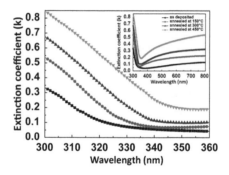

FIGURE 2.7 Extinction coefficient (k) of TiO$_2$ thin film on glass substrate. Inset: Its value for 300–800 nm wavelength range.

The imaginary part of refractive index, i.e. extinction coefficient (k) is shown in Figure 2.7 in the wavelength range of 300–360 nm; the entire measurement range (300–800 nm) is shown in inset.

Extinction coefficient (k) can be seen increasing in the UV region (<400 nm). As-deposited sample exhibits the lowest extinction coefficient, which further increases with the rise in annealing temperatures, because annealing makes TiO$_2$ film denser [29,38]. Extinction coefficient and refractive index at 550 nm are listed in Table2.2. Comparison of measured refractive index at 550 nm with other reported data are listed in Table 2.3.

2.3.3.4 Porosity (P)

A study on refractive index and thickness reveals that the highest temperature (450°C) annealed sample has the densest TiO$_2$ film. So, porosity (P) of other samples is calculated with respect to 450°C annealed sample by using the following formula [42]:

$$P = 1 - \frac{n^2 - 1}{n_{450}^2 - 1} \tag{2.4}$$

wheren_{450} is the refractive index of 450°C annealed sample and n represents the refractive index of the measuring sample. % porous vs. photon energy for the as-deposited and annealed samples (150°C and 300°C) with respect to 450°C annealed

TABLE 2.2

Refractive Index and Extinction Coefficient

Sample	Refractive Index (RI) at 2.255 eV (550 nm)	Extinction Coefficient (k) at 550 nm
As deposited	2.12	0.081
Annealed at 150°C	2.25	0.188
Annealed at 300°C	2.47	0.266
Annealed at 450°C	2.86	0.423

TABLE 2.3

Comparison of Measured Refractive Index with Other Reported Results

Process	Substrate	Refractive Index at 2.255 eV (550 nm)
ALD [40]	Si	2.65
Pulsed bias arc ion plating [28]	Glass	2.51
FCVA [13]	Si	2.56
RF sputtering [41]	Quartz	2.73
RF-magnetron sputtering [39]	Glass	~ 2.76
Spin coating, annealed at 450°C (present work)	Glass	2.86

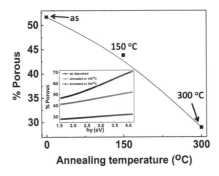

FIGURE 2.8 Porosity of as deposited and annealed (150°C and 300°C) samples with respect to 450°C annealed sample at 2.255 eV (550 nm wavelength). Inset: Entire measurement range (1.550 – 4.133 eV).

sample is shown in Figure 2.8. At 2.255 eV (550 nm wavelength), the as-deposited sample reveals the highest % of porosity of 51.71%, whereas it reduces with rise in annealing temperature as the TiO_2 film becomes denser.

Porosity reduces with decrease in photon energy for all the three samples. Change in porosity with photon energy is highest for the as-deposited sample and it reduces with rise in annealing temperatures. Porosity % at 2.255 eV for all the samples with respect to 450°C annealed sample is listed in Table 2.4.

2.3.3.5 Optical Bandgap

Tauc and Menth [31] method is employed to compute optical bandgap of TiO_2 thin films. The following expression is used to determine the optical bandgap:

$$(\alpha h\gamma)^x = C(h\gamma - E_g)$$ (2.5)

where

TABLE 2.4

Porosity

Sample type	% Porous with Respect to 450°C Annealed Sample at 2.255 eV (550 nm)
As deposited	51.71
Annealed at 150°C	43.8
Annealed at 300°C	28.95

α = absorption coefficient,

h = Plank's constant (4.136×10^{-15}eV.s),

γ = frequency of light (s^{-1}),

C = constant independent of photon energy,

E_g = optical bandgap (eV) and

$x = \dfrac{1}{2}$ for indirect and 2 for direct bandgap.

Most of the reported data indicate that anatase phase of TiO_2 exhibits indirect bandgap, whereas rutile and brookite are direct bandgap materials [43]. Some researchers calculated both direct and indirect bandgap as hγ versus $(\alpha h\gamma)^x$ plot has linear fit over a broad range of photon energy [44]. In amorphous material, long-range periodicity is not present; therefore, destruction of Brillouin zone arises and hence only a particular band structure may not be present. So, to compute optical bandgap of amorphous as-deposited and annealed TiO_2 thin films, both direct and indirect optical bandgaps are calculated by employing Equation (2.5). hγ versus $(\alpha h\gamma)^x$ plot with considering $x = \dfrac{1}{2}$ and 2 is found to be linear fit over a wide stretch of photon energy for all the four samples.

Plot of hγ versus $(\alpha h\gamma)^{1/2}$ to find out the indirect bandgap is shown in Figure 2.9. Linear extrapolation on the x-axis (hγ) gives the value of indirect bandgap (E_g) for all the four samples. Bandgaps for all the samples are listed in Table 2.5, from where it is evident that with increase in annealing temperature bandgap decreases. This change in indirect bandgap with annealing is due to the fact that TiO_2 thin film is

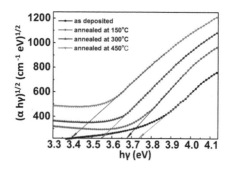

FIGURE 2.9 Plot of hγ versus $(\alpha h\gamma)^{1/2}$.

TABLE 2.5
Optical Bandgap

Sample type	Optical Indirect Bandgap (eV)	Optical Direct Bandgap (ev)
As deposited	3.737	3.875
annealed at 150°C	3.637	3.841
annealed at 300°C	3.548	3.791
annealed at 450°C	3.394	3.726

attaining higher density with larger grain size [45,46], which is confirmed with the SEM images shown in Figure 2.2.

Optical indirect bandgap discussed in this chapter is in line with other reported data [2,24,45], which also suggests decrease in bandgap due to increase in annealing temperature. *Naceur et al.* [45] reported 3.65 eV optical bandgap for amorphous TiO_2 film, which after annealing at 800°C changes to anatase phase with 3.22 eV. *Dang et al.*[2] reported optical bandgap of 3.7 eV for the as deposited and 3.6 eV for annealed (500°C) crystalline TiO_2 film.

Plot of hγ versus $(\alpha h\gamma)^2$ to calculate direct optical bandgap is shown in Figure 2.10. Linear extrapolation on the *x*-axis reveals direct optical bandgap. Similar to indirect bandgap, direct bandgap too reduces with rise in annealing temperatures, as annealing makes TiO_2 film denser with larger grain size. So, a tunable bandgap can be obtained by changing the annealing temperature. Indirect and direct optical bandgap is listed in Table 2.5.

2.3.3.6 Optical Dielectric Constant

Optical dielectric constant (ε) consists of real and imaginary part and is calculated by using the following formulae [39]:

$$\varepsilon_r = n^2 - k^2 \qquad (2.6)$$

$$\varepsilon_i = 2nk \qquad (2.7)$$

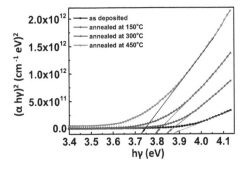

FIGURE 2.10   Plot of hγ versus $(\alpha h\gamma)^2$.

where ε_r is the real part and ε_i is the imaginary part, n is the refractive index and k represents the extinction coefficient. The real and imaginary part of dielectric constant is shown in Figures 2.11 and 2.12, respectively. The calculated dielectric constant using Equations (2.6) and (2.7) at 2.255 eV (550 nm wavelength) is listed in Table 2.6.

2.3.3.7 Optical Conductivity

Optical conductivity (σ_o) for all the four samples is calculated with the help of the following expression [39]:

$$\sigma_o = \frac{\alpha n c}{4\pi} \tag{2.8}$$

where α is the absorption coefficient, n is the refractive index and c represents the velocity of light. Optical conductivity is shown in Figure 2.13 over the wavelength stretch of 300–800 nm. It can be seen from Figure 2.13 that with rise in annealing temperature optical conductivity also increases, which is due to the increased absorption. In the UV region, a significant increase in optical conductivity is observed owing to high absorbance, which in turn increases conduction band electron density [29].

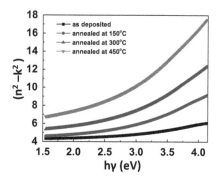

FIGURE 2.11   Real part of dielectric constant.

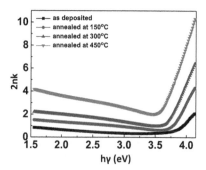

FIGURE 2.12   Imaginary part of dielectric constant.

TABLE 2.6
Optical Dielectric Constant

Sample type	Dielectric constant at 2.255 eV (550 nm)	
	Real part	Imaginary part
As deposited	4.461	0.555
Annealed at 150°C	4.973	1.233
Annealed at 300°C	5.979	1.842
Annealed at 450°C	7.854	3.382

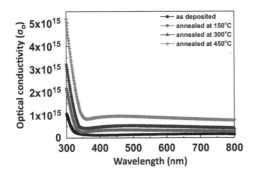

FIGURE 2.13 Plot of optical conductivity versus wavelength.

2.4 CONCLUSIONS

As-deposited and annealed (30 minutes) amorphous TiO_2 thin films at different temperatures (150°C, 300°C and 450°C) are prepared on glass substrate by employing sol–gel spin coating method. XRD pattern reveals amorphous nature for all the samples. Intensity of XRD increases with rise in annealing temperatures. Thickness measurement reveals significant film thickness shrinkage due to annealing. Thickness for as-deposited sample is 86.24 nm and it decreases further to 58.69, 50.10 and 43.12 nm for 150°C, 300°C and 450°C annealed samples, respectively. SEM images show that with rise in annealing temperature the grain size also increases. Density of TiO_2 film increases with higher annealing temperature; thus, the refractive index is highest for the 450°C sample and lowest for the as-deposited sample. Refractive indexes for the as-deposited and annealed (150°C, 300°C and 450°C) samples are 2.12, 2.25, 2.47 and 2.86 at 2.255 eV. In the UV region, as-deposited sample shows the lowest value of extinction coefficient and it increases with increase in annealing temperatures. Porosity is highest for the as-deposited sample with 51.71 % and it reduces with rise in annealing temperatures to 43.8% and 28.95% for 150°C and 300°C annealed samples with respect to the 450°C annealed sample at 2.255 eV. Optical indirect bandgap study exhibits 3.737 eV for the as-deposited sample and it reduces further to 3.673, 3.548 and 3.394 eV for the annealed samples, whereas optical direct bandgap for as-deposited sample reveals 3.875 eV and it decreases to 3.841, 3.791 and 3.726 eV after annealing, as TiO_2 thin film is attaining higher density with larger grain size. Owing to the highest refractive index of 450°C annealed sample, the optical dielectric constant is also highest which reduces further with decrease in annealing temperature and is lowest for the as-deposited sample. Optical conductivity increases due to annealing as the absorption increases. In the UV region, optical conductivity increases sharply as absorption becomes high. Thus, TiO_2 thin film with tunable optical properties is discussed in this chapter, which can be used over a wide variety of optical applications.

REFERENCES

1. Acrivos, J.V., Mott, N.F. and Joffe, A.D. eds., 2012. *Physics and Chemistry of electrons and ions in condensed matter* (Vol. 130). Springer Science & Business Media.
2. Dang, V.S., Parala, H., Kim, J.H., Xu, K., Srinivasan, N.B., Edengeiser, E., Havenith, M., Wieck, A.D., de los Arcos, T., Fischer, R.A. and Devi, A., 2014. Electrical and optical properties of TiO2 thin films prepared by plasma-enhanced atomic layer deposition. *Physica Status Solidi (a)*, 211(2), pp.416–424.
3. Mondal, S. and Kumar, A., 2016. Tunable dielectric properties of TiO2 thin film based MOS systems for application in microelectronics. *Superlattices and Microstructures*, 100, pp.876–885.
4. Yoldas, B.E. and O'Keeffe, T.W., 1979. Antireflective coatings applied from metal–organic derived liquid precursors. *Applied Optics*, 18(18), pp.3133–3138.
5. Yoldas, B.E., 1982. Deposition and properties of optical oxide coatings from polymerized solutions. *Applied Optics*, 21(16), pp.2960–2964.
6. Serpone, N., Lawless, D. and Khairutdinov, R., 1995. Size effects on the photophysical properties of colloidal anatase TiO2 particles: size quantization versus direct transitions in this indirect semiconductor? *The Journal of Physical Chemistry*, 99(45), pp.16646–16654.
7. Su, P.G. and Huang, L.N., 2007. Humidity sensors based on TiO2 nanoparticles/polypyrrole composite thin films. *Sensors and Actuators B: Chemical*, 123(1), pp.501–507.
8. Al-Homoudi, I.A., Thakur, J.S., Naik, R., Auner, G.W. and Newaz, G., 2007. Anatase TiO2 films based CO gas sensor: film thickness, substrate and temperature effects. *Applied Surface Science*, 253(21), pp.8607–8614.
9. San Vicente, G., Morales, A. and Gutierrez, M.T., 2002. Sol–gel TiO2 antireflective films for textured monocrystalline silicon solar cells. *Thin Solid Films*, 403, pp.335–338.
10. Sopyan, I., Watanabe, M., Murasawa, S., Hashimoto, K. and Fujishima, A., 1996. An efficient TiO2 thin-film photocatalyst: photocatalytic properties in gas-phase acetaldehyde degradation. *Journal of Photochemistry and Photobiology A: Chemistry*, 98(1–2), pp.79–86..
11. O'Regan, B. and Grätzel, M., 1991. A low-cost, high-efficiency solar cell based on dye-sensitized colloidal TiO2 films, *Nature*, 353, 737–740. Solar cells–New aspects and solutions, 208.
12. DeLoach, J.D. and Aita, C.R., 1998. Thickness-dependent crystallinity of sputter-deposited titania. *Journal of Vacuum Science & Technology A: Vacuum, Surfaces, and Films*, 16(3), pp.1963–1968.
13. Zhao, Z., Tay, B.K. and Yu, G., 2004. Room-temperature deposition of amorphous titanium dioxide thin film with high refractive index by a filtered cathodic vacuum arc technique. *Applied Optics*, 43(6), pp.1281–1285.

14. Singh, R.S., Rangari, V.K., Sanagapalli, S., Jayaraman, V., Mahendra, S. and Singh, V.P., 2004. Nano-structured CdTe, CdS and TiO2 for thin film solar cell applications. *Solar Energy Materials and Solar Cells*, 82(1–2), pp.315–330.

15. Boyadjiev, S., Georgieva, V., Vergov, L., Baji, Z., Gáber, F. and Szilágyi, I.M., 2014. Gas sensing properties of very thin TiO2 films prepared by atomic layer deposition (ALD). *Journal of Physics: Conference Series*, 559(1), p.012013.

16. Calnan, S., Upadhyaya, H.M., Dann, S.E., Thwaites, M.J. and Tiwari, A.N., 2007. Effects of target bias voltage on indium tin oxide films deposited by high target utilisation sputtering. *Thin Solid Films*, 515(24), pp.8500–8504.

17. Mayabadi, A.H., Waman, V.S., Kamble, M.M., Ghosh, S.S., Gabhale, B.B., Rondiya, S.R., Rokade, A.V., Khadtare, S.S., Sathe, V.G., Pathan, H.M. and Gosavi, S.W., 2014. Evolution of structural and optical properties of rutile TiO2 thin films synthesized at room temperature by chemical bath deposition method. *Journal of Physics and Chemistry of Solids*, 75(2), pp.182-187.

18. Wang, X., Shi, F., Gao, X., Fan, C., Huang, W. and Feng, X., 2013. A sol–gel dip/spin coating method to prepare titanium oxide films. *Thin Solid Films*, 548, pp.34–39.

19. Yang, C., Fan, H., Xi, Y., Chen, J. and Li, Z., 2008. Effects of depositing temperatures on structure and optical properties of TiO2 film deposited by ion beam assisted electron beam evaporation. *Applied Surface Science*, 254(9), pp.2685–2689.

20. Murakami, M., Matsumoto, Y., Nakajima, K., Makino, T., Segawa, Y., Chikyow, T., Ahmet, P., Kawasaki, M. and Koinuma, H., 2001. Anatase TiO2 thin films grown on lattice-matched LaAlO3 substrate by laser molecular-beam epitaxy. *Applied Physics Letters*, 78(18), pp.2664–2666.

21. Karuppuchamy, S., Amalnerkar, D.P., Yamaguchi, K., Yoshida, T., Sugiura, T. and Minoura, H., 2001. Cathodic electrodeposition of TiO2 thin films for dye-sensitized photoelectrochemical applications. *Chemistry Letters*, 30(1), pp.78–79.

22. Yu, J.G., Yu, H.G., Cheng, B., Zhao, X.J., Yu, J.C. and Ho, W.K., 2003. The effect of calcination temperature on the surface microstructure and photocatalytic activity of TiO2 thin films prepared by liquid phase deposition. *The Journal of Physical Chemistry B*, 107(50), pp.13871–13879.

23. Aarik, J., Karlis, J., Mändar, H., Uustare, T. and Sammelselg, V., 2001. Influence of structure development on atomic layer deposition of TiO2 thin films. *Applied Surface Science*, 181(3–4), pp.339–348.

24. Shi, Y.J., Zhang, R.J., Zheng, H., Li, D.H., Wei, W., Chen, X., Sun, Y., Wei, Y.F., Lu, H.L., Dai, N. and Chen, L.Y., 2017. Optical constants and bandgap evolution with phase transition in sub-20-nm-thick TiO2 films prepared by ALD. *Nanoscale Research Letters*, 12(1), p.243.

25. Hasan, M.M., Haseeb, A.S.M.A., Saidur, R. and Masjuki, H.H., 2008. Effects of annealing treatment on optical properties of anatase TiO2 thin films. *International Journal of Chemical and Biological Engineering*, 1(2), pp.92–96.

26. Yoo, D., Kim, I., Kim, S., Hahn, C.H., Lee, C. and Cho, S., 2007. Effects of annealing temperature and method on structural and optical properties of TiO2 films prepared by RF magnetron sputtering at room temperature. *Applied Surface Science*, 253(8), pp.3888–3892.

27. Shao, P., Tian, J., Zhao, Z., Shi, W., Gao, S. and Cui, F., 2015. Amorphous TiO2 doped with carbon for visible light photodegradation of rhodamine B and 4-chlorophenol. *Applied Surface Science*, 324, pp.35–43.

28. Zhang, M., Lin, G., Dong, C. and Wen, L., 2007. Amorphous TiO2 films with high refractive index deposited by pulsed bias arc ion plating. *Surface and Coatings Technology*, 201(16–17), pp.7252–7258.

29. Guang-Lei, T., Hong-Bo, H. and Jian-Da, S., 2005. Effect of microstructure of TiO2 thin films on optical bandgap energy. *Chinese Physics Letters*, 22(7), p.1787.
30. Cras, J.J., Rowe-Taitt, C.A., Nivens, D.A. and Ligler, F.S., 1999. Comparison of chemical cleaning methods of glass in preparation for silanization. *Biosensors and Bioelectronics*, 14(8–9), pp.683–688.
31. Tauc J, Menth A, 1972. States in the gap. *J Non-Cryst Solids*, 8, pp.569–585.
32. Liu, X., Yin, J., Liu, Z.G., Yin, X.B., Chen, G.X. and Wang, M., 2001. Structural characterization of TiO2 thin films prepared by pulsed laser deposition on GaAs (1 0 0) substrates. *Applied Surface Science*, 174(1), pp.35–39.
33. Kajitvichyanukul, P., Ananpattarachai, J. and Pongpom, S., 2005. Sol–gel preparation and properties study of TiO2 thin film for photocatalytic reduction of chromium (VI) in photocatalysis process. *Science and Technology of Advanced Materials*, 6(3–4), p.352.
34. Naceur, J.B., Gaidi, M., Bousbih, F., Mechiakh, R. and Chtourou, R., 2012. Annealing effects on microstructural and optical properties of nanostructured-TiO2 thin films prepared by sol–gel technique. *Current Applied Physics*, 12(2), pp.422–428.
35. Di Paola, A., Bellardita, M. and Palmisano, L., 2013. Brookite, the least known TiO$_2$ photocatalyst. *Catalysts*, 3(1), pp.36–73.
36. İcli, K.C., Güllü, H.H. and Hosseini, A., 2013. Preparation and characterization of porous TiO2 thin films by sol-gel method for extremely thin absorber-ETA solar cell applications.
37. Ayieko, C.O., Musembi, R.J., Waita, S.M., Aduda, B.O. and Jain, P.K., 2012. Structural and optical characterization of nitrogen-doped TiO2 thin films deposited by spray pyrolysis on fluorine doped tin oxide (FTO) coated glass slides. *International Journal of Energy Engineering*, 2(3), pp.67–72.
38. Rantala, J.T. and Kärkkäinen, A.H.O., 2003. Optical properties of spin-on deposited low temperature titanium oxide thin films. *Optics Express*, 11(12), pp.1406–1410.
39. Astinchap, B., Moradian, R. and Gholami, K., 2017. Effect of sputtering power on optical properties of prepared TiO2 thin films by thermal oxidation of sputtered Ti layers. *Materials Science in Semiconductor Processing*, 63, pp.169–175.
40. Aarik, J., Aidla, A., Kiisler, A.A., Uustare, T. and Sammelselg, V., 1997. Effect of crystal structure on optical properties of TiO2 films grown by atomic layer deposition. *Thin Solid Films*, 305(1–2), pp.270–273.
41. Nair, P.B., Justinvictor, V.B., Daniel, G.P., Joy, K., Raju, K.J., Kumar, D.D. and Thomas, P.V., 2014. Optical parameters induced by phase transformation in RF magnetron sputtered TiO2 nanostructured thin films. *Progress in Natural Science: Materials International*, 24(3), pp.218–225.
42. Yoldas, B.E., 1980. Investigations of porous oxides as an antireflective coating for glass surfaces. *Applied Optics*, 19(9), pp.1425–1429.
43. Zhang, J., Zhou, P., Liu, J. and Yu, J., 2014. New understanding of the difference of photocatalytic activity among anatase, rutile and brookite TiO2. *Physical Chemistry Chemical Physics*, 16(38), pp.20382–20386.
44. Slav, A., 2011. Optical characterization of TiO2-Ge nanocomposite films obtained by reactive magnetron sputtering. *Digest Journal of Nanomaterials and Biostructures*, 6(3), pp.915–920.
45. Naceur, J.B., Gaidi, M., Bousbih, F., Mechiakh, R. and Chtourou, R., 2012. Annealing effects on microstructural and optical properties of nanostructured-TiO$_2$ thin films prepared by sol–gel technique. *Current Applied Physics*, 12(2), pp.422–428.
46. Yu, J., Zhao, X. and Zhao, Q., 2001. Photocatalytic activity of nanometer TiO$_2$ thin films prepared by the sol–gel method. *Materials Chemistry and Physics*, 69(1–3), pp.25–29.

3 Metamaterials and Metasurfaces for High-Frequency Applications

Somak Bhattacharyya

CONTENTS

3.1 INTRODUCTION TO METAMATERIALS

Electromagnetic metamaterials, also popularly known as 'metamaterials', are structural material exhibiting artificial electromagnetic properties, in general, not exhibited by nature under electromagnetic excitation [1]. The phrase 'meta' implies *beyond*, and hence the term 'metamaterials' attributes to the composite materials with structures where unusual electromagnetic properties have been observed which cannot be found in naturally available materials when the electromagnetic wave is excited under certain conditions [2]. The electromagnetic properties associated with the metamaterials can be considered to possess effective medium parameters instead of the bulk parameters. These effective parameters can vary over an extremely wide range, ranging from extremely high value to a very low value even with a negative sign which has not been feasible in any naturally available materials. As any construction can be electromagnetically characterized in terms of permeability and permittivity, the metamaterials are characterized as 'effective' media possessing effective permittivity and effective permeability values. The variations in the values of effective

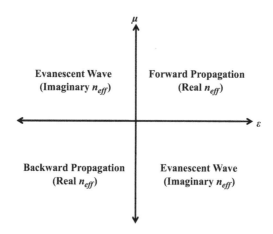

FIGURE 3.1   Permittivity-permeability co-ordinate system.

permeability and permittivity lead to the formation of a co-ordinate plane consisting of effective permeability and effective permittivity as provided in Figure 3.1.

In a conventional material, the bulk permittivity and permeability lies on the first quadrant. The extreme high values of either or both of them cannot be achieved in a conventional material (like permittivity of 2000 and permeability of 300). However, these can be realizable in the case of metamaterial-based structures under required excitation. The effective properties of the electromagnetic parameters come into picture in the case of metamaterials as the metamaterial-based structure consists of an array of periodic unit cells where the dimension of the unit cell itself (a) has been designed much smaller than the incident wavelength in the guided medium (λg). Thus, the incident electromagnetic wave interacts with the 'effective' medium instead of the bulk medium consisting of a single unit cell. Materials whose effective constitutive electromagnetic properties lie in the first quadrant are also known as Double Positive (DPS) materials.

In the second quadrant, effective permeability is positive, whereas effective permittivity becomes negative. On the contrary, effective permeability is negative while effective permittivity is positive. Materials belonging to either second or fourth quadrant are known as Single NeGative (SNG) material. Specifically, materials in the second quadrant are of the class Epsilon NeGatine (ENG) while the materials lying in the fourth quadrant are Mu NeGative (MNG) in nature.

The 'effective' refractive index n_{eff} of the medium can be defined as Equation (3.1) where ε_{eff} and μ_{eff} are the effective permittivity and effective permeability, respectively,

$$n_{eff} = \sqrt{\varepsilon_{eff} \mu_{eff}} \qquad (3.1)$$

Further, it is to be noted that the effective constitutive electromagnetic parameters are expressed as provided in Equations (3.2a) and (3.2b), respectively,

$$-\varepsilon_{eff} = |\varepsilon_{eff}| e^{j.\pi} \qquad (3.2a)$$

$$-\mu_{eff} = \left|\mu_{eff}\right|e^{j.\pi} \tag{3.2b}$$

For second and fourth quadrant, n_{eff} can be derived and found to have an imaginary value. This leads to an evanescent nature of the electromagnetic wave as the imaginary value of n_{eff} results in the absorption in the medium:

$$n_{eff} = \sqrt{-|\varepsilon||\mu|}e^{j\frac{\pi}{2}} \tag{3.3}$$

It was experimentally found that an array of periodically arranged metallic cylinders provide negative effective permittivity provided electric field of the incident electromagnetic wave becomes parallel to the axis of the cylinder [3]. If the metallic cylinders are arranged with a periodicity of p provided p lies in sub-wavelength region ($p \leq \lambda/4$), and the incident electric field is oriented along z-direction as evident from Figure 3.2(a), the effective permittivity can be defined as given in Equation (3.4a) while the plasma frequency ω_p is defined in Equation (3.4b) where r is the radius of the metallic rod and σ is the conductivity of the metal. For an aluminum array ($\sigma = 3.4 \times 10^7 S/m$) of cylinders with radii 5 μm and periodicity of 5 mm, the frequency response of the complex effective permittivity are illustrated in Figure 3.2(b).

$$\varepsilon_{eff} = 1 - \frac{\omega_p^2}{\omega\left(\omega + j\dfrac{\varepsilon_0 p^2 \omega_p^2}{\pi r^2 \sigma}\right)} \tag{3.4a}$$

$$\omega_p^2 = \frac{2\pi c^2}{p^2 \ln\left(\dfrac{p}{r}\right)} \tag{3.4b}$$

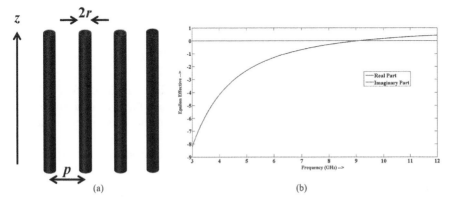

(a)   (b)

FIGURE 3.2 (a) Experimental set-up of realization of negative permittivity of an array of periodic metallic cylinders where the electric field is oriented along z-direction and (b) the corresponding frequency response of the retrieved effective permittivity.

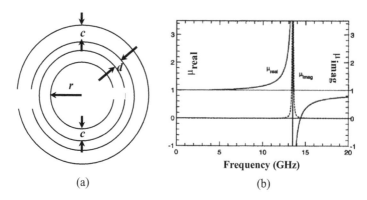

(a)   (b)

FIGURE 3.3 (a) Experimental set-up of realization of negative permeability of an array of periodic split-ring resonators where the magnetic field is incident normal to the plane and (b) the corresponding frequency response of the retrieved effective permeability.

Similarly, under the excitation of the magnetic field directed perpendicularly to the plane of an array of split-ring resonators results in the negative permeability over a narrow frequency range [4]. If the magnetic field is oriented along the perpendicular direction in the plane of the metallic split-ring resonators, the effective permittivity can be given as shown in Equation (3.5) where the geometrical dimensions are shown in Figure 3.3(a). It can be observed that the application of the magnetic field results in the generation of circulating current within the metallic region, thereby forming an effective inductance. On the other hand, the gap between the two rings as well as the mutual separation forms equivalent capacitance; thereby generating a LC resonance. This, in turn, yields a resonance over a narrow frequency band due to the high selectivity:

$$\mu_{eff} = 1 - \frac{\dfrac{\pi r^2}{a^2}}{1 + j\dfrac{2l\sigma_1}{\omega r \mu_0} - \dfrac{3lc_0^2}{\pi \omega^2 \ln\left(\dfrac{2c}{d}r^3\right)}} \tag{3.5}$$

The effective permeability is calculated and shown in Figure 3.3(b) for $c = 1$ mm, $a = 10$ mm, $d = 0.1$ mm, $r = l = 2$ mm (separate between the two consecutive split-ring resonators) and $\sigma_1 = 200$ Ω/m. The resonating nature has been observed close to 13.5 GHz.

When the material properties are lying in the third quadrant viz., for negative values of both ε_{eff} and μ_{eff}, the effective refractive index can be derived in Equation (3.6) using Equations (3.1) and (2):

$$n_{eff} = \sqrt{(-\varepsilon)(-\mu)} = \sqrt{|\varepsilon||\mu|e^{j.2\pi}} = \sqrt{|\varepsilon||\mu|e^{j.\pi}} = -\sqrt{|\varepsilon||\mu|} \tag{3.6}$$

The negative value of n_{eff} suggests that the propagation of the electromagnetic wave takes place. This concept was first mathematically proposed by Vaselago [5]. The energy propagates in the forward direction while the wave is considered to be propagating along the reverse direction. This means the Poynting vector $\vec{S}$ and wave vector $\vec{k}$ are anti-parallel in nature. The direction of $\vec{S}$ is determined by the direction of the cross-product $\vec{E} \times \vec{H}$ while $\vec{k}$ is oriented along $-\vec{E} \times \vec{H}$. Materials lying in the third quadrant are more commonly known as Double NeGative (DNG) materials as both ε_{eff} and μ_{eff} possess negative values.

Initially, the conception was popular among the researchers that metamaterials lie in the fourth quadrant only as the simultaneous realization of negative values of both ε_{eff} and μ_{eff} are not possible in practice. However, the metamaterials can offer any values of ε_{eff} and μ_{eff} as the structure is excited under the appropriate electromagnetic field. So, it can be considered that all DNG materials are necessarily metamaterials, but the reverse is not true.

The conventional Snell's law gets modified when the wave propagates from one DPS medium to a DNG medium. If the refractive indices of the DPS and DNG media are n_{DPS} and n_{DNG}, respectively, the Snell's law is given by Equation (3.7) where θ_i and θ_r are the respective angles of incidence and refraction:

$$n_{DPS} \sin \theta_i = n_{DNG} \sin \theta_r \qquad (3.7)$$

Since n_{DNG} assumes negative value, so to satisfy the positive sign of the right-hand side of Equation (3.7), θ_r must be negative. This signifies that the wave in the DNG medium flows away from the normal in the other direction. This has been clearly explained in Figure 3.4. When both the media are DPS in nature, Snell's law has been followed using the conventional case as shown in Figure 3.4(a). However, for the later one, the wave propagates from one DPS medium to a DNG one as depicted from Figure 3.4(b). This leads to the fact that the wave diverges its direction from the interface AB toward the other end. This can be treated as reversal of Snell's law.

If a DPS medium of refractive index 1.3 is inserted in an air medium with refractive index unity, it is observed that the incident wave from a point source gets diverged following conventional Snell's law. On the contrary, the introduction of a DNG medium with refractive index -1 in an air medium results in focusing of the beam in

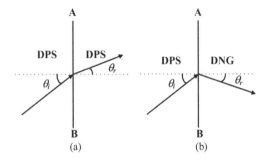

FIGURE 3.4   Snell's law of refraction at the interface AB for (a) two DPS media and (b) one DPS and one DNG medium.

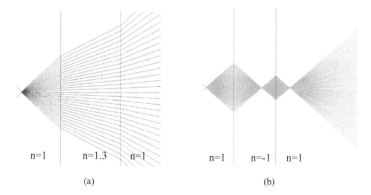

n=1   n=1.3   n=1   n=1   n=-1   n=1

(a)   (b)

FIGURE 3.5 (a) Divergence and (b) focusing of an electromagnetic beam in the interface.

the LH media as well as in the air medium in the propagating direction following Snell's law as observed from Figure 3.5(b)[6].

The first experimental realization of the DNG medium was carried out by Smith *et al.* [7], where he had used both ENG and MNG medium by creating a three-dimensional plane consisting of an array of metallic cylinders and split-ring resonator structures embedded on the dielectric. Initially, the metallic lines were not inserted and only the split-ring resonators were aligned and excited by a perpendicularly applied magnetic field, thereby creating an MNG medium and subsequently no transmission of wave takes place as illustrated by the dip close to 5 GHz. Simultaneously, ENG medium has been designed at the same frequency (5 GHz) by optimizing the geometrical dimensions of the metallic rods. This, in turn, results in formation of a DNG medium. The transmission of the electromagnetic wave through the DNG medium takes place at 5 GHz.

Consequently, the first experimental realization of the negative refractive index was reported by Shelby *et al.* [8]. A microwave detector has been put in the receiver side while the power has been launched from the sample side as illustrated in the schematic shown in Figure 3.6. When the sample is a DPS medium consisting of Teflon, power has been received at a positive value of angle (θ), obeying Snell's law. Later, the Teflon sample has been replaced by an LHM one to receive power at a negative angle, thereby following reversal of Snell's law.

3.2 METASURFACES

Metasurfaces are considered as the two-dimensional representation of a metamaterial structure in which the third dimension viz., the thickness is limited to the

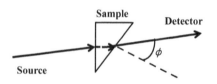

Sample   Detector

ϕ

Source

FIGURE 3.6 Schematic of the first experimental realization of the negative refractive index.

sub-wavelength dimension. The structure is confined in the two-dimensional plane, consisting of an array with sub-wavelength periodicity. A frequency selective surface (FSS) is also a 2D configuration comprising the metallic patches arranged in a periodic manner [9]. However, when the periodicity falls below the sub-wavelength dimension, it can be termed as metasurface [10]. The periodicity of the unit cell is considered to be lesser than the quarter-wavelength of the incident radiation. This is known as the effective homogeneity limit [1]. The incident beam under this condition can "visualize" the complete surface instead of interacting with a single unit cell. Several geometrical parameters involved in the design of the metasurface structure under the excitation of incident electromagnetic wave generate effective inductance and capacitance to the structure; thereby forming a high-Q resonator structure.

The metasurface has opened out the horizons of electromagnetic engineering covering a broad electromagnetic spectrum in recent years including ultra-thin absorbers, polarization-converting structures, antennas, filters, etc.

3.3 METASURFACE ABSORBERS

In absorbing applications, initially Salisbury screen was proposed [9] to absorb over a single frequency using destructive interference theory as shown in Figure 3.7. The incident wave striking on an interface comprising a thin resistive sheet is reflected back partly as the wave '1' while propagating as the wave '2'. A thin metal sheet is kept at a distance $\lambda/4$ away from the resistive sheet. The wave '2' has now struck the metal interface and reflected back completely as the wave '3' and incident again at the resistive interface. The wave '4' is again coming out of the interface as evident from Figure 3.8. It is noted that the wave '1' and '4' are parallel to each other while there is an additional path difference of $\lambda/2$ (= $\lambda/4$ due to wave '2' + $\lambda/4$ due to wave '3') between the two emergent waves, which in turn results in a phase change of 180°. Thus, the wave corresponding to the incident wavelength λ does not come out of the system or in other words, gets absorbed within the system.

The concept of Salisbury screen was extended to develop Jaumann layer as well as Dalenbach layer, where a number of resistive sheets are used to realize absorption of multiple bands [11]. Consequently, other FSS-based absorbers have been reported too [9,11–13]. With the advent of time, the electromagnetic absorbers have also been constructed implementing lossy dielectric and magnetic materials [14,15]. Lossy

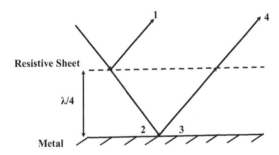

FIGURE 3.7 Schematic diagram of a Salisbury screen.

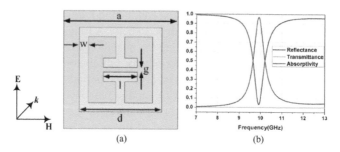

FIGURE 3.8 (a) Top view of the unit cell of the ELC structure with complete metal backing with (b) absorptivity responses (Reproduced from [26]).

dielectric absorbers have been fabricated using carbon black [16] while lossy magnetic absorbers are made up of ferrite-based composites [17,18]. But, the bottleneck of these structures are their bulky nature. Over a broader frequency range, carbon foam-based pyramidal absorbers [9] have been reported finding applications in anechoic chamber [19]. These pyramidal absorbers are also bulky in nature as their thicknesses are in the order of wavelength (~ λ). The large thickness accompanied by the bulky nature of the absorbers limits their applications for all practical considerations like thermal detector, RFID system, radar cross-section reduction in stealth technology etc. [20–23]. This leads to the research of the absorbers using metasurface structures.

When the wave is incident on an interface, a part of the wave is getting reflected while the rest is transmitted into the second medium. When the reflected as well as transmitted wave undergoes minimization simultaneously, the wave can be considered as absorbed in the second medium, provided there is absence of any scattering effect due to the surface roughness (surface roughness <<λ). The absorption in the medium has been characterized by the absorptivity or the absorbed power $A(f)$ defined in Equation (3.8). Here, the reflectivity $R(f)$ and the transmittance $T(f)$ are defined in Equations (3.9a) and (3.9b), respectively. Both the $R(f)$ and $T(f)$ consist of both cross-polarized and co-polarized components:

$$A(f) = 1 - |R(f)| - |T(f)| \tag{3.8}$$

$$R(f) = |r_{co}(f)|^2 + |r_{cross}(f)|^2 \tag{3.9a}$$

$$T(f) = |t_{co}(f)|^2 + |t_{cross}(f)|^2 \tag{3.9b}$$

In 2006, magnetically excited split-ring resonator-based arrangement has been proposed to achieve absorption at 2 GHz [24]. The magnetic field is incident perpendicularly to the axis of the split rings. The frequency responses of insertion loss as well as insertion loss show that both are minimized at 2 GHz, thereby providing absorption at 2 GHz. However, this arrangement is very difficult to realize in practice.

A couple of years later, the first metasurface-based structure has been proposed [25] where the ultra-thin nature was realized implementing an electric field driven LC (ELC) resonating structure. The configuration was designed over FR-4 dielectric where both the top and bottom sides of the unit cells were designed using metallic patches. Both the reflectivity and transmittance was minimized near 11.7 GHz implying that absorption occurs at 11.7 GHz. This structure is considered as the first experimentally realized metasurface absorber structure.

The transmittance through the structure can be further minimized by incorporating complete metallic ground instead of the partial one [26]. The top view of the unit cell along with the direction of the incident electromagnetic field is shown in Figure 3.8(a). The corresponding responses are shown in Figure 3.8(b) where it has been observed clearly that the structure offers zero transmission. The reflectance becomes minimum at 9.92 GHz to realize absorption at the said frequency.

For a completely backed metallic structure, complete restriction of e.m wave transmission has been taken place and hence the Equation (3.8) has been reduced to Equation (3.10a) where the reflectance $R(f)$ can be described in the Equation (3.10b) for the microwave frequency region when there is no reflection due to cross-polarized effect:

$$A(f) = 1 - |R(f)| \qquad\qquad (3.10a)$$

$$R(f) = |S_{11}(f)|^{22} \qquad\qquad (3.10b)$$

When the electromagnetic wave is incident from air medium to an interface of another medium, it can be shown that the wave impedance $Z(f)$ can be expressed in Equation (3.11a), where η_0 is the free space impedance. For completely metal-backed metasurface-based structure, it can be reduced further to the form given in Equation (3.11b)[27]:

$$Z(f) = \eta_0 \sqrt{\frac{(1+S_{11})^2 - S_{21}^2}{(1-S_{11})^2 - S_{21}^2}} \qquad\qquad (3.11a)$$

$$Z(f) = \eta_0 \frac{1+S_{11}}{1-S_{11}} \qquad\qquad (3.11b)$$

For a perfect absorber, $S_{11} = 0$, i.e. the minimization of the reflection coefficient from the top side of the sample takes place. This implies that the wave and the intrinsic impedances should be equal, i.e. $Z(f) = \eta_0$. Further, the wave impedance can be written as given in Equation (3.12a) while the intrinsic impedance is expressed in Equation (3.12b). The wave impedance has been considered to be consisting of an efficient medium with permeability and permittivity of μ_{eff} and ε_{eff}, respectively,

$$Z(f) = \sqrt{\frac{\mu_{eff}\mu_0}{\varepsilon_{eff}\varepsilon_0}} \qquad\qquad (3.12a)$$

$$\eta_0 = \sqrt{\frac{\mu_0}{\varepsilon_0}} \qquad (3.12b)$$

Since both of them are equal for zero reflection from the top surface, the effective permeability and permittivity are equal to each other as evident from Equation (3.13). This signifies that the real parts of effective permeability and permittivity are equal as well as imaginary parts of the respective effective parameters are also equal at the frequency of absorption:

$$\frac{\mu_{eff}}{\varepsilon_{eff}} = 1$$

$$\Rightarrow \mu_{eff} = \varepsilon_{eff} \qquad (3.13)$$

$$\Rightarrow Re\left(\mu_{eff}\right) = Re\left(\varepsilon_{eff}\right) \& \& Im\left(\mu_{eff}\right) = Im\left(\varepsilon_{eff}\right)$$

3.3.1 SINGLE-BAND METASURFACE ABSORBER

The top view of the unit cell of a single-band metasurface absorber is shown in Figure 3.9(a) where the completely copper laminated bottom side [28]. The metallic patches of the unit cell are made of 0.035 mm thin copper printed on FR-4 dielectric possessing 1 mm thickness. The geometrical dimensions are considered as $d = 3.6$ mm, $a = 5$ mm, $w = 0.4$ mm and $g = 0.2$ mm. The absorptivity response of the structure is shown in Figure 3.9(b) where absorption has been achieved at 7.46 GHz.

From the study of the surface currents at the top and bottom surfaces at 7.46 GHz provided in Figure 3.10(a) and Figure 3.10(b), respectively, it can be seen that the directions are anti-parallel to each other, creating a circulating current loop within the dielectric. The currents are further found to be concentrated on the metallic region perpendicular to the magnetic field. The electric field distribution study of the metasurface structure implies that the electric field is fairly strong near the gap of the unit cell at 7.46 GHz as seen from Figure 3.10(c). These two phenomena generate magnetic and electric excitations simultaneously at 7.46 GHz, thereby creating strong electromagnetic absorption at 7.46 GHz.

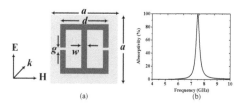

(a)   (b)

FIGURE 3.9 (a) Top view of the unit cell of the single-band absorber along with (b) absorptivity response.

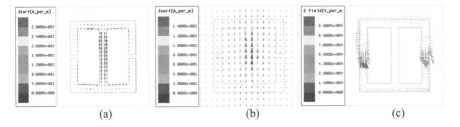

FIGURE 3.10 Surface current distributions at the (a) top and (b) bottom surfaces of the single-band absorber structure along with (c) electric field distribution at the top metallic side at 7.46 GHz.

The absorption within the metasurface-based structure takes place owing to the presence of both ohmic loss and the dielectric loss. The absorptivity responses for different loss tangent values are provided in Figure 3.11(a). For zero loss tangent, the absorptivity falls to 40% at the same frequency of 7.46 GHz. This is caused because of copper ohmic loss. From Figure 3.11(b), it can be seen that the peak absorptivity enhances rapidly with initial increase of tanδ, reaches maximum for loss tangent of 0.023 and thereafter, gradually decreases. The peak absorptivity again falls ~40% for loss tangent of 0.2. It has also been observed from Figure 3.14(b) that the peak absorptivity is high ($\geq$ 90%) for the range $0.01 \leq \tan\delta \leq 0.04$, while the other geometrical parameters remaining constant. For the higher values of loss tangent, the dipoles present in the system (in dielectric) undergo enough friction to orient along the direction of applied electric field. With further increase of the loss tangent, the applied electric field fails to provide the proper excitation (because of high friction) to achieve resonance within the structure [29].

The refractive index n corresponding to the effective medium can be computed employing Equation (3.14), d being the distance propagated by the incident electromagnetic wave and k_0 is the free space wave vector while m is an integer [27,30]:

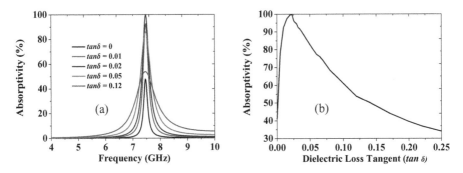

FIGURE 3.11 (a) Absorptivity response and (b) absorptivity variation with dielectric loss tangent of the FR-4 substrate for the single-band absorber structure.

$$n = \frac{1}{k_0 d}\left\{\left[Im\left[\ln\left(e^{jnk_0 d}\right)\right] + 2m\pi\right] - j\,Re\left[\ln\left(e^{jnk_0 d}\right)\right]\right\}$$

$$e^{jnk_0 d} = \frac{S_{21}}{1 - S_{11}\dfrac{z-1}{z+1}} \tag{3.14}$$

From Equations (3.11a) and (3.14), the effective permittivity and permeability are computed and provided in Equations (3.15a) and (3.15b), respectively,

$$\varepsilon_{eff} = \frac{n}{Z} \tag{3.15a}$$

$$\mu_{eff} = nZ \tag{3.15b}$$

For the single-band structure shown in Figure 3.9, there is complete metal backing and hence complete restriction of electromagnetic wave transmission takes place. However, the equation containing computation of refractive index consists of the term involving the transmission coefficient. To realize a finite transmission, small cuts at the metallic bottom surface are implemented maintaining the top surface unaltered so that the reflection minima should occur at the same frequency. The top and bottom views of the unit cell of the single-band absorber are shown in Figure 3.12(a) and Figure 3.12(b), respectively. The simulated responses are shown in Figure 3.12(c) from which it has been observed clearly that the reflection minima occurs at 7.50 GHz, very close to the original frequency of the reflection minima while negligibly small finite transmission takes place through the back surface.

The effective electromagnetic parameters are retrieved and shown in Figure 3.13 where the real parts of effective permittivity and permeability are compared in Figure 3.13(a) while the imaginary parts of the same are shown in Figure 3.13(b). It can be observed that the real and imaginary values of the retrieved parameters are nearly equal at 7.50 GHz evident from Table 3.1, thereby satisfying the condition of reflection minima.

The dimension of the basic unit cell of the single-band absorber discussed in Figure 3.9(a) is scaled to change the frequency of absorption. All the geometrical

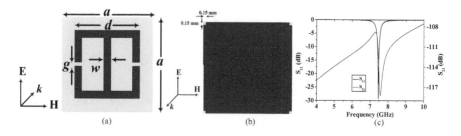

(a)   (b)   (c)

FIGURE 3.12   (a) Top view and (b) bottom view of the structure with cuts and (c) simulated response.

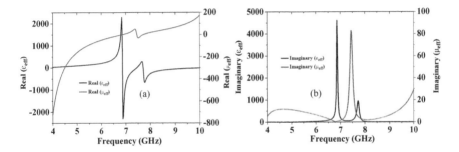

FIGURE 3.13 Comparisons of (a) real and (b) imaginary parts of the constitutive parameters.

TABLE 3.1

Comparison of the Electromagnetic Parameters Retrieved for the Single-Band Absorber

Frequency of Absorption (GHz)	Real (ε_{eff})	Real (μ_{eff})	Imaginary (ε_{eff})	Imaginary (μ_{eff})
7.50	1.5	1.1	80.7	81.5

---

dimensions (l, w, g, d) are scaled up by the factor k, keeping the periodicity of the unit cell size a constant. The scaling factor greater than 1 results in decrease of absorption frequency while the same less than 1 yields increase of absorption frequency e.g., scale factors of 1.1 and 0.9 generate absorption corresponding to frequencies 6.84 GHz and 8.06 GHz, respectively.

3.3.2 DUAL-BAND METASURFACE ABSORBER

The unit cell of the single-band absorber discussed in Figure 3.9(a) has been scaled up by two different factors so that the design can exhibit absorption at two distinct frequencies. This process has been implemented in the dual-band absorber whose unit cell is shown in Figure 3.14(a). The sub-cells '1' and '2' are identical in nature and placed diagonally to constitute the complete unit cell [31]. The sub-cell '1' has been multiplied by a factor of 1.2 while the sub-cell '2' has been scaled up by a factor of 1.3. The designed dual-band absorber demonstrates absorption at two dissimilar frequencies 5.64 GHz and 6.22 GHz as seen from Figure 3.14(b). The surface current distributions at the bottom and the top surfaces at 5.64 GHz and 6.22 GHz are shown in Figure 3.15. It is clear from Figure 3.15 that the surface currents are concentrated at the sub-cell '1' at 6.22 GHz while the sub-cell '2' is responsible for absorption at 5.64 GHz. Moreover, the anti-parallel surface currents at the bottom and top surfaces at the two frequencies support magnetic excitation.

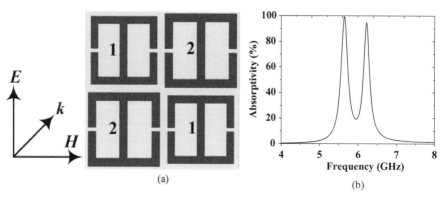

(a)

(b)

FIGURE 3.14   (a) Design of the unit cell of a dual-band metasurface absorber along with (b) absorptivity response (Reproduced from [31]).

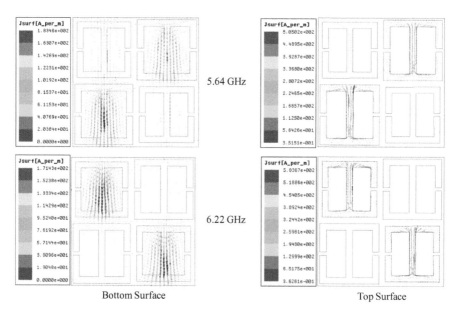

FIGURE 3.15   Surface currents at the bottom and top metallic surfaces at 5.64 GHz and 6.22 GHz (Reproduced from [31]).

The dual-band metasurface structure has been fabricated with dimension 2.4 cm × 2.4 cm as illustrated in Figure 3.16(a). The enlarged view of the fabricated sample is shown in Figure 3.16(b). The complete set-up to determine the reflection coefficient of the sample is provided in Figure 3.16(c). The sample has been designed on the top of a dielectric layer where the other side of the layer is completely metal-backed. The sample has been tested in the anechoic chamber by a pair of identical horn antennas, connected to VNA. The electromagnetic wave is launched using one of the antennas acing as

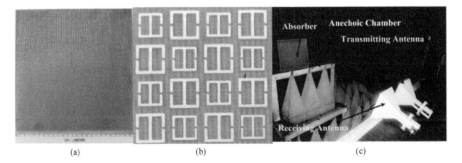

(a)   (b)   (c)

FIGURE 3.16 (a) Prototype of the fabricated dual-band metasurface with (b) enlarged view and (c) experimental set-up within the anechoic chamber.

transmitting antenna while the other one is behaving as the receiving antenna which receives the electromagnetic wave reflected from the sample.

The measured responses from the sample are recorded in the VNA and shown in Figure 3.17. Initially, the absorber sample has been removed and the background reflection has been measured as shown in Figure 3.17(a). Then, a larger metallic sheet made of copper with 1 m × 1 m dimension has been taken to measure the reflection from the sample. The large size of the sample has been considered to ensure that the electromagnetic beam launched from the antenna can illuminate the sample properly. Thereafter, a sample with the exact dimension of the absorber prototype (240 mm × 240 mm) has been considered. The reflection coefficients of both the larger and smaller metallic sheets are nearly identical as indicated in Figure 3.17(a). Then, the absorber sample has been illuminated with the beam and compared with the reflection coefficient of the copper sheet having identical dimension to that of the metal sheet as provided in Figure 3.17(b). The difference between the two provides the reflection coefficient from the absorber sample. This, in turn, generates the absorptivity response which has been compared with the simulated absorptivity response of the dual-band absorber as depicted in Figure 3.17(c). The measured and simulated absorptivity responses are in good agreement as seen from Figure 3.17(c).

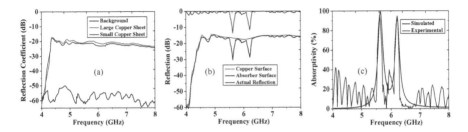

FIGURE 3.17 (a) Reflection coefficient of the sample from a large and small copper sheet with background response; (b) reflection coefficient of the absorber sample and the copper sample for actual reflection from the absorber and (c) comparison of measured and simulated absorptivity responses of the dual-band absorber prototype (Reproduced from [31]).

TABLE 3.2

Comparison of Retrieved Electromagnetic Parameters for the Dual-Band Absorber

Frequency of absorption (GHz)	Real (ε_{eff})	Real (μ_{eff})	Imaginary (ε_{eff})	Imaginary (μ_{eff})
5.64	−10.59	−6.57	62.56	50.39
6.22	−17.28	−9.23	44.61	60.35

The real and imaginary parts of the retrieved effective parameters of the dual-band absorber are provided in Table 3.2.

3.3.3 Triple-Band Metasurface Absorber

Dual-band absorber structure has also been constructed by combining two separate unit cells as shown in the unit cell design SHOWN in Figure 3.21(a). The unit cell has been designed on a copper-backed 1 mm thick FR-4 dielectric substrate using 0.035 mm thick copper lines. The geometrical dimensions of the unit cells are a = 7.5 mm, d = 6.2 mm, d_1 = 1.6 mm, l = 2.6 mm, w = 0.7 mm, w_1 = 0.35 mm, g = 0.35 mm and g_1 = 0.3 mm. When the incident electric field is oriented along x-direction as shown in Figure 3.18(a), it exhibits dual-band absorption at 5.46 GHz and 9.54 GHz illustrated in Figure 3.18(b)[32]. The orientation of the electric field along x-direction offers 0° polarization of the electromagnetic wave. Interestingly, when the incident electric field is orthogonally polarized (90° polarization angle), absorption has been achieved at 7.40 GHz as observed from Figure 3.18(c).

This idea has been implemented to design a structure exhibiting triple-band absorption. The unit cell of the design of the absorber is shown in Figure 3.19(a). It is noted that the sub-cells '1' and '2' are identical in nature; however, oriented orthogonally to each other [33]. The geometrical dimensions of the unit cells are a = 7.5 mm, d = 6.2 mm, d_1 = 1.6 mm, l = 2.6 mm, w = 0.7 mm, w_1 = 0.4 mm, g = 0.4 mm and g_1 = 0.3 mm designed over a FR-4

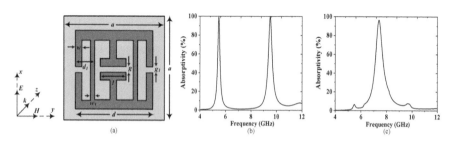

FIGURE 3.18 (a) Unit cell design of a dual-band absorber with absorptivity responses under polarization angles of (b) 0° and (c) 90° (Reproduced from [33]).

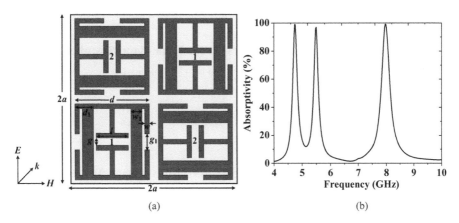

(a)   (b)

FIGURE 3.19 (a) Unit cell design of the triple-band metasurface absorber along with (b) absorptivity response (Reproduced from [33]).

dielectric with 1 mm thickness. The structure offers high absorption at 4.74 GHz, 5.50 GHz and 7.98 GHz as evident from the absorptivity response in Figure 3.19(b).

The triple-band absorber structure is four-fold symmetric around its axis offers polarization insensitivity of the structure. The set-up for the variation of the polarization angle of the incident electromagnetic wave is shown in Figure 3.20(a). It can be observed that the electric field has been confined in the xy-plane only; thereby making normal incidence along the z-direction. The change in the electric field by an angle ϕ with respect to the x-axis results in the rotation of the magnetic field too by an angle ϕ with respect to the y-axis. The simulated and measured responses under polarization angle variation are provided in Figure 3.20(b) and Figure 3.20(c), respectively. Experimentally, the sample itself has been rotated around its axis to generate various polarization angles. The absorptivity responses in both simulated and experimental ones remain invariant to the change of polarization angle.

The complex effective parameters of the triple-band absorber whose unit cell is shown in Figure 3.18(a) has been computed and listed in Table 3.3.

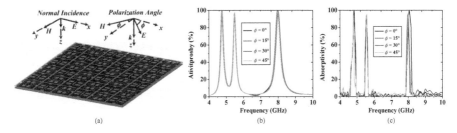

(a)   (b)   (c)

FIGURE 3.20 (a) Set-up of the polarization angle variation of the triple-band metasurface structure along with (b) simulated and (c) experimental responses (Reproduced from [33]).

TABLE 3.3

Computed Electromagnetic Parameters for the Triple-Band Absorber

Frequency of absorption (GHz)	Real (ε_{eff})	Real (μ_{eff})	Imaginary (ε_{eff})	Imaginary (μ_{eff})
4.74	−2.557	−3.478	18.44	17.97
5.50	−1.778	−1.738	22.53	11.52
7.98	−0.271	−0.446	12.45	10.72

3.3.4 Multiband and Bandwidth-Enhanced Metasurface Absorber

Several metasurface absorbers have been reported working over dual as well as triple frequency bands covering a wide range of electromagnetic spectrum [34–36]. Some of the reported structures are polarization-insensitive in nature too owing to the inherent four-fold symmetry of the basic unit cell design [37–50]. Furthermore, multiband (quad-band, penta-band *et al.*) metasurface absorbers with polarization-insensitive behavior have also been reported [51–58].

Subsequently, bandwidth-enhanced absorbers using metamaterials have also been reported using multiple scaling factors, varying geometrical dimensions, using multiple layers, etc. [59–72]. Moreover, broadband metasurface absorbers have been introduced using lumped elements where the loss can be significantly enhanced to increase the absorption bandwidth [73–77]. Most of the absorber applications in microwave domain are designed for stealth technological applications in FCC specified spectrum [78].

Elements such as varactor diode, liquid crystal and graphene have also been incorporated to realize the tunability in absorption bandwidth [79–85].

3.4 METASURFACE POLARIZATION-CONVERTING STRUCTURE

The asymmetric split-ring resonators are designed along with S-shaped resonator in the unit cell to realize polarization conversion in the terahertz range [86]. The unit cell of the construction is shown in Figure 3.21(a). The transmitted power along x-direction for an incident electromagnetic wave polarized along x-direction has been found to be significantly reduced over 0.48–1.04 THz. On the other hand, the transmitted power through the structure along y-direction has been found to be significantly enhanced in the same frequency band. The polarization conversion ratio (PCR) of the transmitted wave is given in Equation (3.16), where t_{xx} is the co-polarized transmission coefficient, i.e. the transmission coefficient along x-direction when the incident wave is polarized along x-direction too and where t_{yx} is the cross-polarized transmission coefficient, i.e. the transmission coefficient along y-direction when the incident wave is polarized along x-direction:

$$PCR = \frac{t_{yx}^2}{t_{xx}^2 + t_{yx}^2} \tag{3.16}$$

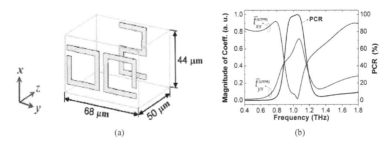

FIGURE 3.21 (a) Unit cell design of the polarization-converting structure employing metasurface with (b) frequency response. (Reproduced from [86]).

The *PCR* has been found to be maximum over the frequency range 0.48–1.04 THz as seen from Figure 3.21(b). This implies the plane of the polarization of the incident electromagnetic wave undergoes orthogonal rotation of plane of polarization after the transmission.

Consequently, a few transmittive type cross-polarization converters using metasurface designs have been proposed [87–90]. Recently, a non-dispersive transmittive type cross-polarization converter (CPC) has been reported in terahertz domain [91]. The perspective view of the unit cell of the CPC structure along with the orientation of electromagnetic fields is shown in Figure 3.22(a). The middle layer consists of a fractal design shown in Figure 3.22(b), while the top and bottom layers consist of the metal strips shown in Figure 3.22(c) and Figure 3.22(d), respectively. Gold has been used as metal in the design where Drude modelling has been incorporated and ZnSe has been chosen for good chemical properties [92,93]. The CPC structure offers minimization of co-polarized transmitted wave and maximization of cross-polarized transmitted wave simultaneously in the terahertz spectrum. The polarization conversion occurs between 10.25 THz to 22.7 THz covering a bandwidth of 12.45 THz, where the PCR value is greater than 90% (vide Figure 3.23(a) and Figure 3.23(b)). The same can be verified by observing the rotation of the electric field at 17 THz as evident from Figure 3.23(c).

Similarly, the reflective type cross-polarization converters have also been reported where the reflected cross-polarized component achieves maximum

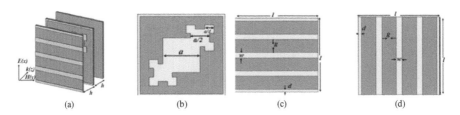

FIGURE 3.22 (a) 3D perspective view of cross-polarization converter (CPC) structure using metasurface along with incident electromagnetic wave directions where (b) middle layer as third-order modified T-square fractal, (c) vertical strips arranged in top layer and (d) bottom layer consisting of horizontal strips (Reproduced from [91]).

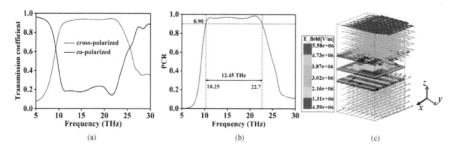

FIGURE 3.23 (a) Co- and cross-polarized transmission coefficient responses along with (b) PCR response of the CPC structure. (c) Electric field of incident and transmitted waves of the CPC structure at 17 THz (Reproduced from [91]).

over a certain band where the cross-polarized component is very small [94–103]. In this case, the bottom side of the unit cell is metal-backed like absorbing structures [94]. The co-polarized reflection coefficients are also minimized; however, cross-polarized reflection coefficients become maximized over the same band. The geometrical parameters of a typical unit cell of the reflective-type polarization converter shown in Figure 3.24(a) are a = 9 mm, r = 3.95 mm, w = 0.3 mm and ψ = 35.4°. The minimization of co-polarized reflection coefficient and maximization of cross-polarized reflection coefficient occur over a large microwave frequency band as provided in Figure 3.24(b). This leads to an enhancement of the PCR bandwidth as illustrated in Figure 3.24(c).

The tunability characteristics of the CPC structures can be achieved by incorporation of graphene structures [104–112].

3.5 METASURFACE ANTENNA

In modern days, the wireless communication systems become compact and low cost as the requirement of dual-band or multiband antennas have been enhanced due to frequency versatility and elimination capabilities of undesirable coupling between the antennas [113].

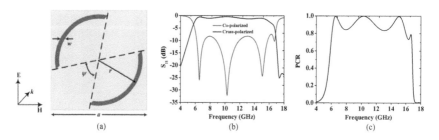

FIGURE 3.24 (a) Top view of the unit cell of the reflective type polarization converting structure with metasurface along with (b) co-polarized and cross-polarized reflection coefficients and (c) polarization conversion ratio (Reproduced from [94]).

There are distinct techniques proposed for miniaturization of antennas by incorporation of high dielectric substrates, resistive or reactive loading as well as modification of the electrical length [113,114]. Metamaterials take part an important role due to involvement of the latest methods, concepts, information or design styles for various microwave devices [114,115]. Various microwave devices such as artificial magnetic conductors (AMC), high impedance surfaces and electromagnetic bandgap (EBG) structures have been reported by utilizing the properties of MS [116–118]. Further, the surface properties of MS are characterized by inserting the lumped circuit elements or by reshaping the structural geometries or manipulating the electromagnetic fields [117,118]. Another important feature of the MS that they have found their applications in enhancement of antenna parameters due to their multi-functional properties [116–118]. Recently, antenna engineers focus on the development of MS antennas owing to low profile, wide bandwidth, and high gain features [116–120].

Metasurface antennas provide large antenna aperture and consequently high value of directivity and gain as comparable to conventional antenna arrays [113,116–118]. However, as the most important and unique feature of this new type of antennas, the antenna aperture and consequently antenna gain can incrementally change if the number of resonator elements are arranged in a periodic manner [121]. Due to this reason, it enhances the bandwidth of the antenna as well as gain. Also, metasurface is formed in terms of transmission-reflection-selective fashion that exhibits desired electromagnetic (EM) functionalities both in reflection and transmission mode controlled by the exciting polarizations [122–124].

For high-frequency range application with proper antenna miniaturization, microstrip patch antennas with fractal geometry and periodic metasurface at the bottom of the substrate have been reported [122,125,126]. The defected ground structure leads to the surface wave reduction which further reduces the cross-polarized radiation [127]. The introduction of the slot in the patch results in the enhancement of the co-polarized radiation characteristics in the broadside direction maintaining wide impedance bandwidth [128]. The metasurface design in the ground plane blocks the spurious surface waves, thereby reducing the antenna dimension designed on the same dielectric [120,121,129,130].

The top and bottom views of a dual-band antenna design using metasurface pattern at the ground plane has been shown in Figure 3.25(a) and Figure 3.25(b), respectively [116]. The geometrical dimensions of the design are $W_s = 28$ mm, $L_s = 28$ mm, $w_p = 16$ mm, $l_p = 16$ mm, $m = 4$ mm, $n = 2$ mm, $s = 3.6$ mm, $t = 1.5$ mm, $f_p = 6$ mm,

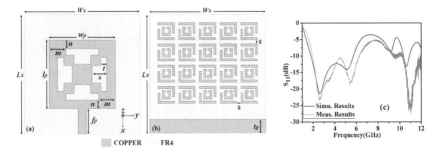

FIGURE 3.25 (a) Top and (b) bottom views of a dual-band metasurface-loaded antenna with (c) simulated and measured return losses. (Reproduced from [116]).

$g = 1$ mm and$t_g = 1$ mm. The antenna has been designed on a 1.6 mm thick FR-4 dielectric. The simulated result of the return loss of the antenna along with the experimental result is provided in Figure 3.25(c). It can be seen from Figure 3.25(c) that the antenna operates over dual wideband frequencies ranging from 1.8 to 5.7 GHz and 10.38 to 10.92 GHz with respective fractional bandwidths of 150% and 5.09%. It has been revealed from the measured results in Figure 3.25(c) that the antenna operates over 1.93–6.62 GHz and 10.32–11.90 GHz.

The polar plots at the several operating frequencies are depicted in Figure 3.26. It has been observed that the reasonable gain has been achieved at 10.92 GHz with a maximum gain of 7.16 dBi. Furthermore, the E-plane ($\varphi = 0°$) and H-plane ($\varphi = 90°$) radiation characteristics of the antenna are shown in Figure 3.27(a) and Figure 3.27(b), respectively. Nearly omnidirectional radiation characteristics have been observed at 2.60 GHz and 5.10 GHz. On the other hand, the directional radiation characteristics have been observed along E-plane at 10.92 GHz. The cross-polarization levels of both E and H-planes are reduced by 15 dB whereas the co-pol levels offer large value at a few frequencies [116].

The performances of a few recently reported metasurface-based antennas have been compared as shown in Table 3.4.

The use of metasurface as superstrate configuration in the fractal-shaped slotted patch has been very recently reported in which the bandwidth has been significantly enhanced where the high gain as well as good impedance matching has been successively achieved [141]. The MS layer in the superstrate is designed on FR4 dielectric having 0.8 mm thickness. The patch has also been designed on a separate 1.6 mm thick FR-4 dielectric. The conventional patch and the MS layer are separated by a Teflon dielectric (loss tangent = 0.001 and relative permittivity = 2.1) having 6 mm

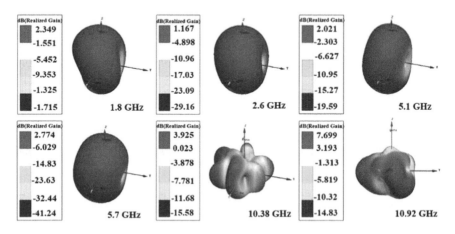

FIGURE 3.26 The three-dimensional polar plot of the dual-band metasurface-loaded antenna (Reproduced from [116]).

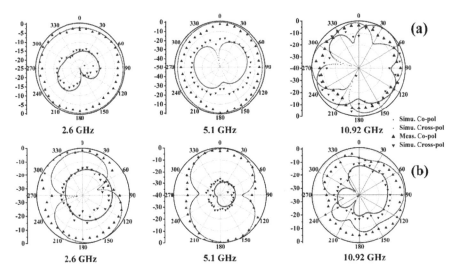

FIGURE 3.27 (a) E-plane and (b) H-plane radiation patterns at different frequencies of the dual-band antenna (Reproduced from [116]).

TABLE 3.4
Performance Comparison of a Few Recently Reported Metasurface-Based Antennas

Antenna literature	Dimension (mm²)	Operating frequency (GHz)	Fractional BW (%)	Gain (dBi)
Nasser et al. [114]	80 × 80	5.62	33.6	8.5
Li et al. [115]	63 × 108	3.75	101.3	7.32
Samantaray et al. [116]	28 × 28	2.6, 10.60	150, 5.09	1.05, 1.91, 7.16
Xu et al. [131]	88 × 88	0.72, 0.79, 0.92	2.3, 0.8, 18	3.7, 2, 4.8
Alibakhshi-Kenari et al. [132]	40 × 35	2.76	165.84	4.45
Zhai et al. [133]	104 × 104	2.4, 5.2	15.6, 9.3	7.2, 7.3
Cai et al. [134]	40 × 45	3.5	3.7	4.9
Li et al. [135]	32 × 28	7.7	46.37	7.2
Mitra et al. [136]	60 × 60	8.66	10.74	4.5
Huang et al. [137]	39.2 × 40	5.6	32	7.8
Zhu et al.[138]	120 × 120	2.3, 2.6	15.5, 17.6	5.8
Nasimuddin et al. [139]	60 × 88	4	34.56	7.4
Xu et al.[140]	43.5 × 43.5	3.11	1.61	4.15

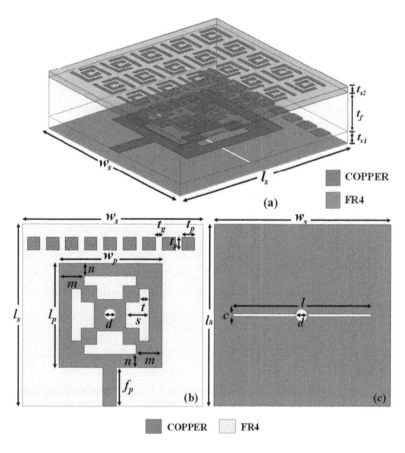

FIGURE 3.28 (a) Three-dimensional geometry of the proposed structure with (b) top and (c) rear views of conventional patch (Reproduced from [141]).

thickness. The optimized geometrical dimensions of conventional antenna are represented in Figure 3.28.

The top and the rear views of the fabricated prototype are provided in Figure 3.29(a) and Figure 3.29(b), respectively. The top view of the fabricated 5 × 5 order MS layer has been shown in Figure 3.29(c). The three-dimensional view of the sample is shown in Figure 3.29(d). The return loss of the antenna prototype has been tested by Anritsu Vector Network Analyzer (VNA) with model MS 2037C as shown in Figure 3.29(e). Figure 3.29(f) compares the experimentally measured return loss from which it can be depicted that the antenna offers wide impedance matching over the frequency range 9.24 GHz–11.25 GHz in X-band. The as-discussed superstrate antenna is of low cost as well as electrically thin in nature as it involves low-cost dielectric structures [118,142–145].

Metasurface-inspired reconfigurable antennas have been reported too employing tuning mechanism [146]. By employing metamaterial-loading in electrically small antennas, the reactive impedance has been nullified accordingly [147]. The suitable

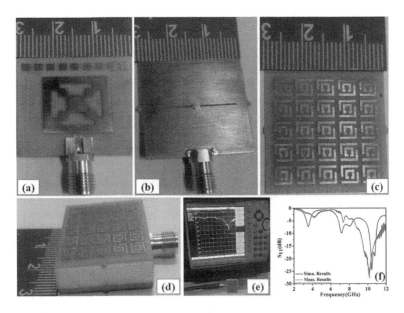

FIGURE 3.29 (a) Top view, (b) rear view, (c) 5 × 5 order metasurface, (d) three-dimensional view of the superstrate-based metasurface antenna, (e) S-parameter measurement setup using VNA and (f) simulated and measured S_{11} (dB) (Reproduced from [141]).

loading of the antenna by a metamaterial lens has also been used to enhance the gain of the antennas [148]. Antenna structure with metametrial-inspired designs have also been reported for applications like cognitive radio, terahertz sensing and wearable device etc. [149–153].

Metasurface structures have also found its applications like filter, phase-shifter, etc. which has not been covered here [154–157].

3.6 CONCLUSION

This chapter focuses on three applications of metamaterials and metasurfaces viz., absorber, polarization converter and antennas. Initially, a single-band metasurface-based absorbing structure in C-band has been discussed where the structure consists of an array of sub-wavelength unit cells. The physics behind the absorption has been discussed in detail with the help of electromagnetic resonance as well as the retrieval of electromagnetic parameters. The metallic and dielectric losses incurred in the absorber structure play a significant role for absorption of the incident electromagnetic wave. The absorption frequency in a metasurface design can be changed by varying the scaling factor of the dimension of the unit cell. Two different scaling factors of the dimension of the unit cell have been incorporated to form a super-cell to constitute a dual-band absorber. Thereafter, the experimental set-up for measuring the absorption of such a structure has been discussed step-by-step. Then, a single unit cell has been discussed exhibiting dual-band absorption for horizontal polarization while the same unit cell offers single-band absorption for vertical polarization of the incident electromagnetic wave. The super-cell constituting two separate unit cells having the identical dimensions but with orthogonal orientation. The super-cell of the metasurface design shows triple-band absorber. The retrieved effective permittivity and permeability support the theory of absorption in dual-band and triple-band metasurface absorber designs. The same concept has been used for the enhancement of the bandwidth. Further enhancement of bandwidth can be achieved by increasing the loss within the structure by embedding resistance in the metasurface unit cell.

Next, the metasurface for polarization conversion application has been designed. Several transmittive and reflective type polarization conversion structures have been discussed. In the case of transmittive type polarization converting devices, the co-polarized transmission coefficient gets minimized while the cross-polarized transmission coefficient gets maximized. This leads to the high polarization conversion ratio so that the polarization of the incident wave gets converted into the orthogonal polarization. The electric fields at the input and output surfaces offer the change of orientation of polarization of the electromagnetic wave in an orthogonal direction. On the contrary, in the case of reflective type polarization conversion, the co-polarized reflection coefficient gets minimized as well as the maximization of cross-polarized reflection coefficient occurs at the same frequency to achieve the highest polarization conversion ratio. The tunability at high frequency can be achieved by using graphene in the metasurface design as the chemical potential of the graphene can be controlled externally by applied electrical bias as well as mechanical strain as well as the doping of the device.

Finally, metasurface designs have also been incorporated to achieve enhanced bandwidth as well as high directivity of the antenna. The enhanced bandwidth of the antenna can be achieved by incorporating metasurface design in the ground plane of the antenna. The gain of the antenna can be enhanced significantly when a superstrate configuration on a patch antenna has been realized.

In summary, the use of metasurface in the designs makes the structures ultra-thin and more compact in nature. This is extremely useful in modern-day 5G and higher data-rate communication system where a number of devices are embedded in a single chip to constitute the complete system.

ACKNOWLEDGEMENT

The author wants to acknowledge the Science and Engineering Research Board, Govt. of India for partial funding of the research work on metasurface.

REFERENCES

1. CalozC, ItohT, *Electromagnetic Metamaterials: Transmission Line Theory and Microwave Applications*, John Wiley & Sons, 2006, Inc.
2. R. Marques, F. Martin, and M. Sorolla, "*Metamaterials with Negative Parameters: Theory, Design, and Microwave Applications*," John Wiley, 2008.
3. J.B. Pendry, A.J. Holden, W.J. Stewart, and I. Youngs, "Extremely Low Frequency Plasmons in Metallic Mesostructure," *Physical Review Letters*, Vol. 76, No. 25, pp. 4773–4776, June 1996.

4. J.B. Pendry, A.J. Holden, D.J. Robbins, and W.J. Stewart, "Magnetism from conductors and enhanced nonlinear phenomena," *IEEE Transactions on Microwave Theory & Techniques*, Vol. 47, No. 11, pp. 2075–2084, Nov. 1999.

5. V. Vaselago, "The electrodynamics of substances with simultaneously negative values of μ and ε," *Soviet physics Uspekhi*, vol. 10, no. 4, pp. 509–514, 1968.

6. A.O. Pinchuk, and G.C. Schtz, "Focussing a beam of light with left-handed metamaterials," *Solid State Electronics*, Vol. 51, Issue 10, pp. 1381–1387, October 2007.

7. D.R. Smith, W.J. Padilla, D.C. Vier, S.C. Nemat-Nasser, and S. Schultz, "Composite medium with simultaneously negative permeability and permittivity," *Physical Review Letters*, Vol. 84, No. 18, pp. 4184–4187, May 2000.

8. R.A. Shelby, D.R. Smith, and S. Schultz, "Experimental verification of a negative index of refraction," *Science*, Vol. 292,pp. 77–79, April 2001.

9. B. A. Munk, "Frequency Selective Surfaces Theory and Design," *John Willey & Sons*, pp. 315–335, 2000.

10. H. T. Chen, A. J. Taylor, and N. Yu, "A review on metasurfaces: physics and applications," *Reports on Progress in Physics*, Vol. 79, pp. 076401, 2016.

11. P. Saville, "Review of radar absorbing materials," Defense R and D Canada-Atlantic, January, 2005.

12. B. Chambers, and A. Tennant, "Active Dallenbach radar absorber," *IEEE International Symposium on Antennas and Propagation*, Albuquerque, New Mexico, USA, pp. 381–384, 9–14 July, 2006.

13. G. I. Kiani, A. R. Weily, K. P. Esselle, "Frequency selective surface absorber using resistive cross-dipoles," *IEEE Antennas and Propagation Society International Symposium 2006*, New Mexico, USA, pp. 4199–4202, 9–14 July, 2006.

14. P. Zhihua, P. Jingcui, P. Yangfeng, O. Yangyu, and N. Yantao, "Complex permittivity and microwave absorption properties of carbon nanotubes/polymer composites: a numerical study," *Physics Letters A*, Vol. 372, Issue 20, pp. 3714–3718, 2008.

15. K. Matous, and G. J. Dvorak, "Optimization of electromagnetic absorption in laminated composite plates," *IEEE Transactions on Magnetics*, Vol. 39, No. 3, pp. 1827–1835, May 2003.

16. J. H. Oh, K. Oh, C. G. Kim, and C. S. Hong, "Design of radar absorbing structures using glass/epoxy composite containing carbon black in X-band frequency ranges," *Composites Part B: Engineering*, Vol. 35, Issue 1, pp. 49–56, January 2004.

17. R. C. Parida, D. Singh, and N. K. Agarwal, "Implementation of multilayer ferrite radar absorber coating with genetic algorithm for radar cross-section reduction at Xband," *Indian Journal of Radio and Space Physics*, Vol. 36, pp. 145–152, April 2007.

18. P. Singh, V. K. Babbar, A. Razdan, R. K. Puri, and T. C. Goel, "Complex permittivity, permeability, and X-band microwave absorption of CaCoTi ferrite composites," *Journal of Applied Physics*, Vol. 87, Issue 9, pp. 4362, 2000.

19. K. Ishihara, and Y. Tomiyama, "Electromagnetically anechoic chamber and shield structures therefor," *US Patent* 5134405, filed 27 February, 1989, granted 28 July, 1992.

20. H. Tao, N. I. Landy, C. M. Bingham, X. Zhang, R. D. Averitt, and W. J. Padilla, "A metamaterial absorber for the terahertz regime: design, fabrication and characterization," *Optics Express*, Vol. 16, Issue 10, pp. 7181–7188, 12 May, 2008.

21. F. Costa, S. Genovesi, and A. Monorchio, "A chipless RFID based on multiresonant high-impedance surface," *IEEE Transactions on Microwave Theory and Techniques*, Vol. 61, No. 1, pp. 146–153, 2013.

22. W. F. Bahret, "The beginning of stealth technology," *IEEE Transactions on Aerospace and Electronics system*, Vol. 29, Issue 4, pp. 1377–1385, October 1993.

23. M. R. Meshram, N. K. Agrawal, B. Sinha, and P. S. Misra, "Characterization of M-type barium hexagonal ferrite-based wide band microwave absorber," *Journal of Magnetism and Magnetic Materials*, Vol. 271, Issues 23, pp. 207–214, May 2004.

24. F. Bilotti, L. Nucci and L. Vegni, "An SRR-based microwave absorber," *Microwave and Optical Technology Letters*, Vol. 48, No. 11, pp. 2171–2175, Nov. 2006.

25. N. I. Landy, S. Sajuyigbe, J. J. Mock, D. R. Smith, and W. J. Padilla, "Perfect Metamaterial Absorber," *Physical Review Letters*, vol. 100, pp. 207402, 2008.

26. M. H. Li, L. Hua Yang, B. Zhou, X. Peng Shen, Q. Cheng, and T. J. Cui, "Ultrathin multiband gigahertz metamaterial absorbers," *Journal of Applied Physics*, vol. 110, pp. 014909, 2011.

27. D. R. Smith, D. C. Vier, Th. Koschny, and C. M. Soukoulis, "Electromagnetic parameter retrieval from inhomogeneous metamaterials," *Physical Review E*,Vol. 71, pp. 036617, 2005.

28. S. Bhattacharyya, H. Baradiya, R. K. Chaudhary, and K. V. Srivastava, "An Electric Field Driven LC Resonator Structure as Ultra Thin Metamaterial Absorber," *5th Annual Conference, Antenna Test and Measurement Society*, Mumbai, India, 2–3 February, 2012.

29. C. G. Hu, X. Li, Q. Feng, X. Chen, and X. G. Luo, "Investigation on the role of the dielectric loss in metamaterial absorber," *Optics Express*, Vol. 18, No. 7, pp. 6598–6603, 29 March, 2010.

30. X. Chen, T. M. Grzegorczyk, B.-I. Wu, J. Pacheco, and J. A. Kong, "Robust method to retrieve the constitutive effective parameters of metamaterials," *Physical Review E*, Vol. 70, pp. 016608, 2004.

31. S. Bhattacharyya, S. Ghosh, H. Baradiya, and K. V. Srivastava, "Study on Ultra-thin Dual Frequency Metamaterial Absorber with Retrieval of Electromagnetic Parameters," *IEEE National Conference on Communication (NCC 2014), IIT Kanpur*, India, 28 February – 2 March, 2014.

32. S. Bhattacharyya, and K. V. Srivastava,"An Ultra Thin Electric Field Driven LC Resonator Structure as Meta-material Absorber for Dual Band Applications," *2013 URSI International Symposium on Electromagnetic Theory (EMTS)*, Hiroshima, Japan, May 20–24, 2013.

33. S. Bhattacharyya, and K. V. Srivastava, "Triple band polarization-Independent ultra-thin metamaterial absorber using electric field-driven LC Resonator," *Journal of Applied Physics*, vol. 115, pp. 064508, 2014.

34. H. Tao, C. M. Bingham, D. Pilon, K. Fan, A. C. Strikwerda, D. Shrekenhamer, W. J. Padilla, "A dual band terahertz metamaterial absorber," *Journal of Physics D: Applied Physics*, Vol. 43, No. 22, pp. 225102, May 2010.

35. Q. Y. Wen, H. W. Zhang, Y. S. Xie, Q. H. Yang, and Y. L. Liu, "Dual band terahertz meta-materials absorber: design, fabrication and characterization," *Applied Physics Letters*, Vol. 95, Issue 24, pp. 241111, 2009.

36. Y. Q. Xu, P. H. Zhou, H. B. Zhang, L. Chen, and L. J. Deng, "A wide-angle planar meta-material absorber based on split-ring resonator coupling," *Journal of Applied Physics*, Vol. 110, Issue 4, pp. 044102, 2011.

37. L. Huang, and H. Chen, "Multi-band and polarization insensitive metamaterial absorber," *Progress In Electromagnetics Research*, Vol. 113, pp. 103–110, 2011.

38. B. Zhang, Y. Zhao, Q. Hao, B. Kiraly, I.-C. Khoo, S. Chen, and T. J. Huang, "Polarization-independent dual-band infrared perfect absorber based on a metaldielectric-metal ellip-tical nanodisk array," *Optics Express*, Vol. 19, Issue 16, pp. 15221–15228, August 2011.

39. R. Feng, W. Ding, L. Liu, L. Chen, J. Qiu, and G. Chen, "Dual-band infrared perfect absorber based on asymmetric T-shaped plasmonic array," *Optics Express*, Vol. 22, Issue S2, pp. A335–A343, March 2014.

40. S. D. Campbell, and R.W. Ziolkowski, "Lightweight, flexible, polarization-insensitive, highly absorbing meta-films," *IEEE Transactions on Antennas and Propagation*, Vol. 61, Issue 3, pp. 1191–1200, March 2013.

41. X. Shen, T. J. Cui, J. Zhao, H. F. Ma, W. X. Jiang, and H. Li, "Polarization-independent wide-angle triple band metametrial absorber," *Optics Express*, Vol. 19, Issue 10, pp. 9401–9407, May 2011.

42. P. V. Tuong, J. W. Park, J. Y. Rhee, K. W. Kim, W. H. Jang, H. Cheng, and Y. P. Lee, "Polarization-insensitive and polarization-controlled dual-band absorption in metamaterials," *Applied Physics Letters*, Vol. 102, No. 8, pp. 081122, 2013.

43. X.-R. Guo, Z. Zhang, J.-H. Wang, and J.-J. Zhang, "The design of a triple-band wide-angle metamaterial absorber based on rectangular pentagon close ring," *Journal of Electromagnetic Waves and Applications*, Vol. 27, Issue 5, pp. 629–637, 2013.

44. G.-D. Wang, J.-F. Chen, X.-W. Hu, Z.-Q. Chen, and M.-H. Liu, "Polarization-insensitive triple band microwave metamaterial absorber based on rotated square rings," *Progress In Electromagnetic Research*, Vol. 145, pp. 175–183, 2014.

45. J. Lee, and S. Lim, "Bandwidth-enhanced and polarization-insensitive metamaterial absorber using double resonance," *Electronics Letters*, Vol. 47, No. 1, pp. 8–9, 6 Jan 2011.

46. B. Bian, S. Liu, S. Wang, X. Kong, H. Zhang, B. Ma, and H. Yang, "Novel triple-band polarization-insensitive wide-angle ultra-thin microwave metamaterial absorber," *Journal of Applied Physics*, Vol. 114, No. 19, pp. 194511, 2013.

47. P. Pitchappa, C. P. Ho, P. Kropelnicki, N. Singh, D. -L. Kuong, and C. Lee, "Dual band complementary metamaterial absorber in near infrared region," *Journal of Applied Physics*, Vol. 115, Issue 19, pp. 193109, 2014.

48. Y. Kaiprath, S. Ghosh, S. Bhattacharyya and Kumar Vaibhav Srivastava, "An Ultrathin polarization-independent wide-angle metamaterial absorber for dual-band applications," in*IEEE Applied Electromagnetics Conference 2013*, 18–20December, 2013, KIIT University, Bhubaneshwar, India.

49. S. Bhattacharyya, S. Ghosh, and K. V. Srivastava, "An ultra-thin Polarization-Independent Metamaterial Absorber for Triple Band Applications," in*IEEE Applied Electromagnetics Conference 2013*, pp. 1–2, 18–20 December, 2013, KIIT University, Bhubaneshwar, India.

50. D. Chaursaiya, S. Ghosh, S. Bhattacharyya, and K. V. Srivastava, "The Design of Dual band Polarization-Independent Wide-angle Circular-Shaped Metamaterial Absorber," in *IEEE International Microwave & RF Conference (IMaRC 2014)*, pp. 96–99, Bangalore, India, 15–17 December, 2014.

51. D. Chaurasiya, S. Ghosh, S. Bhattacharyya, and K. V. Srivastava, "An ultra-thin quad-band polarization-insensitive wide-angle metamaterial absorber," *Microwave and Optical Technology Letters*, Vol. 57, Issue 3, pp. 697–702, March 2015.

52. D. Chaurasiya, S. Ghosh, S. Bhattacharyya, A. Bhattacharya, and K. V. Srivastava, "A compact multi-band polarization-insensitive metamaterial absorber," *IET Microwaves, Antennas and Propagation*, Vol. 10, Issue 1, pp. 94–101, 2016.

53. A. Sarkhel and S. R. Bhadra Chaudhuri, "Compact quad-band polarization-insensitive ultrathin metamaterial absorber with wide angle stability," *IEEE Antennas and Wireless Propagation Letters*, vol. 16, pp. 3240–3244, 2017, doi: 10.1109/LAWP.2017.2768077.

54. D. Sood, and C. C. Tripathi, "A polarization insensitive compact ultrathin wide-angle penta-band metamaterial absorber,"*Journal of Electromagnetic Waves and Applications*, vol. 31, issue 4, pp. 394–404,2017, DOI:10.1080/09205071.2017.1288172

55. A. Bhattacharya, S. Bhattacharyya, S. Ghosh, D. Chaurasiya, and K. V. Srivastava, "An ultra-thin penta-band polarization-insensitive compact metamaterial absorber for airborne radar application," *Microwave and Optical Technology Letters*, Vol. 57, No. 11, pp. 2519–2524, 2015.

56. P. Munaga, S. Bhattacharyya, S. Ghosh, and K. V. Srivastava, "An ultra-thin compact polarization-independent hexa-band metamaterial absorber," *Applied Physics A*, Vol. 124, Issue 4, pp. 331, April 2018.

57. R. Dutta, S. C. Bakshi and D. Mitra, "An ultrathin compact polarization-insensitive hepta-band absorber," *2018IEEE MTT-S International Microwave and RF Conference (IMaRC)*, Kolkata, India, 2018, pp. 1–4, doi:10.1109/IMaRC.2018.8877104.

58. Verma, V.K., Mishra, S.K., Kaushal, K.K. et al. An uctaband polarization insensitive terahertz metamaterial absorber using orthogonal elliptical ring resonators. *Plasmonics* vol. 15, pp. 75–81, 2020.

59. S. Ghosh, S. Bhattacharyya, and K. V. Srivastava, "Bandwidth-enhancement of an ultra-thin polarization insensitive metamaterial absorber," *Microwave and Optical Technology Letters*, Vol. 56, Issue 2, pp. 350–355, 2014.

60. S. Bhattacharyya,S. Ghosh, and K. V. Srivastava, "Triple band polarization-independent metamaterial absorber with bandwidth enhancement at X-band," *Journal of Applied Physics*, Vol. 114, Issue 9, pp. 094514, 2013.

61. S.Bhattacharyya,S.Ghosh,and K.V.Srivastava, "Bandwidth enhanced metamaterial absorber using electric field driven LC resonator for airborne radar applications," *Microwave and Optical Technology Letters*, Vol. 55, Issue 9, pp. 2131–2137, September 2013.

62. S. Ghosh, S. Bhattacharyya,Y. Kaiprath, and K. V. Srivastava, "Bandwidth-enhanced polarization insensitive microwave metamaterial absorber and its equivalent circuit model," *Journal of Applied Physics*, Vol. 115, Issue 10, pp. 104503, 2014.

63. S. Bhattacharyya,S. Ghosh, D. Chaurasiya, and K. V. Srivastava, "Bandwidth-enhanced dual-band dual-layer polarization-insensitive ultra-thin metamaterial absorber," *Applied Physics A*, Vol. 118, Issue 1, pp. 207–215, 2015.

64. S. Bhattacharyya, and K. V. Srivastava, "Dual layer polarization insensitive dual band metamaterial absorber with enhanced bandwidths," in *IEEEAsia Pacific Microwave Conference (APMC) 2014*, pp. 816–818, Sendai, Japan, 4–7 November, 2014.

65. G. Dayal, and S. A. Ramakrishna, "Design of multi-band metamaterial perfect absorbers with stacked metaldielectric disks," *Journal of Optics*, Vol. 15, Issue 5, pp. 055106, 2013.

66. F. Ding, Y. Cui, X. Ge, Y. Jin, and S. He, "Ultra-broadband microwave metamaterial absorber," *Applied Physics Letters*, Vol. 100, Issue 10, pp. 103506, 2012.

67. T. H. Nguyen, S. T. Bui, T. T. Nguyen, T. T. Nguyen, Y. P. Lee, M. A. Nguyen, and D. M. Vu, "Metamaterial-based perfect absorber: polraization insensitivity and broadband," *Advances in Natural Sciences: Nanoscience and Nanotechnology*, Vol. 5, No. 2, pp. 025013, 2014.

68. D. Wen, H. Yang, Q. Ye, M. Li, J. Guo, and J. Zhang, "Broadband metamaterial absorber based on a multi-layer structure," *Physica Scripta*, Vol. 88, pp. 015402, 2013.

69. J. Zhu, Z. Ma, W. Sun, F. Ding, Q. He, L. Zhou, and Y. Ma, "Ultra-broadband terahertz metamaterial absorber," *Applied Physics Letters*, Vol. 105, Issue 2, pp. 021102, 2014.

70. B.-X. Wang, L.-L. Wang, G.-Z. Wang, W.-Q. Huang, X.-F. Li, and X. Zhai, "Theoretical investigation of broadband and wide-angle terahertz metamaterial absorber," *IEEE Photonics Technology Letters*, Vol. 26, No. 2, pp. 111–114, 15 January, 2014.

71. V. T. Pham, J. W. Park, D. L. Vu, H. Y. Zheng, J. Y. Rhee, K. W. Kim, and Y. P. Lee, "THz-metamaterial absorbers," *Advances in Natural Sciences: Nanoscience and Nanotechnology*, Vol. 4, Issue 1, pp. 015001, 2013.

72. J. A. Bossard, L. Lin, S. Yun, L. Liu, D. H. Werner, and T. S. Meyer, "Near-ideal optical metamaterial absorbers with super octave bandwidth," *ACS Nano*, Vol. 8, No. 2, pp. 1517–1524, 2014.

73. S. Ghosh, S. Bhattacharyya, and K. V. Srivastava, "Design, characterization and fabrication of a broadband polarization-insensitive multi-layer circuit analog absorber," *IET Microwaves, Antennas and Propagation*, Vol. 10, Issue 8, pp. 850–855, 2016.

74. P. Munaga, S. Ghosh, S. Bhattacharyya, and K. V. Srivastava, "A Fractal based Compact Broadband Polarization Insensitive Metamaterial Absorber using Lumped Resistors," *Microwave and Optical Technology Letters*, Vol. 58, No. 2, pp. 343–347, February 2016.

75. D. Kundu, A. Mohan and A. Chakrabarty, "Single-layer wideband microwave absorber using array of crossed dipoles," *IEEE Antennas and Wireless Propagation Letters*, vol. 15, pp. 1589–1592, 2016, doi:10.1109/LAWP.2016.2517663.

76. Khalid Saeed Lateef Al-badri, "Electromagnetic broad band absorber based on metamaterial and lumped resistance," *Journal of King Saud University-Science*, Vol. 32, Issue 1, pp. 501–506, 2020.

77. Y. Z. Cheng, Y. Wang, Y. Nie, R. Z. Gong, X. Xiong, and X. Wang, "Design, fabrication and measurement of a broadband polarization-insensitive metamaterial absorber based on lumped elements," *Journal of Applied Physics*, vol. 111, issue 4, pp. 044902, 2012.

78. *Federal Radar Spectrum Requiremnt*, U.S. Department of Commerce, May2000.

79. S. Ghosh, and K. V. Srivastava, "Polarization-insensitive single–/dual-band tunable absorber with independent tuning in wide frequency range," *IEEE Transactions on Antennas and Propagation*, vol. 65, no. 9, pp. 4903–4908, 2017.

80. S. Ghosh, and K. V. Srivastava, "A polarization-independent broadband multi-layer switchable absorber using active frequency selective surface," *IEEE Antennas and Wireless Propagation Letters*, vol. 16, pp. 3147–3150, 2017.

81. S. Ghosh, and S. Lim, "Fluidically-reconfigurable multifunctional frequency selective surface with miniaturization characteristic," *IEEE Transactions on Microwave Theory and Techniques*, vol. 66, no. 8, pp. 3857–3865, 2018.

82. S. Ghosh, and S. Lim, "A multifunctional reconfigurable frequency selective surface using liquid metal alloy," *IEEE Transactions on Antennas and Propagation*, vol. 66, no. 9, pp. 4953–4957, 2018.

83. S. K. Ghosh,S. Bhattacharyya, and S. Das, "A Graphene Based Metasurface with Wideband Absorption in the Lower Mid Infrared Region," in *IEEE International Microwave & RF Conference (IMaRC 2018)*, Kolkata, India, 28–30 November, 2018.

84. S. K. Ghosh, V. S. Yadav, S. Das, and S. Bhattacharyya, "Tunable graphene based metasurface for polarization-independent broadband absorption in lower mid infrared (mir) range," *IEEE Transactions on Electromagnetic Compatibility*, vol. 62, no. 2, pp. 346–354, April 2020.

85. S. K. Ghosh, S. Das, andS. Bhattacharyya, "A Graphene Based Broadband Metasurface Absorber in the Terahertz Region," in *2019 URSI Asia Pacific Radio Science Conference (AP-RASC 2019)*, New Delhi, India, 9–15 March, 2019.

86. Y.-J. Chiang, and T.-J. Yen, "A composite-metamaterial-based terahertz-wave polarization rotator with an ultrathin thickness, an excellent conversion ratio, and enhanced transmission,"*Applied Physics Letters*, Vol. 102, pp. 011129, 2013.

87. M. Saikia, S. Ghosh, S. Bhattacharyya and K. V. Srivastava, "Broadband Polarization Rotator using Multilayered Metasurfaces," 2015 IEEE Applied Electromagnetics Conference, 2015.

88. M. Saikia, S. Ghosh and K. V. Srivastava, "Design and analysis of ultra-thin polarization rotating frequency selective surface using v-shaped slots," *IEEE Antenna and Wireless Propagation Letters*, vol. 16, no. 1, pp-2022–2025, December 2017.

89. M. R. Akram, M. Q. Mehmood, X. Bai, R. Jin, M. Premaratne, and W. Zhu, "High efficiency ultrathin transmissive metasurfaces," *Advanced Optical Materials*, vol. 7, issue 11, pp. 1801628, June 2019.

90. R. Nandi, Nilotpal, and S. Bhattacharyya, "a transmittive type broadband cross polarization converter for mid wavelength infrared region," in *2019 URSI Asia Pacific Radio Science Conference (AP-RASC 2019)*, New Delhi, India, 9–15 March, 2019.

91. Nilotpal, L.Nama, S. Bhattacharyya, and P. Chakrabarti, "A metasurface-based broadband quasi non-dispersive cross polarization converter for far infrared region," *International Journal of RF and Microwave Computer-Aided Engineering*, Vol. 29, Issue 10, Article No. e21889, October 2019.

92. M. A. Ordal, L. L. Long, R. J. Bell, S. E. Bell, R. R. Bell, R. W. Alexander, Jr, and C. A. Ward, "Optical properties of the metals Al, Co, Cu, Au, Fe, Pb, Ni, Pd, Pt, Ag, Ti, and W in the infrared and far-infrared," *Applied Optics*, vol. 22, pp. 1099–1119, 1983.

93. G. Hawkins and R. Hunneman, "The temperature-dependent spectral properties of filter substrate materials in the far-infrared (640 μm)," *Infrared Physics & Technology*, vol. 45, pp. 69–79, 2004.

94. S. Bhattacharyya, S. Ghosh, and K. V. Srivastava, "A wideband cross polarization conversion using metasurface," *Radio Science*, vol. 52, pp. 1395–1404, November 2017.

95. L. Nama, S. Bhattacharyya, and P. K. Jain, "Ultra-thin wideband reflective polarization converter using elliptical split ring structures," *Eleventh Annual Conference, Antenna Test and Measurement Society*, February 2018.

96. G. Zhou, X. Tao, Z. Shen, G. Zhu, B. Jin, L. Kang, W. Xu, J. Chen&P. Wu, "Designing perfect linear polarization converters using perfect electric and magnetic conducting surfaces," *Scientific Reports*, vol. 6, pp. 38925, 2016.

97. Y. Zhi Cheng, W. Withayachumnankul, A. Upadhyay, D. Headland, Y. Nie, R. Zhou Gong, M. Bhaskaran, S. Sriram, and D. Abbott, "Ultra broadband reflective polarization convertor for terahertz waves," *Applied Physics Letters*, vol. 105, pp. 181111, 2014.

98. L. Cong, W. Cao, X. Zhang, Z. Tian, J. Gu, R. Singh, J. Han, and W. Zhang, "A perfect Metamaterial polarization rotator," *Applied Physics Letters*, vol. 103, pp. 171107, 2013.

99. N. K. Grady, J. E. Heyes, D. Roy Chowdhury, Y. Zeng, M. T. Reiten, A. K. Azad, "Terahertz metamaterials for linear polarization conversion and anomalous refraction," *Science*, vol. 340, pp. 1304–1307, 2013.

100. X. Zheng, Z. Xiao, and X. Ling, "A tunable hybrid metamaterial reflective polarization converter based on vanadium oxide film," *Plasmonics*, vol. 13, pp. 287–291, 2018.

101. Z. Xiao, H. Zou, X. Zheng, X. Ling, and L. Wang, "A tunable reflective polarization converter based on hybrid metamaterial," *Opt. Quant. Electron.* vol. 49, 2017.

102. C. Yang, Y. Luo, J. Guo, Y. Pu, D. He, Y. Jiang, J. Xu, and Z. Liu, "Wideband tunable mid-infrared cross polarization converter using rectangle-shape perforated graphene," *Optics Express*, vol. 24, pp. 16913–16922, 2016.

103. Khan, M. I., Fraz, Q., &Tahir, F. A. , "Ultra-wideband cross polarization conversion metasurface insensitive to incident angle," *Journal of Applied Physics*, vol. 121, issue 4, pp. 045103, 2017.

104. C. Yang, Y. Luo, J. Guo, Y. Pu, D. He, Y. Jiang, J. Xu and Z. Liu, "Wideband tunable mid-infrared cross polarization converter using rectangle-shape perforated graphene," *Optics Express*, vol. 24, pp. 16913–16922, 2016.

105. M. Chen, L. Chang, X. Gao, H. Chen, C. Wang, X. Xiao and D. Zhao, "Wideband tunable cross polarization converter based on a graphene metasurface with a hollow-carved 'H' array," *IEEE Photonics Journal*, vol. 9, pp. 5, 2017.

106. S. Luo, B. Li, A. Yu, J. Gao, X. Wang and D. Zuo, "Broadband tunable terahertz polarization converter based on graphene metamaterial," *Optics Communication*, vol. 413, pp. 184–189, 2018.

107. X. Yu, X. Gao, W. Qiao, L. Wen, and W. Yang, "Broadband tunable polarization converter realized by graphene-based metamaterial," *IEEE Photonics Technology Letters*, 28(21), 2399–2402 (2016).

108. J. Zhu, S. Li, L. Deng, C. Zhang, Y. Yang, and H. Zhu, "Broadband tunable terahertz polarization converter based on a sinusoidally-slotted graphene metamaterial," *Optical Materials Express*, vol. 8, pp. 1164–1173, 2018.

109. J. Ding, B. Arigong, H. Ren, J. Shao, M. Zhou, Y. Lin, H. Zhang, "Mid-Infrared tunable dual-frequency cross polarization converters using graphene-based l-shaped nanoslot array," *Plasmonics*, vol. 10, pp. 351–356, 2015.

110. Y. Zhang, Y. Feng, T. Jiang, J. Cao, J. Zhao, B. Zhu, "Tunable broadband polarization rotator in terahertz frequency based on graphene metamaterial," *Carbon*, vol. 133, pp. 170–175, 2018.

111. V. S. Yadav, S. K. Ghosh, S. Das, and S. Bhattacharyya, "Wideband tunable mid-infra-red cross-polarization converter using monolayered graphene-based metasurface over a wide angle of incidence," *IET Microwaves, Antennas and Propagation*, Vol. 13, Issue 1, pp. 82–87, January 2019.

112. V. S. Yadav, S. K. Ghosh, S. Bhattacharyya, and S. Das, "Graphene based metasurface for tunable broadband terahertz cross polarization converter over wide angle of incidence," *Applied Optics*, Vol. 57, Issue 29, pp. 8720–8726, October 2018.

113. BadaweME, AlmoneefT S, RamahiOM, "A true metasurface antenna," *Scientific Reports*, Jan. 2016; 6: 1–8.

114. Nasser S S Syed, LiuW, ChenZ N, "Wide bandwidth and enhanced gain of a low-profile dipole antenna achieved by integrated suspended metasurface," *IEEE Transactions on Antennas and Propagation*, March 2018; 66: 1540–1544.

115. LiH, WangG, GaoX, LiangJ, HouH, "A novel metasurface for dual-mode and dual-band flat high-gain antenna application," *IEEE Transactions on Antennas and Propagation*, July 2018; 66:3706–3711.

116. D. Samantaray, S. Bhattacharyya, and K. V. Srinivas, "A modified fractal-shaped slot-ted patch antenna with defected ground using metasurface for dual band applications", *International Journal of RF and Microwave Computer-Aided Engineering*, vol. 29, issue 12, pp. e21932, August 2019.

117. R. M. Hashmi, B. A. Zeb and K. P. Esselle, "Wideband high-gain ebg resonator anten-nas with small footprints and all-dielectric superstructures," *IEEE Transactions on Antennas and Propagation*, vol. 62, issue 6, pp. 2970–2977, June 2014, doi:10.1109/TAP.2014.2314534.

118. Y. Zheng et al., "Wideband gain enhancement and rcs reduction of fabry–perot resonator antenna with chessboard arranged metamaterial superstrate," in *IEEE Transactions on Antennas and Propagation*, vol. 66, issue 2, pp. 590–599, February 2018, doi:10.1109/TAP.2017.2780896.

119. H. Attia, L. Yousefi and O. M. Ramahi, "Analytical model for calculating the radiation field of microstrip antennas with artificial magnetic superstrates: Theory and experi-ment," *IEEE Transaction on Antennas and Propagation*, vol. 59, issue 5, pp. 1438–1445, May 2011, doi:10.1109/TAP.2011.2122295

120. YangX, LiuY, GongS, "Design of a wideband omnidirectional antenna with charac-teristic mode analysis," *IEEE Antennas and Wireless Propagation Letters*, vol. 17, pp. 993–997, June 2018.

121. BalanisC A, *Antenna Theory Analysis and Design*, Third Edition, John Willy and Sons, Inc. Wiley Reprint, India, 2010.

122. LockerC, VaupelT, EibertT F, "Radiation efficient unidirectional low-profile slot antenna elements for x-band application," *IEEE Transactions on Antennas and Propagation*, August 2005; 53: 2765–2768.

123. T. Cai, S. W. Tang, G. M. Wang, H. X. Xu, S. L. Sun, Q. He, and L. Zhou, "High-performance bi-functional metasurfaces in transmission and reflection geometries," *Advanced Optical Materials*, vol. 5, pp. 1600506, 2017.

124. L. Zhang, R. Y. Wu, et al., "Transmission-reflection-integrated multifunctional cod-ing metasurface for full-space controls of electromagnetic waves," *Advanced Optical Materials*, vol. 28, no. 33, pp. 1802205, 2018.

125. J. K. Ali, Z. A. AL-Hussain, A. A. Osman, and A. J. Salim, "A new compact size fractal based microstrip slot antenna for GPS applications," *Progress in Electromagnetics Research Symposium, PIERS*, pp. 700–703, Kuala Lumpur, Malaysia, March 27–30, 2012.

126. QuarfothR, SievenpiperD, "Artificial tensor impedance surface waveguides,"*IEEE Transactions on Antennas and Propagation*, vol. 61, pp. 3597–3606, 2013.

127. GhoshA, ChakrabortyS, ChattopadhyayS, NandiA, BasuB, "Rectangular microstrip antenna with dumbbell shaped defected ground structure for improved cross polarized radiation in wide elevation angle and its theoretical analysis," in *IET Microwaves, Antennas & Propagation*, vol. 10, issue 1, pp. 68–78, 2016.

128. ChakrabortyS, GhoshA, ChattopadhyayS, SinghL L K, "Improved Cross-Polarized Radiation and Wide Impedance Bandwidth From Rectangular Microstrip Antenna With Dumbbell-Shaped Defected Patch Surface," in *IEEE Antennas and Wireless Propagation Letters*, vol. 15, pp. 84–88, 2016.

129. MitraD, GhoshB, SarkhelA, ChaudhuriS R B, "A miniaturized ring slot antenna design with enhanced radiation characteristics," *IEEE Transactions on Antennas and Propagation*, vol. 64, pp. 300–305, January 2016.

130. PozarD M, SchaubertD H, *Microstrip Antennas*, New York: IEEE press, 1995.

131. XuK, LiuY, DongL, PengL, ChenS, ShenF, YeX, ChenX, Wang G, "Printed multi-band compound meta-loop antenna with hybrid-coupled SRRs," in *IET Microwaves, Antennas & Propagation*, 2018; 12, 8: 1382–1388.

132. Alibakhshi-KenariM, Naser-MoghadasiM, SadeghzadehR A, VirdeeB S, Limiti E, "Periodic array of complementary artificial magnetic conductor metamaterials-based multiband antennas for broadband wireless transceivers," in *IET Microwaves, Antennas & Propagation*, December 2016; 10, 15: 1682–1691.

133. ZhaiH, ZhangK, YangS, FengD, "A low-profile dual-band dual-polarized antenna with an AMC surface for WLAN applications," *IEEE Antennas and Wireless Propagation Letters*.Aug 2017; 16: 2692–2695.

134. CaiT, WangGM, ZhangXF, WangYW, ZongBF, XuHX, "Compact microstrip antenna with enhanced bandwidth by loading magneto-electro-dielectric planar waveguided metamaterials," *IEEE Transactions on Antennas and Propagation*, February 2015; 63, 5: 2306–2311.

135. LiLW, LiYN, YeoTS, MosigJR, MartinOJ, "A broadband and high-gain metamaterial microstrip antenna," *Applied Physics Letters*, April 2010; 96, 16: 164101–164103.

136. MitraD, SarkhelA, KunduO, ChaudhuriS R B, "Design of compact and high directive slot antennas using grounded metamaterial slab," *IEEE Antennas and Wireless Propagation Letters*, 2015; 14: 811–814.

137. HuangY, YangL, LiJ, et al.: 'Polarization conversion of metasurface for the application of wideband low-profile circular polarization', *Applied Physics Letters*, 2016, 109, pp. 054101-1–054101-5.

138. ZhuH L, CheungS W, Chung, K.L., et al.: 'Linear to circular polarization conversion using metasurface', *IEEE Transactions on Antennas and Propagation*, 2013; 61, (9): 4615–4622.

139. NasimuddinN, Chen, Z.N., Qing, X. "Bandwidth enhancement of a single feed circularly polarized antenna using a metasurface," *IEEE. Antennas and Propagation Magazine*, 2016; 58, (2): 39–46.

140. H. Xu, G Wang, J Liang, M. Q. Qi, X Gao, "Compact circularly polarized antennas combining meta-surfaces and strong space-filling meta-resonators," *IEEE Transactions on Antennas and Propagation*, July 2013; 61,(7):3442–3450.

141. D. Samantaray and S. Bhattacharyya, "A gain-enhanced slotted patch antenna using metasurface as superstrate configuration," in *IEEE Transactions on Antennas and Propagation*, doi:10.1109/TAP.2020.2990280.

142. A. Baba, R. M. Hashmi, K. P. Esselle, J. G. Marin and J. Hesselbarth, "Broadband partially reflecting superstrate-based antenna for 60 ghz applications," *IEEE Transactions on Antennas and Propagation*, vol. 67, no. 7, pp. 4854–4859, July 2019.

143. M. A. Al-Tarifi, D. E. Anagnostou, A. K. Amert and K. W. Whites, "The puck antenna: A compact design with wideband, high-gain operation," *IEEE Transactions on Antennas and Propagation*, vol. 63, no. 4, pp. 1868–1873, April 2015.

144. R. M. Hashmi, B. A. Zeb and K. P. Esselle, "Wideband high-gain ebg resonator antennas with small footprints and all-dielectric superstructures," *IEEE Transactions on Antennas and Propagation*, vol. 62, no. 6, pp. 2970–2977, June 2014.

145. Y. Zheng et al., "Metamaterial-based patch antenna with wideband RCS reduction and gain enhancement using improved loading method," *IET Microwaves, Antennas & Propagation*, vol. 11, no. 9, pp. 1183–1189, July 2017.

146. Tawk, Youssef, Joseph Costantine, and Christos G. Christodoulou. "Reconfigurable filtennas and MIMO in cognitive radio applications." *IEEE Transactions on Antennas and Propagation*, vol. 62, no. 3 pp. 1074–1083, 2014.

147. Ziolkowski, Richard W., and Aycan Erentok, "Metamaterial-based efficient electrically small antennas." *IEEE Transactions on Antennas and Propagation*, vol. 54, no. 7, pp. 2113–2130, 2006.

148. Hailiang Zhu, Sing Wai Cheung, Tung Ip Yuk "Enhancing Antenna Boresight Gain using a small metasurface lens,"*IEEE Antennas and Wireless Propagation Magazine*, vol. 10, pp. 35–44, 2016.

149. Shaik, Latheef A., Chinmoy Saha, Yahia MM Antar, and Jawad Y. Siddiqui "An antenna advance for cognitive radio" *IEEE Antennas & Propagation Magazine*1045, no. 9243/18 (2018).

150. Saha, Chinmoy, Jawad Y. Siddiqui, and Y. M. M. Antar, *Multifunctional Ultrawideband Antennas: Trends, Techniques and Applications*. CRC Press, 2019.

151. S. Ghosh, S. Das, D. Samantaray, and S. Bhattacharyya, "Meander line based defected ground microstrip antenna slotted with srr for terahertz range," *Engineering Reports*, Vol. 2, Issue 1, Article No. e12088, January 2020.

152. K. Chandran P LD. Samantaray, A. Mohamed, C. Saha and S. Bhattacharyya, "A AMC substrate backed gain enhanced multi-band wearable yagi antenna," *2019IEEE Indian Conference on Antennas and Propogation (InCAP)*, Ahmedabad, India, 2019, pp. 1–4, doi:10.1109/InCAP47789.2019.9134570.

153. M. J. Anand Krishnan, D. Samantaray, A. Mohamed, C. Saha and S. Bhattacharyya, "Gain Enhanced Quad-Band AMC Backed Printed Antenna with Fractal Geometry," *2020 URSI Regional Conference on Radio Science (URSI-RCRS)*, Varanasi, India, 2020, pp. 1–4, doi: 10.23919/URSIRCRS49211.2020.9113517.

154. Varuna, A B S. Ghosh, S. Bhattacharyya, and K. V. Srivastava, "Design of a Dual-Band Polarization-Insensitive and Angular-Stable Frequency Selective Surface," in *IEEE Applied Electromagnetics Conference 2015 (AEMC 2015)*, IIT Guwahati, India, 18–21 December, 2015.

155. Lin, Y., Dai, J., Zeng, Z. et al. Metasurface color filters using aluminum and lithium niobate configurations. *Nanoscale Research Letters*,15, 77 (2020). doi:10.1186/s11671-020-03310-3.

156. Y. Yun Ji, F. Fan, M. Chen, L. Yang, and S.-J. Chang, "Terahertz artificial birefringence and tunable phase shifter based on dielectric metasurface with compound lattice," *Optics Express*, vol. 25, issue 10, pp. 11405–11413, 2017.

157. Colburn, S., Zhan, A.&Majumdar, A. Tunable metasurfaces via subwavelength phase shifters with uniform amplitude. *Scientific Reports*, vol. 7, art no. 40174, 2017, doi:10.1038/srep40174.

4 Design of Superlens Using 2D Photonic Crystal with Various Geometries under Polarized Incidence

Design of Superlens Using 2D Photonic Crystal

Moumita Banerjee, Pampa Debnath and Arpan Deyasi

CONTENTS

4.1   INTRODUCTION

Photons have several advantages over electrons. One of the greatest advantages is that the speed of photon much greater than speed of electron in metallic wire. Another advantages is that the band width of dielectric materials is larger than metal. Photons are not directly interacting with electrons, which therefore helps to reduce the energy absorption losses in system design.

Periodic band structures have for years played an important role in many branches of physics. Laudon [1] and Yablonovithch [2] were the first to recommend the idea of designing the material which may affect the properties of photons similar to the way that ordinary semiconductor crystal affects the properties of electrons. In a photonic crystal, the periodic potential (Bragg-like diffraction)occurs due to the lattice mismatch of the periodic potential of different dielectric material constant. It is the combination of dielectric materials that contain a periodicity in dielectric constant. The periodic contrast in static dielectric constant can create a range of forbidden frequency called photonic bandgap [3]. Photonic crystal can be design with any lattice constant and equivalent periodic potential, therefore allowing the gap to be controlled and the existence of photonic bandgap is highly dependent on the lattice geometry and dielectric contrast. Electromagnetic mode with energies resides within a photonic bandgap cannot travel through the photonic crystal and are therefore forbidden.

A photonic crystal might therefore engineered in a different manner to process a photonic bandgap for a specific range of frequency spectrum where which electromagnetic waves are forbidden to exists within the crystal [4–6]. The electromagnetic modes would be prohibited in perfectly periodic photonic crystal. However, if there is defect in the other perfect periodic crystal, localized photonic states could exist[7] within the photonic bandgap. In the recent study, it has been a great interest the photonic band structure, also known as dispersion curve [8–10], for electromagnetic wave travelling in various two-dimensional and three-dimensional photonic crystals. The aim of this study has been to determine the photonic band structure formed by the branches of this dispersion curve that is abstracted by any photonic bandgap. To be useful, the photonic bandgap must exist for all wave vectors values in the Brillouin zone for photonic crystal band structure under study.

4.1.1　Two-Dimensional Photonic Crystal

In photonic crystal, the dielectric periodicity creates a photonic bandgap in the electromagnetic dispersion curve for that electromagnetic wave that propagates perpendicular to the layers photonic crystal, therefore, offers the possibility of allowing for unprecedented control and manipulation of light [11].A two-dimensional photonic crystal is periodic along the two of its axis and homogeneous along the third. Photonic crystal is classified in two categories considering the first photonic crystal is formed by a square lattice of dielectric column of dielectric constant 'ε_a' in a different dielectric background 'ε_b'. The second photonic crystal is made by arranging the dielectric column of a triangular lattice.

4.1.2　Square Lattice Photonic Crystal

Figure 4.1 illustrates the first class of two-dimensional photonic crystal to be studied namely: the two-dimensional square lattice of column of high dielectric' ε_a' in the background of low dielectric constant 'ε_b' and the square lattice of column of low dielectric constant 'ε_a' in the background of high dielectric constant 'ε_b'.

The square lattice consists of a periodic array of parallel dielectric column of circular cross-section and dielectric constant ε_a where the intersection with perpendicular plane form a square lattice. The dielectric column is incorporated in a dielectric material whose dielectric constant is ε_b.

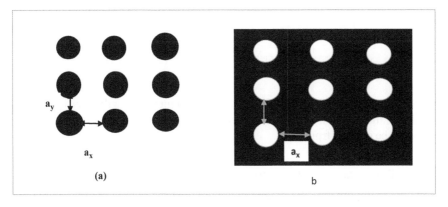

FIGURE 4.1　(a) The two-dimensional square lattice of column of high dielectric ε_a in the background of low dielectric constant ε_a. (b) The square lattice of column of low dielectric constant ε_a in the background of high dielectric constant ε_b.

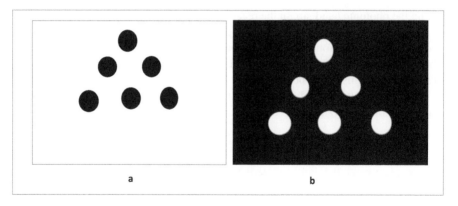

a　　　　　　　　　　　　　　　b

FIGURE 4.2 (a) The two-dimensional triangular lattice of column of high dielectric ε_a in the background of low dielectric constant ε_b. (b) The triangular lattice of column of low dielectric constant ε_a in the background of high dielectric constant ε_b.

4.1.3 TRIANGULAR LATTICE PHOTONIC CRYSTAL

The triangular lattice structure consists of periodic array of parallel dielectric column circular cross-section and dielectric constant ε_a where intersection with a perpendicular plane form a triangular lattice. The dielectric columns are incorporated in a dielectric material in which constant is ε_b.

The square lattice and triangular lattice photonic crystals are the subjects of interest because these are now being used for current experimental purposes.

4.2 MATHEMATICAL FORMULATION

Analogy of the photonic crystal structure is found with semiconductors. Without the presence of external currents and sources, the Maxwell's equation can be modified as the following that is very similar to the Schrödinger equation. The equation of photonic crystal is given by

$$\vec{\nabla} \times \frac{1}{\varepsilon(\vec{r})} \vec{\nabla} \times \vec{H}(\vec{r}) = \frac{\omega^2}{c^2} \vec{H}(\vec{r}) \tag{4.1}$$

where symbols have usual significances. In this context, we will discuss about the solution of the equation with the approach already carried out, and consecutive inferences can be derived from it.

4.2.1 PHOTONIC BAND STRUCTURE

Equation 4.1 represents a linear Hermitian eigenvalue problem. The eigenvalues or eigen-frequencies are determined entirely by the microscopic dielectric function $\varepsilon(\mathbf{r})$. If $\varepsilon(\mathbf{r})$ is completely periodic as in an ideal photonic crystal, then results are denoted by a wave vector k and a band index n. The collection of eigen-frequency in the dispersion curve as a function of wave vector is usually taken under consideration of the first Brillouin zone and is called as photonic band structure.

The photonic band structure depends on lattice constant , the radius of the dielectric column in two dimensions and the dielectric contrast between the cylinders relative to the material making up the background of the photonic crystal. Photonic bandgap may exist for a variety of photonic crystal. The gap is a region of frequency where no electromagnetic modes may subsist within the photonic crystal for any value of wave vector k within the Brillouin zone. The band above the photonic bandgap is commonly known as a air band and the band below the photonic bandgap is commonly known as a dielectric band. Since electromagnetic modes are prohibited, spontaneous emission will be prohibited in cases in which photonic bandgap overlays the electronic bandgap. It is desirable to obtain a photonic band structure with a photonic bandgap, within which the travelling of electromagnetic wave is prohibited for all wave vector k within the Brillouin zone.

Previously, theoretical study of electromagnetic wave in photonic crystal is based on scalar approximation in which vector nature of the electromagnetic field is ignored. These calculations did not give correct result for the band structure. Nowadays, theoretical studies of photonic crystal rely on fully vectored electromagnetic field.

The translation symmetry of photonic crystal can be broken in addition of a dielectric effect. This is essentially a change in the dielectric constant of a specific region within the otherwise perfectly periodic photonic crystal. These dielectric defects may result in defect states within the photonic bandgap, permitting a localized mode to exist about the defect within the crystal. Dielectric defects result in defect states whining the photonic bandgap just below the air band, while air defects result in defect states just above the dielectric band of photonic band structure.

4.2.2 Two-Dimensional Photonic Crystal

In two dimension, owing to the presence of minor symmetry in the plane perpendicular to the dielectric column, it becomes possible to decouple the electromagnetic modes into TE and TM modes with reference to the plane normal to the dielectric column.

The general rule presented by Joannoploulos [12] is that TM bandgaps are recommended in the lattice of outlying region of high dielectric constant and that TE bandgaps are recommended in a connected lattice of high dielectric constant. The design of photonic bandgap of both TE and TM modes' designs are required a photonic crystal that attempts to combine the lattice which favored both TE and TM modes bandgap. Such crystal will have high dielectric constant region that are almost entirely isolated but which are still connected by narrow high veins. The triangular lattice of column of low dielectric constant ε_a in the background of high dielectric constant ε_b is an example of such photonic crystal. The remnants of high dielectric between the columns can be considered to be the isolated region of high dielectric constant required for TM bandgap. If the column is slightly smaller than the maximum packing would allow, then narrow veins of high dielectric constant will connect the mostly isolated region and allow TE bandgap to exits as well.

The dielectric column of dielectric constant 'ε_a' is assumed to be parallel to the z-axis. The convergence of the axis of these columns with the X–Y plane from a two-dimensional lattice (square or triangular) is given by the vector

$$r = l\vec{a}_1 + m\vec{a}_2 \qquad\qquad (4.2)$$

where vector $a_1$and a_2 are the two primitive translation vector of the lattice and l and m are any two integers. For the square lattice of lattice constant , the primitive lattice vector a_1 and a_2 are given by

$$a_1 = a\big[1,0\big]$$
$$a_2 = a\big[1,0\big]$$

For the triangular lattice constant a, the primitive lattice vector a_1 and a_2 are

$$\mathbf{a1} = a\left[\frac{\sqrt{3}}{2}, \frac{1}{2}\right]$$

$$\mathbf{a2} = a\left[-\frac{\sqrt{3}}{2}, \frac{1}{2}\right]$$

The dielectric constant of the this composite system is therefore position dependent, and referred by $\varepsilon(r)$.The area of the primitive unit cell of this lattice is equal to **Au = |a1 × a2|.**

4.2.2.1   Real-Space Representation of Square Lattice

The following Figure 4.3(a) is the real-space depiction of the two-dimensional square lattice column of high dielectric column ε_a in the background of low dielectric constant ε_b.

4.2.2.2   Reciprocal Space Representation of Square Lattice

Figure 4.3(b) is the reciprocal space depiction of two-dimensional square lattice of column. The reciprocal space representation of two-dimensional square lattice is shown in Figure 4.4 with the square Brillouin zone. The triangular irreducible portion of the Brillouin zone is shown in the upper right corner of Brillouin zone.

Figure (4.4) is shown as an expanded view of the Brillouin zone for two-dimensional square lattice. The k-space depiction of high symmetry $\vec{X}$, $\vec{\Gamma}$, $\vec{M}$ defines the irreducible Brillouin zone and is given by

$$\vec{\Gamma} = \frac{2\pi}{a}\big[0,0\big]$$

$$\vec{M} = \frac{2\pi}{a}\left[\frac{1}{2}, \frac{1}{2}\right]$$

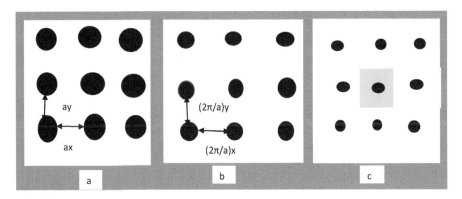

FIGURE 4.3 (a) The real-space depiction of the two-dimensional square lattice of column high dielectric constant in a background of low dielectric constant. (b) The reciprocal depiction of the two-dimensional square lattice column high dielectric constant in a background of low dielectric constant. (c) The reciprocal space depiction of two-dimensional square lattice shown in (b) with the square Brillouin zone. The triangular irreducible portion of the Brillouin zone is shown in the upper right corner of Brillouin zone.

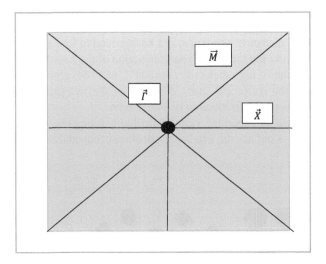

FIGURE 4.4 An expanded view of the Brillouin zone for the two-dimensional square lattice. The triangular irreducible portion of the Brillouin zone is shown in the upper right corner of the Brillouin zone. The **k-space** defines of high symmetry $\vec{X}, \vec{\Gamma}, \vec{M}$.

$$\vec{X} = \frac{2\pi}{a}\left[\frac{1}{2}, 0\right] \quad 39$$

The irreducible portion of the Brillouin zone defines the k–space trajectory to be followed when calculating the photonic band structure. The band structure for all other k – space points outside the irreducible portion of the Brillouin zone can be found by taking advantage of the symmetry of the Brillouin zone.

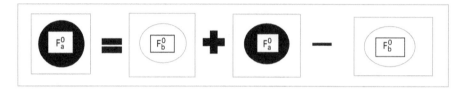

FIGURE 4.5 Schematic illustration of the unit cell construction for 2D square lattice of dielectric column. Shown in the figure are the inverse dielectric function for the unit cell dielectric background ($f_b^o = \dfrac{1}{\varepsilon_a}$) and the dielectric column ($f_a^o = \dfrac{1}{\varepsilon_b}$).

Within each unit cell of the square lattice, we have a single dielectric column (Figure 4.5).

$$f(r) = \frac{1}{\varepsilon(r)}$$

4.2.3 REAL-SPACE REPRESENTATION OF TRIANGULAR LATTICE

Figure 4.6 shows the real-space representation of two-dimensional triangular lattice of column of high dielectric constant ε_a in a background of low dielectric constant ε_b.

Figure 4.6b shows the reciprocal space depiction of the two-dimensional triangular lattice of column of high dielectric constant ε_a in a background of low dielectric constant ε_b with the reciprocal space depiction of the two-dimensional triangular lattice with the hexagonal irreducible Brillouin zone.

Figure4.7 shows an expanded view of Brillouin zone. The *k-space* points high symmetry $\vec{\Gamma}, \vec{M}, \underset{K,}{\rightarrow}$ defines the irreducible Brillouin zone and is given as

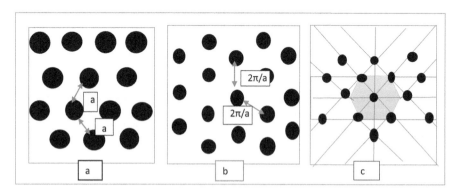

FIGURE 4.6 (a) Real-space depiction of two-dimensional triangular lattice of column of high dielectric constant $\boldsymbol{\varepsilon}_a$ in a background of low dielectric constant $\boldsymbol{\varepsilon}_b$. (b) Reciprocal space depiction of the two-dimensional triangular lattice of column of high dielectric constant $\boldsymbol{\varepsilon}_a$ in a background of low dielectric constant $\boldsymbol{\varepsilon}_b$. (c) Reciprocal space depiction of the two-dimensional triangular lattice with the hexagonal irreducible Brillouin zone.

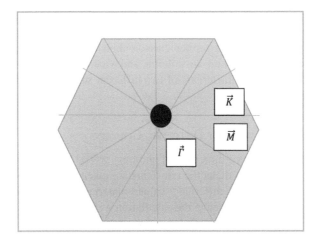

FIGURE 4.7 An expanded view of the Brillouin zone for the two-dimensional triangular lattice. The triangular irreducible portion of the Brillouin zone is shown. The **k-space** points high symmetry $\vec{\Gamma}$, $\vec{M}$, $\underset{K,}{\rightarrow}$ defines the irreducible Brillouin zone.

$$\vec{\Gamma} = \frac{2\pi}{a}\left[0,0\right]$$

$$\vec{M} = \frac{2\pi}{a}\left[\frac{\sqrt{3}}{3},0\right]$$

$$\vec{k} = \frac{2\pi}{a}\left[\frac{\sqrt{3}}{3},\frac{1}{3}\right]$$

The irreducible portion of the Brillouin zone defines *k-space* trajectory to be followed when calculating the photonic band structure. The band structure of all other *k–* space points outside the irreducible portion of the Brillouin zone can be found by taking advantages of the symmetry of the Brillouin zone.

4.3 RESULT ANALYSIS

The computational method was applied in two dimensions for both the square and triangular lattice of dielectric column with dielectric constant 'ε_a', submerged in a background material designated by a dielectric constant 'ε_b' and vice versa.

In the two dimensions, the electromagnetic waves are supposed to be travelling along in plane perpendicular to the columns. Two polarizations of these waves are taken into account. The first was transverse electric or TE polarization where magnetic field vector is parallel to the columns. The second was the transverse magnetic or TM polarization where the electric field vector is parallel to the columns. The eigen-frequency was found for the *k*-dependent matrices for these polarization Θ_{TE} and $\Theta_{TM.}$

Photonic crystal structure can be classified by two different configuring structure which are square lattice structure and triangular lattice structure. Using these two types of configuration, we can further classify their different structure by making lattice of dielectric column and air column and vice versa. The following figure describes the dispersion characteristic of two different structures of photonic crystal (Figures 4.8 and 4.9).

4.3.1 THE SQUARE LATTICE OF DIELECTRIC COLUMNS

The bandgap diagram of two-dimensional photonic crystal of square lattice is shown in the figure. Here different dielectric constants are taken into account. The packing factor P_f and radius vary with different k-valued materials with a varying lattice constant. The reciprocal vectors used in the calculation were chosen from all those reciprocal vectors contained within a circular region in k– space bounded by a maximum $|G|$ that would ensure proper convergence of the eigen frequency. The circular region is a discrete estimate that improves as the number of reciprocal vector is increased. The value of dielectric constant is chosen such a way that might be manufactured for microwave frequency range.

We see from our experimental result that the two-dimensional square lattice of dielectric column has a photonic bandgap for the TM modes but not the TE modes. The TM photonic bandgap exists between the first and second bands and extends throughout the entire irreducible Brillouin zone. The TE modes have a large gap

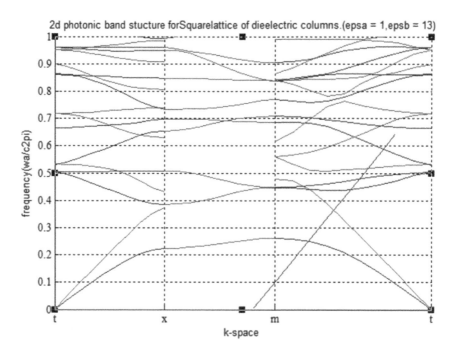

FIGURE 4.8   Dispersion curve of square lattice.

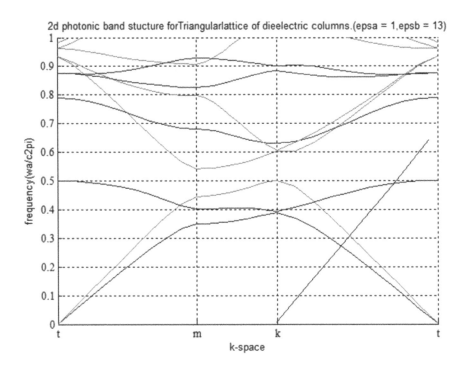

FIGURE 4.9   Dispersion curve of triangular lattice.

throughout the most of the Brillouin zone; however, a photonic bandgap is closed in the *k-space* region from *X* to *M*.

Various dielectric constants with respect to lattice constant and packing factor are taken into account.

1. The simulation is first started by taking air as a background and GaAs (gallium arsenide) as a dielectric column, i.e. $\varepsilon_a = 1$ and $\varepsilon_b = 13$. Here we consider packing factor of.0836 and the radius of the dielectric column is $R = .5a$(where *a* is a lattice constant of materials). The dispersion curve and equal frequency contour are shown in the figure. Only TM modes are visible using the proper photonic bandgap (Figures 4.10 and 4.11).

4.3.2   THE SQUARE LATTICE OF AIR COLUMNS

The photonic band structure for two-dimensional square lattice of air column is shown in the figure. The reciprocal vectors used in the calculation were chosen from all those reciprocal vectors contained within a circular region in *k*– space boundary by a maximum |*G*|that would ensure proper convergence of the eigen frequency. The circular region is a discrete estimate that improves as the number of reciprocal vector is increased. The TM photonic bandgap exists between the first and second bands, and extends throughout most of the irreducible Brillouin zone. Both TM and TE modes larger gap throughout most of the Brillouin zone; however, photonic bandgap for both polarization is closed in the *k* space region from *X* to *M*.

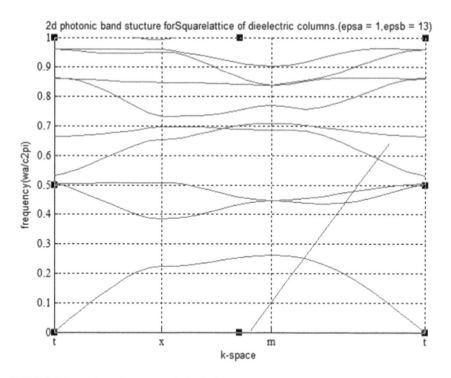

FIGURE 4.10   Dispersion curve of Air-GaAs.

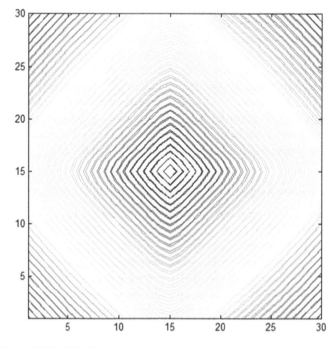

FIGURE 4.11   EFC of Air-GaAs structure.

4.3.3   TRIANGULAR LATTICE OF DIELECTRIC COLUMNS

The two-dimensional triangular lattices of the dielectric column have both photonic bandgap in TE and TM modes. The TM and TE bandgap exists between the first and second bands, and extends throughout the entire irreducible Brillouin zone. Both TE and TM modes have large bandgaps throughout the most Brillouin zone; however, the photonic bandgap for the both polarization is closed in the k-space region from X to M.

The photonic band structure for the two-dimensional triangular lattice of dielectric column is shown in the figure. Different k-valued materials are taken into account and their respective packing factors are used in this structure (Figures 4.12 and 4.13).

4.3.4   TRIANGULAR LATTICE OF AIR COLUMNS

It is desirable to have a photonic crystal with a photonic bandgap in a given range of frequencies for both TM and TE modes. To achieve these requirements, the photonic crystals have high dielectric regions that are practically segregated and connected by narrow links. The photonic crystal of triangular lattice of air columns is shown in the Figures 4.14 and 4.15.

4.3.5   ANALYSIS OF RESULT DIFFERENT K-VALUED MATERIAL IN TRIANGULAR AND SQUARE LATTICES

From the above graph, we get a general formation of photonic crystal structure of square and triangular lattice. Now we analyse the different k-valued material for making photonic crystal for both TE and TM modes, where we get the different λ

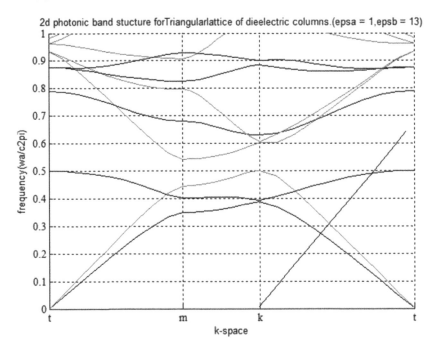

FIGURE 4.12   Dispersion curve of the triangular lattice of dielectric columns.

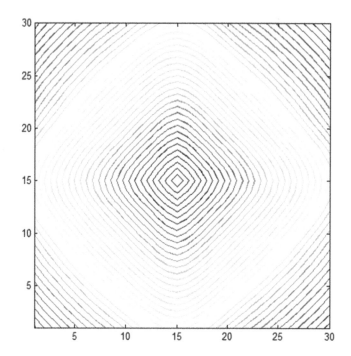

FIGURE 4.13   Equal frequency contour of the triangular lattice of dielectric column.

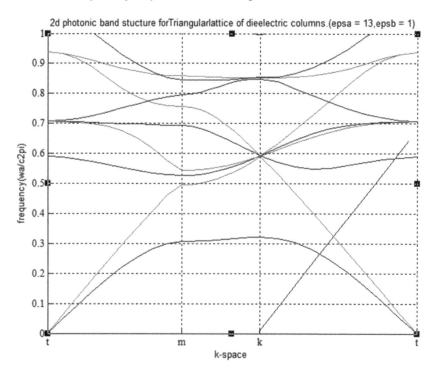

FIGURE 4.14   Dispersion curve of the triangular lattice of air columns.

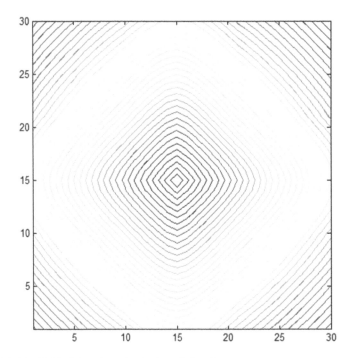

FIGURE 4.15   EFC of the triangular lattice of air columns.

value, i.e. the different wavelengths in which frequency of the photonic crystal can act as a superlens. Here we tried to get the λ value in the visible range for obtaining the super lensing function. To start the project, we choose direct (GaAs, InP) and indirect (Si,Ge) bandgap semiconductor as one of the dielectric materials of photonic crystal and another dielectric material is air for making the photonic crystal structure. From this different structure, we calculate dispersion curve, light line and equal frequency contour to find the frequency or a λ value in which the photonic crystal structure can act as a superlens.

In the below case study, we see the different structure of photonic crystal and different dispersion curve and equal frequency contour. Here we take both square and triangular lattice structure for both TE and TM modes to calculate the cut-off frequency on particular polarization.

So starting with the indirect bandgap semiconductor material followed by direct bandgap material as one dielectric and air is another dielectric on triangular lattice one is air background dielectric column and another is air column with dielectric background. λ value of both the TE and TM modes is calculated from the dispersion curve where the x-axis stands for k-space which is $2\pi/\lambda$, and the y-axis denotes frequency, i.e. $\omega a/2\pi c$.

4.3.6   CASE STUDY 1 (MATERIAL SI AND AIR) IN TE MODES FOR TRIANGULAR LATTICE

Triangular lattice with $\varepsilon_a = 1$ and $\varepsilon_b = 11.7$ (air background and Si dielectric rod) (Figure 4.16 and 4.17)

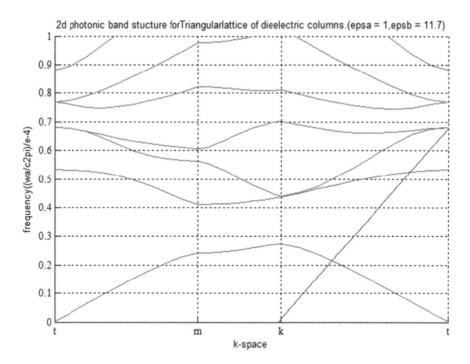

FIGURE 4.16   Dispersion curve of air–Si in TE mode for triangular lattice.

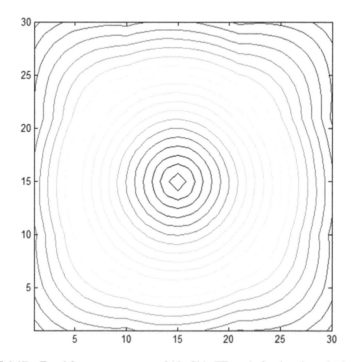

FIGURE 4.17   Equal frequency contour of Air–Si in TE mode for the triangular lattice.

Here the values of different parameters are given as follows: packing factor = 0.34 (because it follows the diamond lattice structure), cube size in *k*-space is (*N* = 2), radius = 0.5*a* and lattice constant is = 0.05μm.

4.3.7 CASE STUDY 2 (MATERIAL SI AND AIR) IN TM MODES FOR TRIANGULAR LATTICE

a) Triangular lattice with $\varepsilon_a = 1$ and $\varepsilon_b = 11.7$ (Air background and Si dielectric column) (Figures 4.18 and 4.19) (Table 4.1)

4.3.8 CASE STUDY 3 (MATERIALS GE AND AIR) IN TM MODES FOR TRIANGULAR LATTICE

Triangular lattice with $\varepsilon_a = 1$ and $\varepsilon_b = 16.2$ (air background and dielectric rod of Ge) (Figures 4.20 and 4.21)

4.3.9 CASE STUDY 4 (MATERIALS GE AND AIR) IN TE MODES FOR TRIANGULAR LATTICE

Triangular lattice with $\varepsilon_a = 1$ and $\varepsilon_b = 16.2$ (air background and dielectric rod of Ge) (Figures 4.22 and 4.23) (Table 4.2)

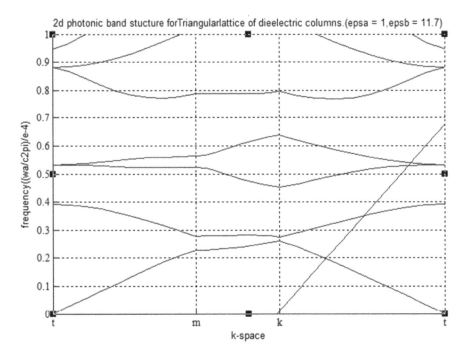

FIGURE 4.18 Dispersion curve of Air–Si in TM mode for triangular lattice.

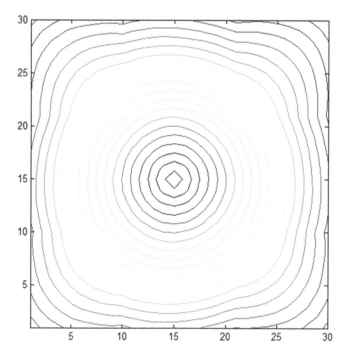

FIGURE 4.19   Equal frequency contour of Air–Si in TM mode for triangular lattice.

TABLE 4.1
Air–Si Triangular Lattice Dataset

	TE Mode			TM Mode		
	x-value	y-value	wavelength (nm)	x-value	y-value	wavelength (nm)
First band	1.137	0.1213	405.012	1.132	0.1181	423.37
Second hand	1.152	0.1363	366.83	1.149	0.1334	374.81
Third band	1.175	0.1571	318.26	1.173	0.1556	321.33
Fourth band	1.209	0.1875	266.66	1.194	0.1731	288.85
Fifth band	1.284	0.2553	195.84	1.213	0.1918	260.68

4.3.10   CASE STUDY 5 (MATERIAL InP AND AIR) IN TM MODES FOR TRIANGULAR LATTICE

Triangular lattice with $\varepsilon_a = 1$ and $\varepsilon_b = 12.5$ (Air background with dielectric column of InP) (Figures 4.24–4.27) (Table 4.3)

4.3.11   CASE STUDY 6 (MATERIAL GaAs AND AIR) IN TM MODES FOR TRIANGULAR LATTICE

Triangular lattice with $\varepsilon_a = 1$ and $\varepsilon_b = 13$ (air background with dielectric column of GaAs) (Figures 4.28–4.31) (Table 4.4)

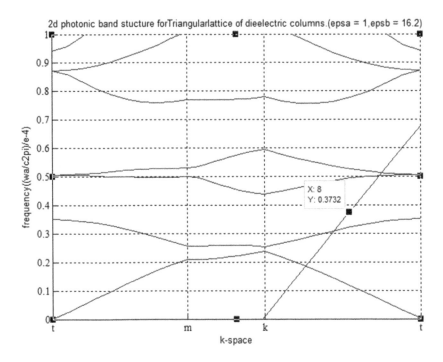

FIGURE 4.20 Dispersion curve of air–Ge in TM mode for triangular lattice.

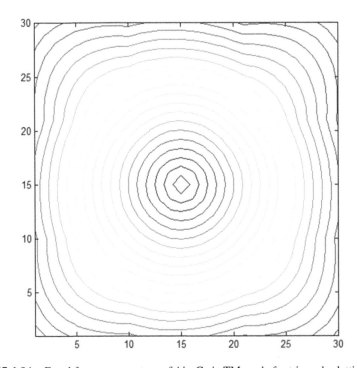

FIGURE 4.21 Equal frequency contour of Air–Ge in TM mode for triangular lattice.

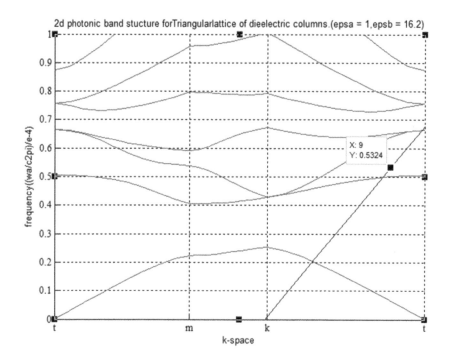

FIGURE 4.22   Dispersion curve of Air–Ge in TE mode for triangular lattice.

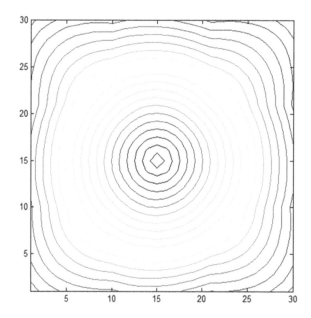

FIGURE 4.23   Equal frequency contour of Air-Ge in TE mode for triangular lattice.

TABLE 4.2

Air–Ge Triangular Lattice Dataset

	TE Mode			TM Mode		
	x-value	*y*-value	Wavelength (nm)	*x*-value	*y*-value	Wavelength (nm)
First band	1.16	0.1086	460.4	1.153	0.1026	487.32
Second band	1.173	0.1198	417.36	1.172	0.119	420.16
Third band	1.197	0.1406	355.61	1.195	0.1387	360.49
Fourth band	1.236	0.1743	286.86	1.22	0.1606	311.33
Fifth band	1.321	0.2468	202.25	1.238	0.1767	282.96

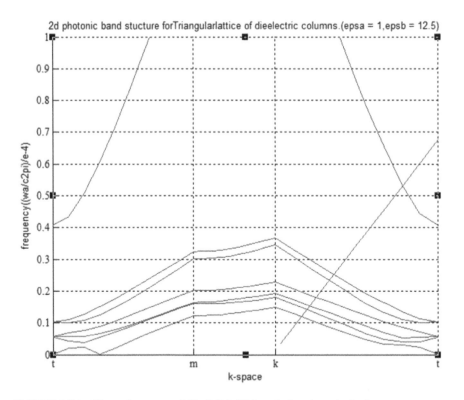

FIGURE 4.24   Dispersion curve of Air–InP in TM mode for triangular lattice.

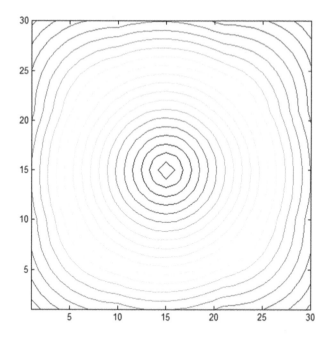

FIGURE 4.25   Equal frequency contour of Air–InP in TM mode for triangular lattice.

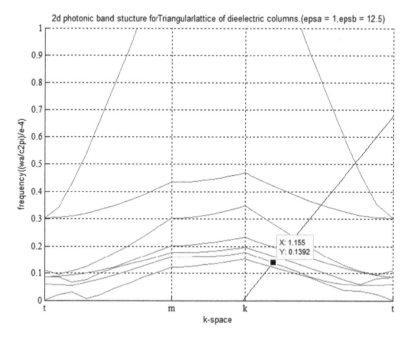

FIGURE 4.26   Dispersion curve of Air–InP in TE mode for triangular lattice.

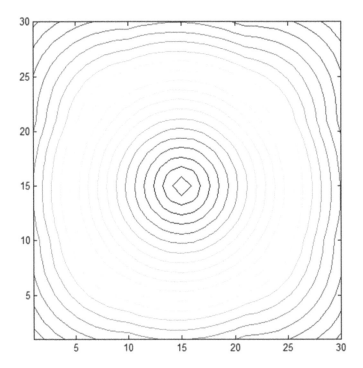

FIGURE 4.27 Equal frequency contour of Air–InP in TE mode for triangular lattice.

TABLE 4.3
Air–InP Triangular Lattice Dataset

	TE Mode			TM Mode		
	x-value	*y*-value	Wavelength (nm)	*x*-value	*y*-value	Wavelength (nm)
First band	1.17	0.1255	398.4	1.168	0.1152	434.02
Second band	1.195	0.139	359.71	1.198	0.1414	353.6
Third band	1.223	0.1637	305.43	1.215	0.1564	319.69
Fourth band	1.261	0.1962	254.84	1.249	0.1856	269.39
Fifth band	1.308	0.2327	214.86	1.31	0.2394	208.85

Now we consider the structure of square lattice with different k-valued material taking dielectric as well as air column structure. The major point shown is that unlike triangular lattice structure only one mode either TE or TM modes is shown in the square lattice structure and also we have seen that square lattice of air column structure can be formed properly. So we consider only the square lattice of dielectric column. We have to calculate the cut-off frequency accordingly. Here are the different case studies sated below:

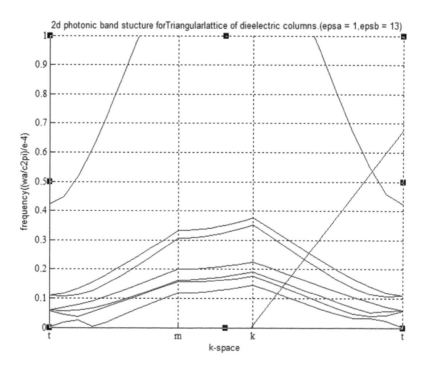

FIGURE 4.28   Dispersion curve of Air–GaAs in TM mode for triangular lattice.

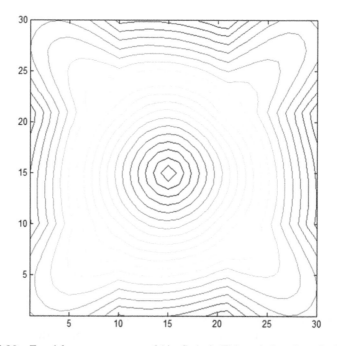

FIGURE 4.29   Equal frequency contour of Air–GaAs in TM mode for triangular lattice.

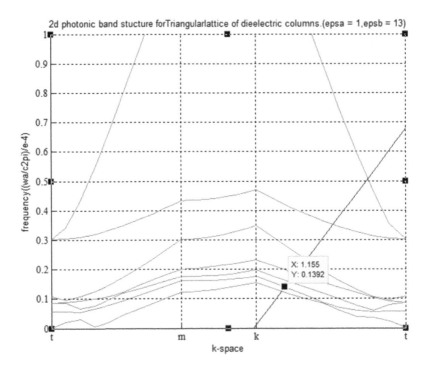

FIGURE 4.30 Dispersion curve of Air–GaAs in TE mode for triangular lattice.

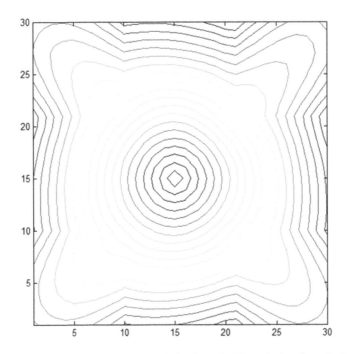

FIGURE 4.31 Equal frequency contour of Air–GaAs in TE mode for triangular lattice.

TABLE 4.4

Air–GaAs Triangular Lattice Dataset

	TE Mode			TM Mode		
	x-value	y-value	Wavelength (nm)	x-value	y-value	Wavelength (nm)
First band	1.178	0.1245	401.6	1.169	0.1165	429.18
Second band	1.195	0.139	359.71	1.194	0.1384	361.27
Third band	1.218	0.159	314.43	1.21	0.1522	328.51
Fourth band	1.252	0.1885	262.84	1.247	0.184	271.73
Fifth band	1.307	0.2368	214.86	1.307	0.2361	211.77

4.3.12   CASE STUDY 7 (MATERIAL AIR AND SI) IN TM MODES IN SQUARE LATTICE

Square lattice with $\varepsilon_a = 1$ and $\varepsilon_b = 11.7$ (Air background and Si dielectric rod) (Figure 32–35)

Here the values of different parameters are given as follows: packing factor = 0.836(because it follows the zinc blend lattice structure), cube size in k-space is ($N = 2$), radios = $0.5a$ and lattice constant is 0.05μm (Table 4.5)

Henceforth, for square lattice, we have always carried out TM mode analysis.

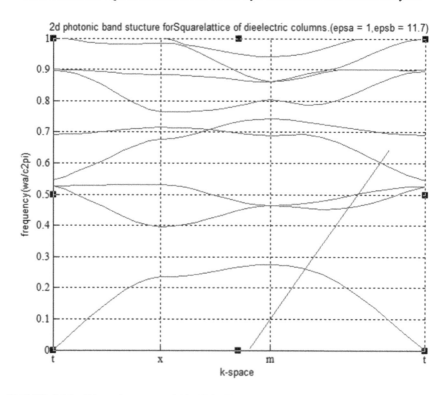

FIGURE 4.32   Dispersion curve of Air–Si in TM mode for square lattice.

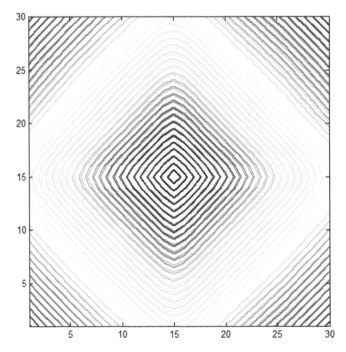

FIGURE 4.33   Equal frequency contour of Air–Si in TM mode for square lattice.

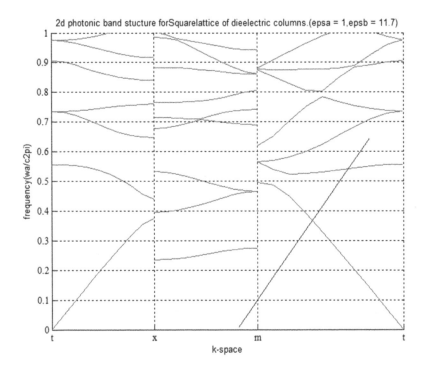

FIGURE 4.34   Dispersion curve of Air–Si in TE mode for square lattice.

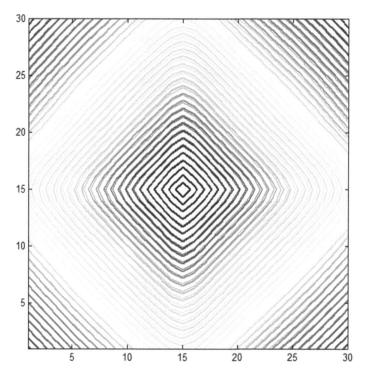

FIGURE 4.35 Equal frequency contour of Air–Si in TE mode for square lattice.

TABLE 4.5
Air–Si Square Lattice Dataset

	TE Mode			TM Mode		
	x-value	*y*-value	Wavelength (nm)	*x*-value	*y*-value	Wavelength (nm)
First band	—	—	—	7.7	0.3255	165.53
Second band	—	—	—	8.7	0.4682	124.97
Third band	—	—	—	9.7	0.6432	89.21

4.3.13 CASE STUDY 8 (MATERIAL AIR AND GE) IN TM MODES IN SQUARE LATTICE

Square lattice with $\varepsilon_a = 1$ and $\varepsilon_b = 16.2$ (Air background and Si dielectric rod) (Figure 4.36 and 4.37) (Table 4.6)

4.3.14 CASE STUDY 9 (MATERIAL AIR AND INP) IN TM MODES IN SQUARE LATTICE

Square lattice with $\varepsilon_a = 1$ and $\varepsilon_b = 12.5$ (Air background and InP dielectric rod) (Figures 4.38 and 4.39) (Table 4.7)

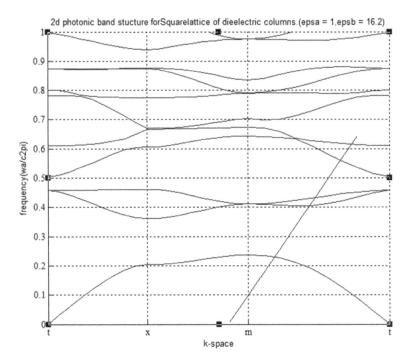

FIGURE 4.36 Dispersion curve of Air–Ge in TM mode for square lattice.

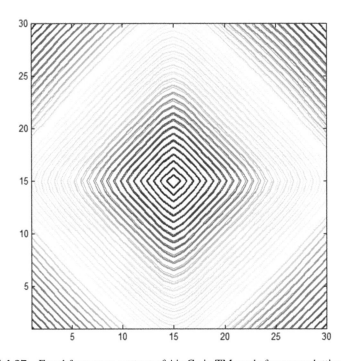

FIGURE 4.37 Equal frequency contour of Air-Ge in TM mode for square lattice.

TABLE 4.6

Air–Ge Square Lattice Dataset

	TE Mode			TM Mode		
	x-value	y-value	Wavelength (nm)	x-value	y-value	Wavelength (nm)
First band	—	—	—	6.7	0.1663	283.83
Second band	—	—	—	8.7	0.4682	124.97
Third band	—	—	—	9.7	0.6432	89.21

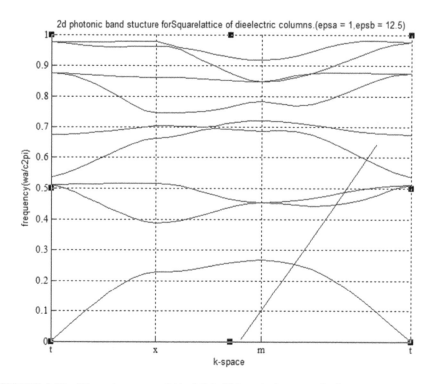

FIGURE 4.38 Dispersion curve of Air–InP in TM mode for square lattice.

4.3.15 CASE STUDY 10 (MATERIAL AIR AND GAAS) IN TM MODES IN SQUARE LATTICE

Square lattice with $\varepsilon_a = 1$ and $\varepsilon_b = 13$(Air background and GaAs dielectric rod) (Figures 4.40 and 4.41) (Table 4.8)

Note that the dataset is the same as we have obtained for Air–InP system.

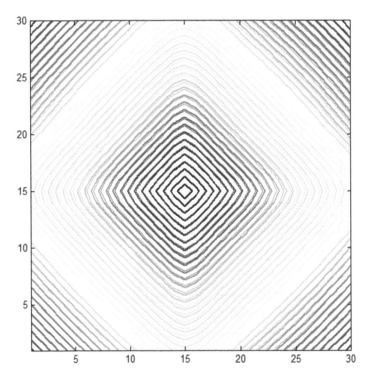

FIGURE 4.39 Equal frequency contour of Air–InP in TM mode for square lattice.

TABLE 4.7
Air–InP Square Lattice Dataset

	TE Mode			TM Mode		
	x-value	y-value	Wavelength (nm)	x-value	y-value	Wavelength (nm)
First band	—	—	—	7.7	0.3255	165.53
Second band	—	—	—	8.7	0.4682	124.97
Third band	—	—	—	9.7	0.6432	89.21

4.4 CONCLUSION

Simulated findings exhibit that square lattice can produce superlens only for TM mode, whereas triangular structure can make it for both the modes. Also, dissimilarity of refractive indices of the elemental materials can affect the wavelengths of different bands of the superlens. A comparative study reveals that triangular lattice can automatically be considered as the candidate for the optical lens design purpose. However, cut-off wavelengths are almost independent for square lattice; however, it is tunable for triangular one. Henceforth, optimization is required from an operational point-of-view.

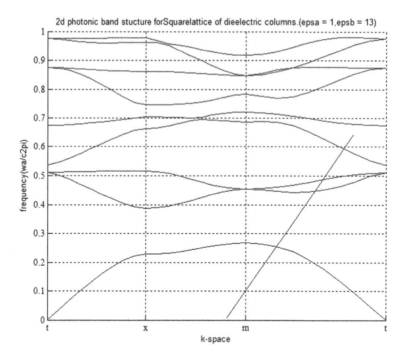

FIGURE 4.40   Dispersion curve of Air–GaAs in TM mode for square lattice.

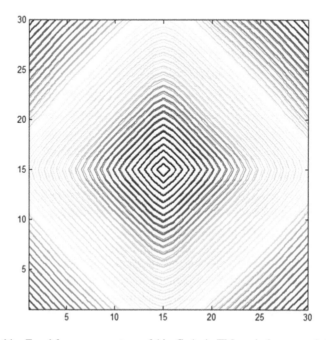

FIGURE 4.41   Equal frequency contour of Air–GaAs in TM mode for square lattice.

TABLE 4.8
Air–GaAs Square Lattice Dataset

	TE Mode			TM Mode		
	x-value	y-value	Wavelength (nm)	x-value	y-value	Wavelength (nm)
First band	—	—	—	7.7	0.3255	165.53
Second band	—	—	—	8.7	0.4682	124.97
Third band	—	—	—	9.7	0.6432	89.21

REFERENCES

1. Loudon, R. The Propagation of Electromagnetic Energy through an Absorbing Dielectric, *Journal of Physics A*, vol. 3, pp. 233-245, 1970

2. Yablonovitch, E. Inhibited Spontaneous Emission in Solid-State Physics and Electronics, *Physical Review Letters*, vol. 58, pp. 2059-2061, 1987

3. Fogel, I. S., Bendickson, J. M., Tocci, M. D., Bloesmer, M. J., Scalora, M., Bowden, C. M., Dowling, J. P. Spontaneous Emission and Nonlinear Effects in Photonic Bandgap Materials, *Pure and Applied Optics: Journal of the European Optical Society Part A*, vol. 7, pp. 393-408, 1998

4. Chakraborty, P., Ghosh, R., Adhikary, A., Deyasi, A., Sarkar, A. Electromagnetic Bandgap Formation in Two-Dimensional Photonic Crystal Structure with DNG Materials under TE Mode, *IEEE International Conference on Devices for Integrated Circuit*, 2019

5. Yablonovitch, E. Photonic bandgap structures, *Journal of Optical Society of America B*, vol. 10, pp. 283-295, 1993

6. Armenise, M. N., Campanella, C. E., Ciminelli, C., Dell'Olio, F., Passaro, V. M. N. Phononic and Photonic Bandgap Structures: Modeling and Applications, *Physics Procedia*, vol. 3(1), pp. 357-364, 2010

7. Painter, O., Srinivasan, K. Localized defect states in two-dimensional photonic crystal slab waveguides: A simple model based upon symmetry analysis, *Physical Review B*, vol. 68, p. 035110, 2003

8. Akbar, F., Syahriar, A., Lubis, A. H. Dispersion Relation of 1-D Photonic Crystal, *IEEE International Conference on Electrical Engineering and Computer Science*, 2014

9. Romo, G., Smy, T. Dispersion Relation Calculation of Photonic Crystals Using the Transmission Line Matrix Method, *International Journal of Numerical Modeling, Electronic Networks, Devices and Fields*, vol. 17(5), pp. 451-459, 2004

10. Guller, F., Inchaussandague, M. E., Depine, R. A. Dispersion Relation and Bandgaps of 3D Photonic Crystals Made of Spheres, Progress in Electromagnetics Research M, vol. 19, pp. 1–12, 2011

11. Lipson, R. H., Lu, C. Photonic Crystals: A Unique Partnership Between Light and Matter, *European Journal of Physics*, vol. 30(4), pp. S17–S32, 2009

5 Investigation on Some Fast Optical/Opto-Electronic Switching Systems for Implementing Different Modulation Schemes

Minakshi Mandal, Suranjan Lakshan,
Debashri Saha, Agnijita Chatterjee and
Sourangshu Mukhopadhyay

CONTENTS

5.1 INTRODUCTION

Communication and data processing is a large and essential part of modern civilization. In the case of communication and data processing, the required speed is very high and it is of the order of THz. The communication and data processing system which are based on electronics show their limitations of speed because of their lower switching speed. Optics can easily reach the speed of THz limit which can overcome the limitation of electronics. Light has a common property of inbuilt parallelism. The storage capacity of light-based optical memory is also found very high. Due to those properties of light, the speed of operation can reach in the superfast level. Also in optical systems, the attenuation and different distortions are seen very low. The bit error rate, crosstalk, signal-to-noise ratio, etc., are found highly improved in optical communication. Use of light as a carrier signal presents unlimited bandwidth. The systems which use optical signals for processing require very low power also. Optical systems offer very low cost.

Optical communication is very much advantageous in modern technology and data processing. In all-optical systems, both communication and data processing are found very fast [1,2]. Optical switching properties show very low response time, which helps the fast operation of all-optical processors.

Optical switches are capable to do channel selection with the fastest possible speed in the domain of all-optical networking. Some of the optical switches convert the optical signal to an electronic signal with tremendous operation speed. Different all-optical and optoelectronic switches are developed for devices related to fast communication and data processing systems [3,4]. They are electro-optic Pockels cell-based switches, Kerr switches, semiconductor optical amplifier based switches, erbium-doped fiber amplifier, etc.

The electro-optic effect is the variation of the refractive index of a medium, caused by an externally applied electric field. When a polarized laser light beam passes through an electro-optic Pockels material, a change in refractive index occurs with the application of an electric field to the material, and this change in refractive index is linearly proportional to the applied field strength. This effect is called the electro-optic pockels effect.

Only the non-centro symmetric materials such as potassium di-deuterium phosphate, potassium di-hydrogen phosphate (KDP), lithium tantalate (LiTaO$_3$), lithium niobate (LiNbO$_3$), etc., can exhibit the linear electro-optic effect very well.

Optical Pockels cells are good candidates for conducting optical circuit operation with very high operation speed. So many all-optical switches are developed by using the optical property of nonlinear electro-optic material. The electro-optic modulator can help the generation of amplitude, phase and polarization modulation very successfully if an optical beam is used as carrier and the biasing electric signal is used as message [5–9]. On the other side, optical tree architectures (OTAs) are also developed using the Pockels material [10]. It is a very good optical system that is capable to organize different Boolean logic operations. The binary arithmetic operation can also be conducted by this type of tree network. It is a multiplying system of breaking a single straight path into several distributed branches and sub-branches. All-optical logic gates are developed using Pockels material and semi-conductor optical amplifier exploiting the phase and polarization encoding technique [11–14].

The optical Kerr effect occurs when an intense beam of light propagates in some materials having second order of nonlinear property. In Kerr electro-optic effect, when an intense beam of laser light propagates through it, the intensity of the light beam modulates itself due to the change of the refractive index of the material. Optical Kerr medium has many applications in optical digital signal processing, optical self-focusing and defocusing and optical switching activities [15–18].

When an intense beam of laser light passes through a Kerr nonlinear medium, the change in refractive index (n) of the medium is expressed as

$$n = n_0 + n_2 I,$$

where n_0 is the refractive index of the material at very low intensity of light, n_2 is Kerr nonlinear coefficient and I is the intensity of the light beam.

The refractive index of the medium depends upon the intensity of the light beam. So when the intensity of the input beam is increased, the refractive index of the material is also increased. Thus by Snell's law, with the increasing of the refractive index of the medium, the refraction angle decreases; hence, the light follows a different path. So by using these phenomena different kinds of Kerr nonlinear optical switches are generated.

Different modulation processes are developed using the Pockels effect and the Kerr effect. The electro-optic effect provides a huge range of applications as it offers a very fast modulation process. So many research works are done by using the characters of optical amplitude modulation phase modulation, intensity modulation, etc., done by electro-optic modulators. Using the phase modulation technique in an electro-optic modulator a higher intensity of the harmonic signals is achieved. Using the optical modulation technique optical switching, all-optical quantum logic gates such as Pauli X, Y, Z gate, SRN gate and SRZ gate are also developed. An optical scheme is developed where the intensity of output light is increased with the help of feedback mechanism in the elctro-optic modulator by phase and amplitude modulation processes. The OTA scheme is developed for the optical conversion system like from binary to decimal or decimal to binary. All-optical tree architecture schemes are developed for using two-input all-optical parallel logic operation, and all-optical data comparison schemes. All-optical half adder/half substructor and full adder/full substructor are also implemented by using the OTA scheme. Optical Kerr medium was successfully used to obtain an ultra-short optical pulse and to convert the frequencies of a Manchester encoded optical data and an optical pulse with saw-tooth intensity pattern. In Section 8 of this chapter, optical Kerr medium has been utilized to generate linear frequency variation of light using Kerr nonlinearity followed by a parabolic intensity-shaped light signal and also using a multi-feedback technique.

In Section 2, a single electro-optic Pockels cell and a single light beam are used for phase modulation by using two biasing voltages parallelly [19]. The biasing voltages are jointly applied to the electro-optic Pockels cell: one is applied parallel to the Z-axis of the crystal and another is applied parallel to the Y-axis. The used electro-optic material is KDP crystal and the biasing signals are two different saw-tooth electronic signals. The ultimate modulation conducted here is phase modulation. In Section 3, a Kerr cell is used to investigate the change of the intensity of light of the central frequency which is second time passed through a Kerr nonlinear material [20]. The used Kerr nonlinear material is fused silica glass. The intensity of the propagating light beam through the material modulates itself. In Section 4, an all-optical X-OR logic gate is designed by using the electro-optic Pockels cell exploiting the optical tree network theory [21]. The phase encoding mechanism is used to design the logic operation of the X-OR gate. The nonlinear material used in the Pockels cell is lithium niobate ($LiNbO_3$). In Section 5, the implementation of intensity-dependent wide-scale phase control technique by the joint effort of Kerr and Pockels materials is discussed. A different idea is proposed behind a large-scale optical phase modulation of light beam after passing through Pockels and Kerr material at a time. In Section 6, an explanation is given about how does the phase of an optical signal propagating through one Pockels and Kerr material is nullified by another second linear electro-optic material to achieve zero phases at final output [22]. The goal of the scheme in Sections 5 and 6 is to design a phase variation scheme of incident optical signal by controlling the intensity and also biasing voltage both by the joint effort of Pockels and Kerr materials in series combination. In Section 7, an alternating scheme is developed to design a theoretical model by using multi-feedback mechanism for increasing the transmission coefficient of an electro-optic modulator [23]. In Section 8, an optical system is developed which can generate the linear frequency variation of light. This scheme is based on Kerr nonlinearity followed by a parabolic light signal and by using the multi-feedback technique. Due to the nonlinear property of optical Kerr medium, this medium can be used to convert the frequency of the light signal having time-dependent intensity pattern passing through the medium. This conversion process is possible only when an optical Kerr medium with very high nonlinear coefficient is triggered by intense laser light.

5.2 MODULATION OF LIGHT BY POCKELS CELL BIASED BY SAW-TOOTH ELECTRONIC PULSE

KDP is normally a uniaxial crystal but when an electric field is applied to it, it becomes bi-axial. When an intense beam of light propagates properly through a KDP crystal, the refractive index of the material is changed linearly with the electric field. This change in refractive index generates different types of modulations such as amplitude modulation, frequency modulation, phase modulation and polarization modulation.

Let a beam of light, polarized in the X-direction, propagates along the Z-direction (optic axis) from one side to another side of the crystal and an electric field (E_z) parallel to the Z-direction is applied to the crystal, then the refractive index (n_x) [5,6] of the crystal is expressed as

$$n_x = n_0 + \frac{1}{2}n_0^3 r_{63} E_z.$$

Similarly, when the electric field is applied parallel to the Y-direction of the crystal, then for the same light beam propagating along the Z-direction of the crystal, the refractive index (n_x') [5,6] of the crystal is expressed as

$$n_x' = n_0 - \frac{1}{2}n_0^3 r_{41} E_y \tan\theta,$$

where n_0 is the constant the refractive index of the material, r_{63} and r_{41} are the electro-optic coefficients of KDP crystal, E_z is the applied field along the Z-direction, and E_y is the field applied along the Y-direction. $\theta\theta$ is a small angle of rotation of the new coordinate with respect to the old coordinate when the applied field is parallel to the Y-direction.

5.2.1 JOINT MODULATION ON A SINGLE LIGHT BEAM

The schematic block diagram of this joint modulation is shown in Figure 5.1. Two different biasing voltages V_y and V_z are applied along the Y- and Z-directions, respectively, to the crystal as shown in Figure 5.1. A light polarized in the X-direction propagates through the crystal along positive Z-direction. The length of the crystal is l and the width is d. Now the expression of the refractive index of the material is

$$n = n_0 + \frac{1}{2}n_0^3 r_{63} E_z - \frac{1}{2}n_0^3 r_{41} E_y \tan\theta.$$

So the expression of the electric field associated with the output beam is given by

$$E_{output} = E_0 e^{i[\omega t - k_0 nl]},$$

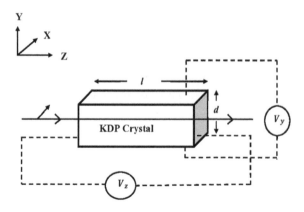

FIGURE 5.1 Joint Modulation of a single light beam.

$$E_{output} = e^{i\left[\omega t - k_0 n_0 l - \frac{1}{2}k_0 n_0^3 r_{63} E_z l + \frac{1}{2}k_0 n_0^3 r_{41} E_y l \tan\theta\right]},$$

where

$$\tan\theta = [(n_0^2 - n_e^2)/r_{41}E_y n_e^2 n_0^2 + r_{41}E_y n_e^2 n_0^2 / (n_0^2 - n_e^2)]$$

$n_0 = 1.5115$, $n_e = 1.4698$, $r_{63} = 10.3 \times 10^{-12} m/V$; $r_{41} = 8.77 \times 10^{-12}$ m/V.

Two different saw-tooth pulses are applied to the system simultaneously as the biasing electric signals. The biasing voltages are taking as $V_z = k_1 t$, which is applied parallel to the Z-axis and $V_y = k_2 t$, which is applied parallel to the Y-axis of the crystal.

Now the expression of the real part of the electric field at the output in terms of V_z and V_y is given by

$$E_{output} = E_0 \cos\left[\omega t - k_0 n_0 l - \frac{1}{2}k_0 n_0^3 r_{63} E_z l + \frac{1}{2}k_0 n_0^3 r_{41} E_y l\right.$$
$$\left. \times \left\{(n_0^2 - n_e^2)/r_{41}E_y n_e^2 n_0^2 + r_{41}E_y n_e^2 n_0^2 / (n_0^2 - n_e^2)\right\}\right]$$

Or,

$$E_{output} = E_0 \cos\left[\omega t - k_0 n_0 l - \frac{1}{2}k_0 n_0^3 r_{63} E_z l + \frac{1}{2}k_0 n_0^3 l (n_0^2 - n_e^2)/n_e^2 n_0^2\right.$$
$$\left. + \frac{1}{2}k_0 n_0^5 n_e^2 E_y^2 r_{41}^2 l / (n_0^2 - n_e^2)\right]$$
$$= E_0 \cos\left[\omega t - k_0 n_0 l - \frac{1}{2}k_0 n_0^3 r_{63} V_z - \frac{1}{2}k_0 n_0^3 l / n_e^2 - \frac{1}{2}k_0 n_0 l + \frac{1}{2}k_0 n_0^5 n_e^2 V_y^2 r_{41}^2 l / (n_0^2 - n_e^2) d^2\right]$$

$$= E_0 \cos\left[\omega t - \frac{3}{2}k_0 n_0 l - \frac{1}{2}k_0 n_0^3 l / n_e^2 - \frac{1}{2}k_0 n_0^3 r_{63} V_z + \frac{1}{2}k_0 n_0^5 n_e^2 V_y^2 r_{41}^2 l / \left(n_0^2 - n_e^2\right) d^2\right]$$

When we put $V_z = k_1 t$ and $V_y = k_2 t$, then the expression of the output beam becomes

$$E_{output} = E_0 \cos\left[\omega t - \frac{3}{2}k_0 n_0 l - \frac{1}{2}k_0 n_0^3 l / n_e^2 - \frac{1}{2}k_0 n_0^3 r_{63} k_1 t \right.$$
$$\left. + \frac{1}{2}k_0 n_0^5 n_e^2 k_2^2 t^2 r_{41}^2 l / \left(n_0^2 - n_e^2\right) d^2\right].$$

5.2.2 ANALYTICAL RESULTS

The voltage-dependent parts take part in the phase modulation of the light beam. So the expression of the phase of the output beam is

$$\varphi = \frac{1}{2}k_0 n_0^5 n_e^2 k_2^2 t^2 r_{41}^2 l / \left(n_0^2 - n_e^2\right) d^2 - \frac{1}{2}k_0 n_0^3 r_{63} k_1 t.$$

Taking, $k_0 = \dfrac{2\pi}{\lambda_0}$; $\lambda_0 = 0.633$ μm; $n_0 = 1.5115$, $n_e = 1.4698$, $r_{41} = 8.77 \times 10^{-12}$ m/V, $r_{63} = 10.3 \times 10^{-12}$ m/V.

The phase difference becomes

$$\varphi = 0.3030 \times 10^{-8} k_2^2 t^2 - 0.2045 \times 10^{-3} k_1 t.$$

Now the time required to get zero phase difference for different values of V_z and V_y is measured. Data to make the zero phase difference between the two saw-tooth signals are shown in Table 5.1.

Here, in the expression of the electric field of the output beam, the phase part contains two voltage-dependent parts, where one is having a positive sign and another is negative. The positive part contains V_y^2 term and the negative part contains V_z term,

TABLE 5.1
Data for Zero Phase Difference

k_1	k_2	t (sec)
1	50	26.96
1	100	6.74
1	150	3
1	200	1.685
1	250	1.078
1	300	0.75
1	350	0.55

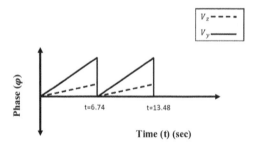

FIGURE 5.2   Variation of phase difference with time.

where V_y^2 appears as $k_2^2 t^2$ and V_z appears as $k_1 t$, so the phase of the modulated beam can be controlled by using two different biasing signals jointly at a time. Taking different biasing signals phase of the modulated beam becomes zero after a few seconds of interval (Figure 5.2).

If $V_z = k_1 t$ and $V_y = k_2 t$, where $k_1 = 1$ and $k_2 = 100$, then the phase difference becomes zero after the interval of 6.74 sec.

The phase part of the modulated output signal contains two voltage-dependent parts, where one is having a positive sign and another is negative. So the phase of the output beam can be controlled by using different biasing voltages.

5.3   CHARACTERISTIC STUDY ON DIFFERENT HARMONICS OF A LIGHT PASSING THROUGH A KERR CELL

An intense light beam is passing through a Kerr nonlinear medium of fused silica of length z. A part of its output is again passed through the medium for second passing by feedback. The schematic block diagram is shown in Figure 5.3.

5.3.1   THEORETICAL ANALYSIS

The expression of the output beam after first passing through the material is

$$E = E_0 \cos(\omega t - kz)$$

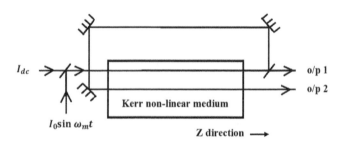

FIGURE 5.3   First passing and second passing of light through a Kerr nonlinear medium.

$$= \left(I_{dc} + I_0 \sin \omega_m t \right)^{1/2} \cos \left(\omega t - k_0 nz \right)$$

$$= \left(I_{dc} + I_0 \sin \omega_m t \right)^{1/2} \cos \left[\omega t - k_0 \left(n_0 + n_2 I \right) z \right]$$

$$= \left(I_{dc} + I_0 \sin \omega_m t \right)^{1/2} \cos \left[\omega t - k_0 n_0 z - k_0 n_2 I z \right]$$

$$= \left(I_{dc} + I_0 \sin \omega_m t \right)^{1/2} \cos \left[\omega t - k_0 n_0 z - k_0 n_2 \left(I_{dc} + I_0 \sin \omega_m t \right) z \right]$$

$$= \left(I_{dc} + I_0 \sin \omega_m t \right)^{1/2} \cos \left[\omega t - k_0 n_0 z - k_0 n_2 I_{dc} z - k_0 n_2 I_0 \sin \omega_m t z \right]$$

$$= \left(I_{dc} + I_0 \sin \omega_m t \right)^{1/2} \cos \left[\omega t - k_0 n_0 z - k_0 n_2 I_{dc} z - \xi \sin \omega_m t \right],$$

where $\xi = k_0 n_2 I_0 z$.
So,

$$E = \left(I_{dc} + I_0 \sin \omega_m t \right)^{1/2} \left[\cos \left(\omega t - k_0 n_0 z - k_0 n_2 I_{dc} z \right) \cos \left(\xi \sin \omega_m t \right) \right.$$
$$\left. - \sin \left(\omega t - k_0 n_0 z - k_0 n_2 I_{dc} z \right) \sin \left(\xi \sin \omega_m t \right) \right]$$

$$= \left(I_{dc} + I_0 \sin \omega_m t \right)^{1/2} \left[\cos \left(\omega t - k_0 n_0 z - k_0 n_2 I_{dc} z \right) \left\{ J_0 \left(\xi \right) + 2 J_2 \left(\xi \right) \sin 2 \omega_m t + \ldots \right\} \right.$$
$$\left. - \sin \left(\omega t - k_0 n_0 z - k_0 n_2 I_{dc} z \right) \left\{ 2 J_1 \left(\xi \right) \sin \omega_m t + 2 J_3 \left(\xi \right) \sin 3 \omega_m t + \ldots \right\} \right]$$

$$= \left(I_{dc} + I_0 \sin \omega_m t \right)^{1/2} \left[J_0 \left(\xi \right) \cos \left(\omega t - k_0 n_0 z - k_0 n_2 I_{dc} z \right) + J_1 \left(\xi \right) \cos \left\{ \left(\omega - \omega_m \right) t \right. \right.$$
$$\left. - k_0 n_0 z - k_0 n_2 I_{dc} z \right\} - J_1 \left(\xi \right) \cos \left\{ \left(\omega + \omega_m \right) t - k_0 n_0 z - k_0 n_2 I_{dc} z \right\} + J_2 \left(\xi \right)$$
$$\times \sin \left\{ \left(\omega + 2 \omega_m \right) t - k_0 n_0 z - k_0 n_2 I_{dc} z \right\} - J_2 \left(\xi \right) \sin \left\{ \left(\omega + 2 \omega_m \right) t - k_0 n_0 z - k_0 n_2 I_{dc} z \right\}$$
$$\left. - J_3 \left(\xi \right) \cos \left\{ \left(\omega - 3 \omega_m \right) t - k_0 n_0 z - k_0 n_2 I_{dc} z \right\} + J_3 \left(\xi \right) \cos \left\{ \left(\omega - 3 \omega_m \right) t - k_0 n_0 z - k_0 n_2 I_{dc} z \right\} \right]$$

The expression of the output beam after second passing through the material is

$$E' = E_0 \cos \left(\omega t - 2 kz \right)$$

$$\left(I_{dc} + I_0 \sin \omega_m t \right)^{1/2} \cos \left(\omega t - 2 k_0 nz \right)$$

$$= \left(I_{dc} + I_0 \sin \omega_m t \right)^{1/2} \cos \left[\omega t - 2 k_0 \left(n_0 + n_2 I \right) z \right]$$

$$= \left(I_{dc} + I_0 \sin \omega_m t \right)^{1/2} \cos \left[\omega t - 2 k_0 n_0 z - 2 k_0 n_2 I z \right]$$

$$= \left(I_{dc} + I_0 \sin \omega_m t \right)^{1/2} \cos \left[\omega t - 2 k_0 n_0 z - 2 k_0 n_2 \left(I_{dc} + I_0 \sin \omega_m t \right) z \right]$$

$$= \left(I_{dc} + I_0 \sin \omega_m t \right)^{1/2} \cos \left[\omega t - 2 k_0 n_0 z - 2 k_0 n_2 I_{dc} z - 2 k_0 n_2 I_0 \sin \omega_m t z \right]$$

$$= \left(I_{dc} + I_0 \sin \omega_m t \right)^{1/2} \cos \left[\omega t - 2 k_0 n_0 z - 2 k_0 n_2 I_{dc} z - \xi' \sin \omega_m t \right]$$

where $\xi' = 2k_0 n_2 I_0 z$.

So,

$$E' = \left(I_{dc} + I_0 \sin \omega_m t\right)^{\frac{1}{2}} \left[\cos\left(\omega t - 2k_0 n_0 z - 2k_0 n_2 I_{dc} z\right)\cos\left(\xi' \sin \omega_m t\right)\right.$$
$$\left. - \sin\left(\omega t - 2k_0 n_0 z - 2k_0 n_2 I_{dc} z\right)\sin\left(\xi' \sin \omega_m t\right)\right]$$

$$= \left(I_{dc} + I_0 \sin \omega_m t\right)^{\frac{1}{2}} \left[\cos\left(\omega t - 2k_0 n_0 z - 2k_0 n_2 I_{dc} z\right)\left\{J_0\left(\xi'\right) + 2J_2\left(\xi'\right)\sin 2\omega_m t + \ldots\right\}\right.$$
$$\left. - \sin\left(\omega t - 2k_0 n_0 z - 2k_0 n_2 I_{dc} z\right)\left\{2J_1\left(\xi'\right)\sin \omega_m t + 2J_3\left(\xi'\right)\sin 3\omega_m t + \ldots\right\}\right]$$

$$= \left(I_{dc} + I_0 \sin \omega_m t\right)^{\frac{1}{2}} \left[J_0\left(\xi'\right)\cos\left(\omega t - 2k_0 n_0 z - 2k_0 n_2 I_{dc} z\right) + J_1\left(\xi'\right)\cos\left\{\left(\omega - \omega_m\right)t\right.\right.$$
$$- 2k_0 n_0 z - 2k_0 n_2 I_{dc} z\right\} - J_1\left(\xi'\right)\cos\left\{\left(\omega + \omega_m\right)t - 2k_0 n_0 z - 2k_0 n_2 I_{dc} z\right\}$$
$$+ J_2\left(\xi'\right)\sin\left\{\left(\omega + 2\omega_m\right)t - 2k_0 n_0 z - 2k_0 n_2 I_{dc} z\right\} - J_2\left(\xi'\right)\sin\left\{\left(\omega + 2\omega_m\right)t\right.$$
$$- 2k_0 n_0 z - 2k_0 n_2 I_{dc} z\right\} - J_3\left(\xi'\right)\cos\left\{\left(\omega - 3\omega_m\right)t - 2k_0 n_0 z - 2k_0 n_2 I_{dc} z\right\}$$
$$\left. + J_3\left(\xi'\right)\cos\left\{\left(\omega - 3\omega_m\right)t - 2k_0 n_0 z - 2k_0 n_2 I_{dc} z\right\}\right]$$

where k_0 is the propagation constant, $k_0 = \dfrac{2\pi}{\lambda_0}$, $\lambda_0 = 800nm$, n_2 is Kerr nonlinear coefficient, $n_2 = 0.35 \times 10^{-19} \dfrac{m^2}{W}$. $z = 10$ cm.

Now the variations of ξ, $J_0(\xi)$ and $J_0^2\left(\xi\right)$ with intensity I_0 are shown in Table 5.2, where ξ depends upon intensity I_0. Here, $J_0(\xi)$ is the amplitude of the central frequency, whereas $J_0^2\left(\xi\right)$ is the intensity of the central frequency for the first passing of the light signal. The variations of ξ', $J_0(\xi')$ and $J_0^2\left(\xi'\right)$ with intensity I_0 are shown in Table 5.3. Here, $J_0(\xi')$ and $J_0^2\left(\xi'\right)$ are the amplitude and intensity of the central frequency for the second passing of the light signal. The variations of $J_0^2\left(\xi\right)$ and $J_1^2\left(\xi\right)$ with intensity are shown in Figure 5.5. Here, $J_1^2\left(\xi\right)$ is the intensity of the first sideband frequency for the first passing of the light signal.

TABLE 5.2
Data for First Passing

$I_0(10^{13}$ W/m²)	ξ	$J_0(\xi)$	$J_0^2(\xi)$
1	0.2748	0.981216	0.962784
2	0.5495	0.925926	0.857337
4	1.099	0.720092	0.518532
6	1.6485	0.427653	0.182887
8	2.198	0.111474	0.012426
10	2.7475	−0.163075	0.026593

TABLE 5.3

Data for Second Passing

$I_0(10^{13}$ W/m²$)$	ξ'	$J_0(\xi')$	$J_0^2(\xi')$
1	0.5495	0.925926	0.857337
2	1.099	0.720092	0.518532
4	2.198	0.111474	0.012426
6	3.297	−0.343632	0.118082
8	4.396	−0.343036	0.118095
10	5.495	−0.008552	0.000073

TABLE 5.4

Data for $J_0^2(\xi)$ and $J_1^2(\xi)$ with Intensity

$I_0(10^{13}$ W/m²$)$	$J_0^2(\xi)$	$J_1^2(\xi)$
1	0.962784	0.018518
2	0.857337	0.069965
4	0.518532	0.221473
6	0.182887	0.329735
8	0.012426	0.309410
10	0.026593	0.182132

5.3.2 ANALYTICAL RESULTS

The variations of intensity with $J_0^2(\xi)$ and $J_0^2(\xi')$ are shown in Figure 5.4. It is seen that the light intensity for the central frequency decreases with I_0 more sharply in case of second passing of the light in contrast to the first passing.

The variations of intensity with $J_0^2(\xi)$ and $J_1^2(\xi)$ are shown in Figure 5.5. It is seen that the intensity of the harmonics is increasing at the cost of decreasing of intensity of the central frequency.

5.4 SOME APPLICATIONS OF OTA USING PHASE ENCODED LIGHT

When an intense light beam passes through an electro-optic Pockels cell with the application of an electric field, then the refractive index of the material used in the Pockels cell is changed. This change of refractive index generates a phase change of the light beam. Using the phase changing property of the Pockels cell, an optical tree network is established.

The two-input optical tree network is shown in Figure 5.6. Here three electro-optic Pockels cells are used in the two-input optical tree network. A single electro-optic Pockels cell is used in each of the lower branches of the optical tree network. Biasing

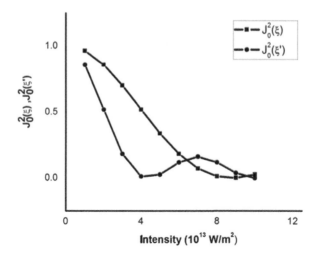

FIGURE 5.4   Variation of intensity with $J_0^2(\xi)$ and $J_0^2(\xi')$.

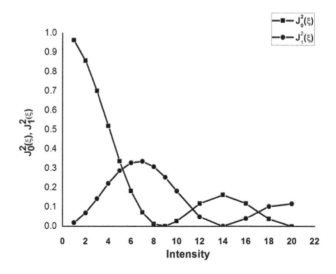

FIGURE 5.5   Variation of intensity with $J_0^2(\xi)$ and $J_1^2(\xi)$.

voltage V_1 is applied to the EOM 1 and biasing voltage V_2 is applied to both EOM 2 and EOM 3.

The three-input optical tree network is shown in Figure 5.7. There are seven electro-optic Pockels cell (EOM 4 to EOM 10) in the lower branch of the three-input optical tree networks. Biasing voltage V_1 is applied to the EOM 4, biasing voltage V_2 is applied to the both EOM 5 and EOM 6, and biasing voltage V_3 is applied to EOM 7 to EOM 10.

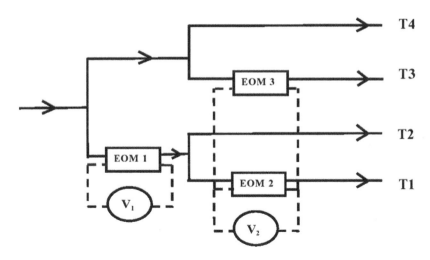

FIGURE 5.6 Optical tree for all-optical X-OR scheme (two-input).

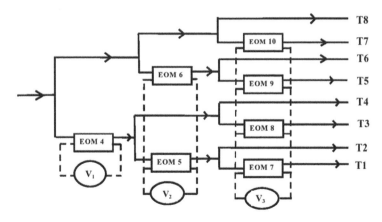

FIGURE 5.7 Optical tree for all-optical X-OR scheme (three-input).

In this scheme, the phase of the light beam is used for encoding and decoding. When V_π voltage is applied to the EOM, a π phase difference is obtained at the output of each EOM is encoded as '1' state. In the absence of applied voltage to the EOM, no phase difference is seen at the output of the EOM. This is considered as '0' state. Also, the even multiple of π is considered as 0 phase difference and odd multiple of π is considered as π phase difference. The phase differences shown at the output channel from T1 to T4 of the two-input optical tree and from T1 to T8 of the three-input optical tree are given in Tables 5.5 and 5.6, respectively. Tables 5.5 and 5.6 represent the truth table of the two-input optical tree network and three-input optical tree network, respectively.

The output of different channels can perform different logic operations. X-OR gate can be constructed by using the phase encoding technique in the OTA. It is seen

TABLE 5.5

Truth Table for Two-Input Optical Tree

V_1	V_2	T_1	T_2	T_3	T_4
0	0	0	0	0	0
0	1	π	0	π	0
1	0	π	π	0	0
1	1	0	π	0	π

TABLE 5.6

Truth Table for Three-Input Optical Tree

V_1	V_2	V_3	T_1	T_2	T_3	T_4	T_5	T_6	T_7	T_8
0	0	0	0	0	0	0	0	0	0	0
0	0	1	π	0	π	0	π	0	π	0
0	1	0	π	π	0	0	π	π	0	0
0	1	1	0	π	π	0	0	π	π	0
1	0	0	π	π	π	π	0	0	0	0
1	0	1	0	π	0	π	π	0	π	0
1	1	0	0	0	π	π	π	π	0	0
1	1	1	π	0	0	π	0	π	π	0

from the two-input scheme that the output from terminal 1 gives the result of an X-OR operation of V_1 and V_2. It is seen from Table 5.1 that the output of terminal 1 gives a three-input X-OR operation of V_1, V_2 and V_3 , terminal 2 gives the X-OR operation of V_1 and V_2, terminal 3 gives the X-OR operation of V_1 and V_3, and terminal 5 gives the X-OR operation of V_2 and V_3.

The electro-optic material used in the Pockels cell is lithium niobate ($LiNbO_3$). When polarized light passes through the Pockels cell with the application of an electric field (E), then the refractive index (n) [3,5,6,13,14] of the material becomes

$$n = n_0 - \frac{1}{2}n_0^3 r_{13}E$$

where n_0 is the constant refractive index of the material and r_{13} is the electro-optic coefficient of the material.

Here the applied electric field is along the direction of light propagation, so this is a longitudinal electro-optic effect. So, if the dimension of the material along the Z-direction is l and the applied biasing voltage is V, the electric field is E, then, $V = El$. The expression of the output beam in terms of the electric field is

$$E_{output} = E_0 e^{i\left[\omega t - k_0 nl\right]}$$

$$E_{output} = E_0 e^{i\left[\omega t - k_0 n_0 l - \frac{1}{2} k_0 n_0^3 n_3 El\right]}.$$

The phase-dependent part is $\varphi = \frac{1}{2} k_0 n_0^3 r_{13} El = \frac{1}{2} k_0 n_0^3 r_{13} V$

To get a π phase difference, the required voltage is 6.16 kV, where $n_0 = 2.286$, $r_{13} = 8.6 \times 10^{-12} m/V$ and $\lambda_0 = 633$ nm.

5.5 CONDUCTION OF WIDE-SCALE PHASE VARIATION BY SIMULTANEOUS USES OF POCKELS AND KERR MATERIALS

Optical phase modulation is required in high-speed and distinct optical communication. The phase of the carrier wave is varied to transmit a lot of information associated with the modulating signal wave. KDP and LiNbO$_3$ are commonly used as an electro-optic modulator for its good switching character in all-optical domain. Pockels material having the first order of nonlinearity shows a phase variation of the emitted light wave due to the change of the optical biasing voltage applied across its two terminals. Again, the Kerr material shows the second order of nonlinearity, i.e. the change of refractive index is directly proportional to the intensity of the incident light beam. The phase of modulating light wave after passing through Kerr material can be controlled by varying the intensity of the carrier wave at the input section only.

5.5.1 ELECTRICALLY CONTROLLED PHASE AND INTENSITY VARIATIONS OF USING POCKELS MATERIAL

The propagation of light beam through the Pockels material in presence of an external biasing voltage induces the birefringence property within it, i.e. re-distribution of the components of the dielectric tensor. The change of refractive index within it depends on the electro-optic coefficient. The basic reason for the electro-optic effect is the variation of the polarizability of the isotropic material in the presence of an external field (E). The expression of polarization (P) [5,6] in terms of the electric field is

$$P = P_0 + \chi^{(1)}E + \chi^{(2)}E.E + \dots$$

The quantity $\chi^{(1)}$ is the measure of the first order of nonlinearity of the material which is responsible for the linear electro-optic effect of the birefringent material. If a light beam propagates through the Pockels material, then it experiences the change of internal property of the material along the particular direction of propagation. An unpolarized light beam initially passed through the polarizer having a particular direction of the pass axis. In the presence of an external biasing voltage, its refractive index will be different along two different pass axes. For this reason, the two perpendicular polarized light waves after traveling a distance through the linear electro-optic crystal experience a phase difference. This amount of phase difference directly

depends on the applied voltage across the material as well as the electro-optic coefficient. One can easily control the phase of the outgoing light wave by the suitable
change of the electrical biasing voltage. Electrically controlled phase modulation
phenomena will be introduced in this process successfully. Suppose a linearly polarized light wave is propagating along z-direction through a linear electro-optic material and it is polarized along x-direction (i.e. 45° with respect to x-axis). The scheme
of the above proposal is shown in Figure 5.8. Along the pass axis, the light wave
passes through the polarizer as well as the crystal without change of state of
polarization.

If the crystal having a length l, then the expression of the electric field vector
emerging from the output terminal becomes

$$E'_x(l) = E'_x(0) \exp i(\omega t - n'_x k_0 l).$$

For KDP crystal, the expression electric field vector emerging from the output
terminal becomes

$$E'_x(l) = E'_x(0) \exp\left\{ i\left(\omega t - n_0 \frac{\omega}{c} l + \frac{\omega}{2c} n_0^3 r_{63} E_z l \right) \right\}$$

where $v_z = E_z l$ is the voltage applied across the crystal and r_{63} is the electro-optic
coefficient of the material.

So,

$$E'_x(l) = E'_x(0) \exp\left\{ i\left(\omega t - n_0 \frac{\omega}{c} l + \frac{\omega}{2c} n_0^3 r_{63} v_z \right) \right\}$$

The phase modulation index is

$$\xi = \frac{\omega}{2c} n_0^3 r_{63} v_z$$

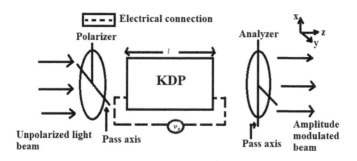

FIGURE 5.8   Schematic diagram for amplitude modulation.

If one applies a sinusoidal voltage across the electro-optic crystal, then it leads to sinusoidal phase variation with its highest peak ξ.

Similarly, the expression of the electric field vector for another polarized light wave after traversing the distance 'l' through the crystal is

$$E'_y(l) = E'_y(0)exp\left\{i\left(\omega t - n_0\frac{\omega}{c}l - \frac{\omega}{2c}n_0^3r_{63}E_zl\right)\right\} = E'_y(0)exp\left\{i\left(\omega t - n_0\frac{\omega}{c}l - \frac{\omega}{2c}n_0^3r_{63}v_z\right)\right\}$$

Now the phase difference developed between two orthogonal components of polarized light waves after completing their journey across the crystal in presence of an applied electric field is

$$\gamma = \frac{\omega}{c}n_0^3r_{63}v_z$$

From the above expression, one can say that the retardation factor directly depends on the amplitude of the electric field applied across the crystal as well as the electro-optic coefficient of the material.

The expression of the electric field vector of the emitted light signal from the analyzer is

$$E = \frac{A}{2}\left[exp\left\{i\left(\omega t - n_0\frac{\omega}{c}l - \frac{\omega}{2c}n_0^3r_{63}v\sin\omega t\right)\right\}\{1 - exp(-i\gamma)\}\right], \text{ where } E'_x(0) = E'_x(0) = \frac{A}{\sqrt{2}}$$

Finally, the intensity of the modulating light wave is

$$I = \frac{1}{2}Re(E.E^*) = \frac{A^2}{2}\sin^2\left(\frac{\gamma}{2}\right)$$

5.5.2   PHASE VARIATION IN KERR MATERIAL BY INTENSITY VARIATION

Optics is very much preferable in data communication and quantum computation due to its inherent properties like its ultra-fast speed, noiseless operation, no crosstalk, etc. For this, all-optical phenomena have much importance to design different optical switches. The optical property like phase can be controlled by varying the strength of the optical signal itself. The Kerr material is very useful in all-optical domains for optical communication. This material has the second order of nonlinearity. The change of refractive index of the light wave in Kerr material is directly proportional to the square of the electric field. A nonlinear polarization character will be introduced when an intense light wave is passing through this type of material. The term $\chi^{(2)}$ is responsible for the intensity-dependent character. The expression of refractive index (n) [5,6,15–17] of the light beam after passing a Kerr material is given by

$$n = n_0 + n_2I$$

where n_0 is the constant refractive index of the material, n_2 is the second-order nonlinear correction term and I is the intensity of the light wave used.

When a highly intense light beam is passed through the Kerr material, then it carries the inherent property of this nonlinear material. The output light wave emitted from the Kerr material develops an extra path difference along the particular direction of propagation which is responsible to generate a phase difference of the light beam. The expression of the electric field of the polarized light beam traversing distance 'l' through the Kerr material along the x'-direction is

$$E'_x(l) = E'_x(0)\exp i(\omega t - n'_x k_0 l)$$
$$= E'_x(0)\exp i(\omega t - n_0 k_0 l - n_2 I k_0 l)$$

The phase difference introduced for the light wave after traveling through the Kerr material is $d\Phi = n_2 I k_0 l$

The phase difference directly depends on the intensity of the light wave used and the second-order nonlinear correction term. So, one can easily control the phase of the light beam by controlling the fine adjustment of the intensity of the light at the input terminal according to our needs. In this case, we can also change the phase of light wave in a wide range.

5.5.3 POCKELS AND KERR MATERIALS SIMULTANEOUSLY FOR MASSIVE PHASE VARIATION

An unpolarized light beam initially passing through the polarizer along the z-direction becomes polarized. The polarized light beam has two components of polarization along x and y-direction. These two components of the polarized light beams have two pass axes along x' and y'-direction within the first electro-optic material in presence of an external biasing voltage v_1. These two polarized light waves are in phase when they touch the input terminal of the first Pockels material. After traveling a finite length through this crystal, a phase difference is introduced between these two components of light wave at the end terminal of the crystal. The expression of the electric field of the polarized light wave along the x'- and y'-directions becomes

$$E'_x(l) = E'_x(0)\exp\left[i\left\{\omega t - n_0\left(\frac{\omega}{c}\right)l + \left(\frac{\omega}{2c}\right)n_0^3 r_{63} E_z l\right\}\right]$$

$$E'_y(l) = E'_y(0)\exp\left[i\left\{\omega t - n_0\left(\frac{\omega}{c}\right)l - \left(\frac{\omega}{2c}\right)n_0^3 r_{63} E_z l\right\}\right]$$

where $v = E_z l$ is the voltage applied across the crystal and r_{63} = electro-optic coefficient of the material.

Here the KDP crystal is used as a Pockels material in this scheme. The scheme is depicted in Figure 5.9, where the components of refractive indices along the x'- and y'-directions are given by

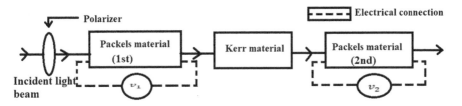

FIGURE 5.9 Schematic diagram for a light wave passing through two Pockels materials and one Kerr material.

$$n'_x = n_0^3 - \frac{1}{2}n_0^3 r_{63}E_z \text{ and } n'_y = n_0^3 + \frac{1}{2}n_0^3 r_{63}E_z.$$

The expression of the electric field of the emitted light from the analyzer is given by

$$E = E'_x(l)\cos\left(\frac{\pi}{4}\right) - E'_y(l)\sin\left(\frac{\pi}{4}\right),$$

$$E = \frac{A}{2}\left[\exp\left\{i\left(\omega t - n_0\frac{\omega}{c}l - \frac{\omega}{2c}n_0^3 r_{63}v_1 \sin \omega t\right)\right\}\{1 - \exp(-i\gamma)\}\right],$$

$$E'_x(0) = E'_y(0) = \frac{A}{\sqrt{2}},$$

where $\gamma = \left(\frac{v_1}{v_\pi}\right)\pi$

The intensity of the signal after crossing the first Pockels material is

$$I = \frac{A^2}{2}\sin^2\left(\frac{v_1}{2v_\pi}\pi\right)$$

Now the intense light beam just emitted from the first Pockels material again passes through a Kerr material. The expression of the electric field of the light beam after emerging from the Kerr material becomes

$$E''_x(l) = \frac{A}{\sqrt{2}}\sin\left(\frac{\gamma}{2}\right)\exp\left\{i\left(\omega t - 2kn_0l + \frac{k}{2}n_0^3 r_{63}v_1 - kn_2\frac{A^2}{8}\frac{v_1^2\pi^2}{v_\pi^2}l\right)\right\},$$

where $k = \frac{\omega}{c}$ and n_2 is the second-order nonlinear correction term.

This amplitude and phase-modulated light beam finally passes again through another linear electro-optic Pockels material in the presence of another external biasing voltage v_2. The expression of the electric field of the light beam after leaving from second Pockels material is

$$E_x'''(l) = \frac{A}{\sqrt{2}}\sin\left(\frac{\gamma}{2}\right)\exp\left\{i\left(\omega t - 2kn_0 l + \frac{k}{2}n_0^3 r_{63}v_1 - kn_2\frac{A^2}{8}\frac{v_1^2\pi^2}{v_\pi^2}l - kn_0 l + \frac{k}{2}n_0^3 r_{63}v_2\right)\right\}.$$

One can evaluate the phase term from the above expression easily. Now the phase term is

$$\Phi = 3kn_0 l - \frac{k}{2}n_0^3 r_{63}\left(v_1 + v_2\right) + kn_2\frac{A^2}{8}\frac{v_1^2\pi^2}{v_\pi^2}l.$$

This phase part includes both the voltages applied across the first and the second Pockels material, respectively, as well as the intensity of the signal wave which is used in this scheme at the input terminal of the first Pockels material.

We take $v_1 = v_2 = v$, then the expression of the phase for successive passing of light signal through two Pockels materials sandwich with a Kerr material (shown in Figure 5.9) is

$$\Phi\text{ with Kerr} = 3kn_0 l - kn_0^3 r_{63}v + kn_2\frac{A^2}{8}\frac{v^2\pi^2}{v_\pi^2}l.$$

In the same manner, if a light beam passes successively through two different Pockels materials biased by two different voltages v_1 and v_2 along their cross-section, then the expression of the electric field after leaving from the second Pockels material depicted in Figure 5.10 becomes

$$E_x''(l) = \frac{A}{\sqrt{2}}\exp\{i(\omega t - 2kn_0 l + \frac{k}{2}n_0^3 r_{63}\left(v_1 + v_2\right)\}.$$

The phase introduced due to simultaneously passing through two Pockels material is

$$\Phi\text{ without Kerr} = 2kn_0 l - \frac{k}{2}n_0^3 r_{63}\left(v_1 + v_2\right).$$

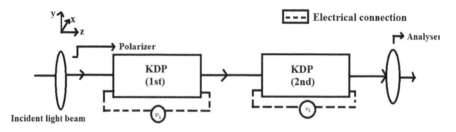

FIGURE 5.10 Schematic diagram for the light wave passing through two Pockels material successively.

When $v_1 = v_2 = v$, $\Phi_{\text{without Kerr}} = 2kn_0l - kn_0^3 r_{63} v$.

5.5.4 THEORETICAL ANALYSIS

From the above theoretical expression of the phase, one can realize that the phase of an outgoing light wave after passing through the Kerr material sandwiched between two Pockels material directly depends upon the two biasing voltages applied across two Pockels material as well as the intensity of the input signal used in our scheme. The phase of the output light wave after passing successively through two Pockels material directly depends on the two electrically applied biasing voltages only. For the wide-scale phase change of a light wave in communication purposes in a sophisticated way, this scheme 1 (shown in Figure 5.9) is very useful. In this scheme, the phase of the light wave is dependent on the intensity of the input light which ensures that without changing any electrically biasing voltage one can control the phase of the outgoing light wave. One can also observe that the fluctuation of the phase happens due to the fluctuation of the applied voltage from a lower value to higher from the analysis of the expressions of $\Phi_{\text{with Kerr}}$ and $\Phi_{\text{without Kerr}}$. It is also observed that $\left(\dfrac{d\Phi}{dv}\right)_{\text{With Kerr}}$ increases linearly with the increase of biasing voltage from a lower value to a higher value and it is given as

$$\left(\frac{d\Phi}{dv}\right)_{\text{With Kerr}} = -kn_0^3 r_{63} + kn_2 \frac{A^2}{4} \frac{v\pi^2}{v_\pi^2} l$$

It suggests that if we increase the phase of the light by controlling the intensity of the light wave at the input terminal keeping the biasing voltage unchanged, the signal will be less distorted for secured and noiseless data communication. In addition, the tuning of the phase will be more perfect according to our requirement. But $\left(\dfrac{d\Phi}{dv}\right)_{\text{without kerr}}$ remains the same due to the variation of the applied voltage and it is given by

$$\left(\frac{d\Phi}{dv}\right)_{\text{With Kerr}} = -kn_0^3 r_{63}$$

5.5.4.1 Efficiency of Modulation after Passing through Only the KDP Material

The intensity of the output light wave is

$$I_0 = \frac{A^2}{2} \sin^2\left(\frac{\gamma}{2}\right),$$

where the intensity of the light wave used at the input terminal is $I_i = \dfrac{A^2}{2}$.

Now the efficiency of the KDP crystal becomes $\dfrac{I_0}{I_i} = \sin^2(\dfrac{\gamma}{2})$, where $\gamma = \dfrac{v\pi}{v_\pi}$.

5.5.4.2 Efficiency of Modulation after Passing through Two Pockels and One Kerr Material

The expression of the electric field of the emerging light wave at the output terminal of the second Pockels material (shown in Figure 5.9) polarized in the x-direction is

$$E_x'''(l) = \frac{A}{\sqrt{2}} \sin\left(\frac{\gamma}{2}\right) \exp\left\{ i\left(\omega t - 2kn_0 l + \frac{k}{2} n_0^3 r_{63} v_1 - kn_2 \frac{A^2}{8} \frac{v_1^2 \pi^2}{v_\pi^2} l - kn_0 l + \frac{k}{2} n_0^3 r_{63} v_2 \right) \right\}$$

$$= \frac{A}{\sqrt{2}} \sin\left(\frac{\gamma}{2}\right) \exp\left\{ i\left(\omega t - 3kn_0 l + \frac{k}{2} n_0^3 r_{63} \left(v_1 + v_2\right) - kn_2 \frac{A^2}{8} \frac{v_1^2 \pi^2}{v_\pi^2} l \right) \right\}.$$

Similarly, the expression of the electric field for the light wave polarized in the y-direction is

$$E_y'''(l) = \frac{A}{\sqrt{2}} \sin\left(\frac{\gamma}{2}\right) \exp\left\{ i\left(\omega t - 2kn_0 l - \frac{k}{2} n_0^3 r_{63} v_1 - kn_2 \frac{A^2}{8} \frac{v_1^2 \pi^2}{v_\pi^2} l - kn_0 l - \frac{k}{2} n_0^3 r_{63} v_2 \right) \right\}$$

$$= \frac{A}{\sqrt{2}} \sin\left(\frac{\gamma}{2}\right) \exp\left\{ i\left(\omega t - 3kn_0 l - \frac{k}{2} n_0^3 r_{63} \left(v_1 + v_2\right) - kn_2 \frac{A^2}{8} \frac{v_1^2 \pi^2}{v_\pi^2} l \right) \right\}.$$

Finally, the beam is passed through an analyzer having a pass axis perpendicular to the analyzer giving the expression of the electric field is

$$E = E_x'''(l) \cos\left(\frac{\pi}{4}\right) - E_y'''(l) \sin\left(\frac{\pi}{4}\right),$$

$$= \frac{A}{2} \sin\left(\frac{\gamma}{2}\right) \exp\left\{ i\left(\omega t - 3kn_0 l + \frac{k}{2} n_0^3 r_{63} \left(v_1 + v_2\right) - kn_2 \frac{A^2}{8} \frac{v_1^2 \pi^2}{v_\pi^2} l \right) \right\} \left\{ 1 - \exp\left(-i\gamma'\right) \right\},$$

where, $\gamma' = kn_0^3 r_{63} \left(v_1 + v_2\right)$.

Now the output intensity of the light wave becomes

$$I_0 = \frac{1}{2} \mathrm{Re}\left(E.E^*\right) = \frac{A^2}{2} \sin^2\left(\frac{\gamma}{2}\right) \sin^2\left(\frac{\gamma'}{2}\right).$$

The input intensity is $I_i = \dfrac{A^2}{2}$

Hence, the efficiency of this modulator design becomes $\dfrac{I_0}{I_i} = \sin^2\left(\dfrac{\gamma}{2}\right) \sin^2\left(\dfrac{\gamma'}{2}\right)$.

5.5.4.3 Efficiency of Modulation after Passing through Two KDP Materials

The expression of the electric field of the light wave after passing successively through two Pockels materials (KDP) becomes

$$E_x''(l) = \frac{A}{\sqrt{2}} \exp i \left\{ \omega t - 2kn_0 l + \frac{k}{2} n_0^3 r_{63} (v_1 + v_2) \right\}.$$

Similarly, the expression of the electric field of the light wave polarized along the y-direction is

$$E_y''(l) = \frac{A}{\sqrt{2}} \exp i \left\{ \omega t - 2kn_0 l - \frac{k}{2} n_0^3 r_{63} (v_1 + v_2) \right\}.$$

The emitted light wave after passing through the analyzer gives the expression of the electric field as

$$E = E_x''(l) \cos \left(\frac{\pi}{4} \right) - E_y''(l) \sin \left(\frac{\pi}{4} \right),$$

$$= \frac{A}{2} \exp i \left\{ \omega t - 2kn_0 l + \frac{k}{2} n_0^3 r_{63} (v_1 + v_2) \right\} \left\{ 1 - \exp(-i\gamma') \right\}.$$

The intensity of the output light wave becomes

$$E = E_x'(l) \cos \left(\frac{\pi}{4} \right) - E_y'(l) \sin \left(\frac{\pi}{4} \right),$$

The input intensity is $I_i = \frac{A^2}{2}$.

So, the efficiency of this modulator design becomes $\frac{I_0}{I_i} = \sin^2(\frac{\gamma'}{2})$.

5.6 NEW METHOD OF CONDUCTION OF OPTICAL PHASE ALGEBRA BY POCKELS AND KERR MATERIALS IN SERIES

Let an incident polarized light beam passes through a linear electro-optic material, a Kerr material and another linear electro-optic material in series (as shown in Figure 5.9). Finally, the emitted light wave from the second Pockels material [Figure 5.9] is amplitude, intensity and phase-modulated. Now we theoretically obtain an expression of the amount of voltage required across second Pockels material to neutralize the phase, which is introduced by the joint effect of the first Pockels material and the Kerr material. In the scheme, the KDP is taken as the Pockels material.

The expression of the electric field of the outgoing light emerging from the output terminal of the second Pockels material is

$$E = \frac{A}{\sqrt{2}} \sin \left(\frac{\gamma}{2} \right) \exp \left\{ i \left(\omega t - 2kn_0 l + \frac{k}{2} n_0^3 r_{63} v_1 - kn_2 \frac{A^2}{8} \frac{v_1^2 \pi^2}{v_\pi^2} l - kn_0 l + \frac{k}{2} n_0^3 r_{63} v_2 \right) \right\}$$

$$= \frac{A}{\sqrt{2}} \sin\left(\frac{\gamma}{2}\right) \exp\left[i\left\{\omega t - 3kn_0 l + \frac{k}{2} n_0^3 r_{63} (v_1 + v_2) - kn_2 \frac{A^2}{8} \frac{v_1^2 \pi^2}{v_\pi^2} l\right\}\right].$$

Here the phase term of the intensity-modulated light wave becomes

$$\Phi = 3kn_0 l - \frac{k}{2} n_0^3 r_{63} (v_1 + v_2) + kn_2 \frac{A^2}{8} \frac{v_1^2 \pi^2}{v_\pi^2} l.$$

The voltage across the second Pockels material to compensate the phase introduced for traveling of the light wave between the first Pockels material and the Kerr material is

$$v_2 = \frac{2}{kn_0^3 r_{63}}\left[3kn_0 l + kn_2 \frac{A^2}{8} \frac{v_1^2 \pi^2}{v_\pi^2} l\right] - v_1.$$

5.7 ALTERNATIVE USE OF MULTI-PASSING FOR INCREASING THE TRANSMISSION COEFFICIENT OF KDP-BASED MODULATOR

A scheme is already developed in which the transmission coefficient of amplitude modulation is increasingly adopting the multi-passing feedback mechanism of light in KDP crystal. [23].

Nowhere, we have explored an alternative and novel process for increasing the transmission coefficient of the same crystal. In our proposed system, an unpolarized optical beam is allowed to propagate through a polarizer having a definite polarization axis. The beam is then made to pass through a KDP crystal of certain dimensions, respectively, with the presence of a biasing voltage applied to the system along the z-direction. The output beam is again modulated and it is passing from the output side to the input side. The re-modulated output beam is again made to feedback through the same modulator from the input side to the output side. Finally, the output beam is made to travel through an analyzer. The input wave, propagating through a polarizer, is polarized along the y-direction and the output wave, propagating through an analyzer, is polarized along the x-direction. The beam is propagating along the z-direction. The schematic diagram is shown in Figure 5.11 [5,6,23].

In this scheme for a wave polarized along the x'-direction, the output wave at z = L for the first time passing through the crystal, the expression of the electric field (E_x') [5,6] is

$$E_x'(z = L) = E_x'(0) \exp\left[i\left\{\omega t - n_0\left(\frac{\omega}{c}\right)L + \left(\frac{\omega}{2c}\right)n_0^3 r_{63} E_z L\right\}\right].$$

When the beam is feedbacking through the crystal from the output side of the crystal (i.e. for the second time passing through the crystal), the expression of the output wave is

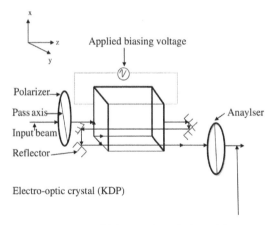

Multi passed amplitude modulated output beam

FIGURE 5.11 Amplitude modulation by feedback mechanism (feedback is done from the output side of the modulator).

$$E_x''(z=L) = E_x''(0)\exp\left[i\left\{\omega t + 2\left(\frac{\omega}{2c}\right)n_0^3 r_{63} E_z L\right\}\right].$$

In the next step, when the beam is again made to pass through the crystal from the input side of the crystal, then the wave emerging from the crystal is

$$E_x'''(z=L) = E_x'''(0)\exp\left[i\left\{\omega t - n_0\left(\frac{\omega}{c}\right)L + 3\left(\frac{\omega}{2c}\right)n_0^3 r_{63} E_z L\right\}\right].$$

Similarly, along the y-direction, we can write

$$E_y'''(z=L) = E_y'''(0)\exp\left[i\left\{\omega t - n_0\left(\frac{\omega}{c}\right)L - 3\left(\frac{\omega}{2c}\right)n_0^3 r_{63} E_z L\right\}\right]$$

Now, we already know that, $E_{output}''' = E_x'''\cos\left(\frac{\pi}{4}\right) - E_y'''\sin\left(\frac{\pi}{4}\right)$

$$E_{output}''' = \frac{1}{\sqrt{2}}\left(E_x''' - E_y'''\right).$$

The final output is expressed as

$$E_{output}''' = \frac{A}{2}\left(\exp\left[i\left\{\omega t - n_0\left(\frac{\omega}{c}\right)L + 3\left(\frac{\omega}{2c}\right)n_0^3 r_{63} E_z L\right\}\right]\right.$$

$$\left. -\exp\left[i\left\{\omega t - n_0\left(\frac{\omega}{c}\right)L - 3\left(\frac{\omega}{2c}\right)n_0^3 r_{63} E_z L\right\}\right]\right)$$

$$I_{output}^{m} = \frac{1}{2} \mathrm{Re}\left[E_{output}^{m} E_{output}^{m*} \right] = \frac{1}{2} A^2 \sin^2\left(\frac{3}{2} \gamma \right).$$

So, after n times passing through the crystal, the expression of the output intensity becomes

$$I_{output}^{m,\dots n} = \frac{1}{2} A^2 \sin^2\left(\frac{n}{2} \gamma \right)$$

So, transmission coefficient $T = \sin^2\left(\dfrac{n}{2} \gamma \right)$

This equation shows that the transmission coefficient of the crystal is increased n times. In this process, the question of traveling the additional paths (what was the case in our earlier work) is removed.

5.8 LINEAR FREQUENCY VARIATION OF LIGHT USING KERR NONLINEARITY BY PARABOLIC LIGHT SIGNAL AND IN A MULTI-PASSING

In optical Kerr medium, the refractive index of the medium alters with the intensity of the input light signal to the medium. The refractive index (n) [5,6,15–17] of the said medium is represented as

$$n = n_0 + n_2 I$$

where n_0 represents a constant refractive index term, n_2 represents the nonlinear correction term and I is the intensity of the applied light signal.

Therefore, the propagation factor for a light beam while passing through the medium can be expressed as

$$\beta = \beta_0 + K_0 n_2 I,$$

where β_0 is a constant factor of propagation and K_0 is the free space propagation constant.

An incident light signal having electric field $E = A exp(i\omega_0 t)$ after passing through a Kerr medium of length d would have the electric field

$$E' = A' \exp\left\{ i\left(\omega_0 t - \beta d \right) \right\},$$

where A and A' represent amplitudes of input and output light signals.

If this output signal is made feedback to the same medium through a different channel by using four mirrors, then the output light after getting first feedback would have the electric field

$$E_1 = A'' \exp\left\{ i\left(\omega_0 t - 2\beta d \right) \right\}.$$

Hence, $E_1 = A'' \exp\{i(\omega_0 t - 2\beta_0 d - 2K_0 n_2 Id)\}$.

The instantaneous phase of this light signal is expressed as

$$\varphi_1 = \omega_0 t - 2\beta_0 d - 2K_0 n_2 Id.$$

So, the instantaneous frequency of this light signal becomes

$$\omega_1(t) = \omega_0 - 2K_0 n_2 d\left(\frac{dI}{dt}\right).$$

Now if we consider that the intensity I is a function of time t, i.e. $I = I(t)$ then a change of output frequency is expected. In this chapter, we propose a new scheme of obtaining linear variation of ω by a special type of $I(t)$.

5.8.1 FREQUENCY RESPONSE OF KERR MEDIUM FOR A PARABOLIC TYPE OF INTENSITY VARYING SIGNAL AFTER THE FIRST FEEDBACK

When an electronic saw-tooth signal $\left\{V(t) = Kt = \left(\dfrac{V_0}{T_0}\right)t, 0 < t < T_0\right\}$ is passed through an electronic integrator circuit and the integrated output is passed to a light emitting diode (LED) or an injection laser diode (ILD), then at the output of the LED/ILD a parabolic type of intensity varying signal is generated whose intensity pattern is expressed as

$$I(t) = \frac{1}{2}Rt^2, 0 < t < T_0,$$

where R is a constant giving the source response factor. The graphical variation of intensity of light from the source with time for a parabolic shaped light signal is shown in Figure 5.12.

A saw-tooth-shaped electronic signal (Figure 5.12b) is passed through an electronic integrator circuit. Integrator output is then passed through an LED or ILD. At the output of LED or ILD, a parabolic-shaped optical signal (Figure 5.12a) is received.

If the parabolic type of intensity varying signal is applied to an optical Kerr medium, then after first passing through this medium the frequency response of the Kerr medium becomes

$$\omega_1(t) = \omega_0 - 2K_0 n_2 dRt.$$

Here, the instantaneous frequency varies with time t.

Thus, the frequency shift (deviation) at $t = T_0$ becomes

$$\Delta\omega_1 = 2K_0 n_2 dRT_0$$

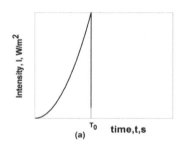

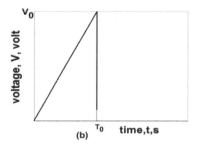

FIGURE 5.12   (a) Parabolic variation of intensity with time for the (b) saw-tooth shaped input electronic signal to the integrator.

This is the amount of frequency shift of a parabolic type of intensity varying signal after first passing through the Kerr medium at time $t = T_0$.

5.8.2   Multi-Passing in Kerr Medium for Change of Frequency

In multi-passing technique, the output-modulated light signal after getting first feedback is sent several times through different channels through the same Kerr medium and a final modulated output light signal can be received at the output terminal of the medium.

If the feedback is done n times to this nonlinear medium, then the instantaneous frequency [17] of the final output signal is expressed as

$$\omega_n(t) = \omega_0 - (1+n) K_0 n_2 d \left(\frac{dI}{dt} \right).$$

For a parabolic-shaped light pulse, the instantaneous frequency becomes

$$\omega_n(t) = \omega_0 - (1+n) K_0 n_2 dRt.$$

So, at $t = T_0$ $\omega_n(t = T_0) = \omega_0 - (1+n) K_0 n_2 dRT_0$.
Thus, the shift of frequency from the original wave is

$$\Delta \omega_n = (1+n) K_0 n_2 dRT_0.$$

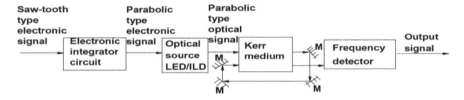

FIGURE 5.13 Schematic diagram of frequency shifting scheme by using Kerr medium, 'M' represents mirror.

In Figure 5.13, the schematic diagram of the frequency shifting scheme for parabolic type signal is shown for a single passing technique. An electronic saw-tooth signal is applied to an electronic integrator circuit and at the output of the integrator circuit one can expect an electronic parabolic signal $\left\{V(t) = \frac{1}{2}Kt^2\right\}$. This signal is then passed to an LED/ILD directly.

The LED/ILD produces a light signal having a parabolic intensity pattern at the output. This signal is then passed through the nonlinear medium having length d. With the help of four mirrors, the output light signal is made feedback to the input of the same medium again through a different channel. After that the final output modulated light signal is applied to a frequency detector to get a light signal with the desired frequency.

The graphical variation of instantaneous frequency with time for different times of passing ($n = 1, n = 2$ and $n = 3$) is shown in Figure 5.14.

From Figure 5.14, one can observe that the variation of instantaneous frequency with time for parabolic-shaped light signal is different for different times of passing through the Kerr medium. Thus, an output light signal with the desired frequency can be received at the output of the medium by selecting proper feedback mechanism.

Here, a physical system is considered to calculate the shift of frequency. A nonlinear material DR1 (disperse red 1 side chain attached) is chosen for its high n_2

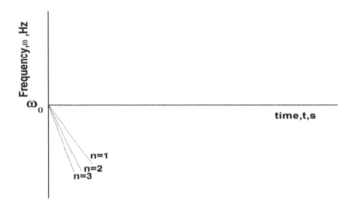

FIGURE 5.14 Variation of instantaneous frequency with time for parabolic light for different times of passing.

$(-1.7 \times 10^{-10}$ m²/w), then for the light having wavelength(λ_0) 640×10^{-9}m the shift of frequency can be calculated.

Again, the length of the material(d), intensity of light (I_0) and time period (T_0) can be considered as

$$d = 10^{-2}\,\text{m}, \quad I_0 = 10^{10}\,\text{w}/\text{m}^2 \quad \text{and}\,T_0 = 2\text{s}.$$

Thus, the shift of frequency for parabolic type light signal from the original light at $t = T_0$ is given as

$$\Delta\omega_n = (1+n)1\cdot 66 \times 10^5\,\text{Hz}.$$

The frequency shift for different times of passing of the parabolic light signal through Kerr medium is shown in Table 5.7.

From Table 5.7, it can be observed that with the increase of the number of passing through the optical Kerr medium the shift of frequency also increases. So, a light signal with the desired frequency can be received at the output of the medium by using this all-optical scheme.

5.9 CONCLUSION

Joint modulation of a single light beam is occurred by using two different saw-tooth pulses at a time. Applying two biasing voltages jointly rather than the application of a single biasing signal to the KDP-based Pockels cell parallel to Y- and Z-directions, the phase part of the light signal can be significantly controlled. One can get a zero phase difference after the operation. One can also organize a good digital phase modulation using the scheme. When an intense light signal is passing through a Kerr type of nonlinear medium multiple times, changes occur at the intensity of the central

---

TABLE 5.7

Frequency Shift vs. Different Times of Passing of Parabolic Type Light Signal through Kerr Medium at Time $=T_0 = 2$ s

Number of Passing Technique (n)	Frequency Shift (10^5 Hz)
1	3 • 32
2	4 • 98
3	6 • 64
4	8 • 3
5	9 • 96
6	11 • 62
7	13 • 28
8	14 • 94
9	16 • 6
a	$(1 + a)1 • 66$

---

frequency of the light signal. The light intensity for the central frequency decreases more sharply in the case of second passing the light in contrast to the first passing. The intensity of the harmonics in the case of second passing increases at the cost of decreases of the intensity of central frequency. This proposed scheme may be useful for the amplification of harmonic powers of light. An all-optical X-OR gate is developed by using the phase encoded mechanisms with OTA. The scheme can easily and successfully be extended and implemented for any higher number of input digits by proper use of an electro-optic modulator and phase encoding technique. A phase-encoded algebraic operation can be conducted with the proposed system. A new method of conduction of optical phase algebra is developed by the use of Pockels and Kerr materials in series. Again a theoretical description is given here against the voltage required across a linear electro-optic material for which one can nullify the introduced phase by the joint effort of Pockels and Kerr cells. A preliminary idea is invoked, how an intensity-controlled phase modulation is useful to get a zero-phase shift for propagating a wave through electro-optic material. Finally, it is realized that one can construct the optical phase algebra by modulation of the intensity of the input light beam as well as the voltage across Pockels materials. An alternative method is developed by using the multi-passing technique to increase the transmission coefficient of a KDP-based electro-optic modulator. In this process, the extra path is removed, which can be finely applied for achieving better modulation action in long haul optical communication. The high nonlinearity of the Kerr medium can cause a high conversion of the frequency of the applied light signal through it. The shift of frequency can be increased by increasing the number of passing through the material. This scheme may be applicable for the study of the molecular spectroscopic systems. Frequency modulation and multiplexing can be more perfectly done by using linear frequency variation with time. Here one important point is to specially mention that the duration of the input light pulse (parabolic) must be greater than the total propagation delay of the light from input to output covering the multi-passing. Here from the above analytical observation, it is seen that the higher number of multi-passing promises higher frequency shifts or conversion.

In the modulation process of the Pockels effect, when the modulator is biased by a single biasing voltage then the phase change occurred at the output is very small. When the biasing voltages are applied jointly to the modulator then the change of phase at the modulated output beam becomes large, the application of the system is more beneficial. The use of joint biasing voltage makes the power consumption lower comparing to the single use of biasing voltage. In the case of Kerr effect when a feedback mechanism is applied to the material, then changes occur at the intensity of central frequency and sideband frequencies. For one time passing of the light, the change of intensity is lower than the multiple time of light passing. Here the intensity of central frequency is decreased and the intensity of sideband frequencies is increased. So by using the feedback mechanism more powers of central frequency is utilized to increase the sideband power. The magnitude of the necessary V_π voltage for a crystal is reduced significantly just by cutting the extreme edge(s) of the crystal obliquely and by passing the light beam through the crystal multiple times. Here in this chapter, we are focused only on the massive variation of the phase of the output modulated light wave by the joint use of Pockels and Kerr materials. The advantages

of using Pockels and Kerr material jointly one can adjust the suitable phase difference between two perpendicular polarized light waves by only varying the intensity
of the input light wave keeping the biasing voltage fixed across two Pockels material.
In this scheme, the nonlinear refractive index of the Kerr material also plays an
important role in creating the finite phase difference. If we trigger a high intense light
beam through the Kerr material then it will produce a massive change of the phase
between the two components of the light waves. It is also noticed that if we use the
Kerr medium with a higher value of nonlinear coefficient then it also provides a significant phase difference. In this scheme, one can control the phase of the output
modulated light beam by suitable variation of the intensity of light wave at the input
end without hampering the biasing voltage at all. In the case of using only two
Pockels materials successively, we have to control the phase of the output light wave
by controlling only on the biasing voltage. So for getting massive phase variation we
have to require large power consumption which is also controlled electrically. High
conversion of the frequency of the applied light signal is achieved when the light is
passing through the high nonlinearity of the Kerr medium. For an applied light signal
having a saw-tooth intensity pattern, the shift of frequency is a constant and it does
not depend on time. On the other hand for a parabolic light signal, the shift of frequency becomes a linear function of time. Hence, a linear frequency variation of light
with time is possible by using this scheme. That's why this scheme is more advantageous than the previous one.

REFERENCES

1. R. Maji and S. Mukhopadhyay, Some analytical investigations of propagation of radiation in electro-optic modulator in connection to optical velocity modulation, *IUP J.
   Phys. IV* (4), 25–29 (2011).
2. C. Zheng, C. Ma, X. Yan, X. Wang, D. Zhang, Optimal design and analysis of a high-
   speed, low-voltage poly-mer Mach–Zehnder interferometer electro-optic switch, *Opt.
   Laser Technol.* 42(3) 457–464 (2010).
3. S. Sen and S. Mukhopadhyay, Reduction of $V\pi$ voltage of an electro-optic modulator
   by jointly using oblique end cutting and multi-rotation, *Opt. Laser Technol*, 59, 19–23,
   (2014).
4. S. Sen, S.K. Pal and S. Mukhopadhyay, A new scheme of using electro-optic crystal
   cut obliquely in both sides for reducing its half wave voltage, *J. Opt.* 43 (2), 154–158,
   (2014).
5. A. Yariv, P. Yeh, *Electro-optic modulation in LASER beams. Photonics – optical electronics in modern communications*. New York: Oxford University Press; 2007.
6. A. Ghatak, K. Thyagarajan, *Optical electronics*. New Delhi: Cambridge University
   Press; 2008.
7. J. F. Diehl, C. E. Sunderman, J. M. Singley, V. J. Urick, K. J. Williams, Control of
   residual amplitude modulation in Lithium Niobate phase modulators. *Optics Express.*
   25(26), 32985–32994(2017).
8. L. Cao, A. Aboketaf, Z. Wang, S. Preble, Hybrid amorphous silicon (a-Si:H)–LiNbO3
   electro-optic modulator. *Optics Communications.* 330, 40–44(2014)
9. Q. Yang , H. Dong, S. Chen, W. Sima, T. Yuan, R. Han, Non-contact overvoltage monitoring sensor based on electro-optic effect. *High Voltage Engineering*, 41(1), 140–145
   (2015).

10. S. Mukhopadhyay, An optical conversion system: from binary to decimal and decimal to binary. *Opt. Commun.* 76(5–6), 309–312 (1990).
11. B. Chakraborty and S. Mukhopadhyay, Alternative approach of conducting phase-modulated all optical logic gates, Opt. Eng., 48, (3), 035201 (2009).
12. S. Saha, S. Biswas and S. Mukhopadhyay, An optical scheme of conversion of a positionally encoded decimal digit to frequency encoded Boolean form using MZI-SOA, *IET Optoelectronic*, 11(5), 201–207 (2017).
13. S. Dey and S. Mukhopadhyay, A new approach of implementing phase encoded quantum SRN gate, *Electronics Letters*, 53(20), 1375–1377 (2017).
14. S. Dey and S. Mukhopadhyay, Implementation of all-optical Pauli-Y Gate by the integrated phase and polarization encoding, *IET Optoelectronics*, 12(4), 176–179 (2018).
15. A. Chatterjee and S. Mukhopadhyay, A New Method of Obtaining an Ultrashort Optical Pulse by the Use of Optical Kerr Material and a Saw-tooth Optical Pulse, *International Journal of Electronics & Communication Technology*, 6(1), 42–43 (2015).
16. A. Chatterjee, S. Biswas and S. Mukhopadhyay, Method of frequency conversion of Manchester encoded data from a Kerr type of nonlinear medium, *J. Opt.*, 46(4), 415–419 (2017).
17. A. Chatterjee and S. Mukhopadhyay, Kerr Medium for Change of Light Frequency in Saw-Tooth Intensity Pattern Accommodating Multi-Passing Technique, International Journal of Computer Sciences and *Engineering*, 7(1), 105–108 (2019)
18. J. Leuuthold and C. S. Bres, All-optical pulse shaping for highest spectral efficiency, Springer Ser. *Opt.sci.*, 194, pp. 217–260 (2015).
19. M. Mandal and S. Mukhopadhyay, A New Study *Of Using Two Different Saw-tooth Pulses For Joint Modulation Of A Single Light Beam, International Conference On Recent Developments in Nonlinear Dynamics and its Applications (CRDNDA 2019)*, March 12–14, 2019, Durgapur Govt. College, Durgapur, 713214, WB, India.
20. M. Mandal and S. Mukhopadhyay, *A new study for the investigation of the change of the intensity of the light for the central frequency after 2nd passing through a Kerr nonlinear material, National Seminar on Recent Trends in Condensed Matter Physics including Laser Applications (NSCMPLA-2019)* January 16–18, 2019, The University Of Burdwan, Golapbag, Burdwan, 713104, WB, India.
21. M. Mandal and S. Mukhopadhyay, 4th Regional Science & Technology Congress (Western Region), 2019 (December 12–13, 2019), The University Of Burdwan, Golapbag, Burdwan, 713104, WB, India.
22. S. Lakshan and S. Mukhopadhyay, A new method of conduction of optical phase algebra by use of Pockels and Kerr materials in series, 4th Regional Science & Technology Congress (Western Region), 2019 (December 12–13, 2019), The University Of Burdwan, Golapbag, Burdwan, 713104, WB, India.
23. S. Lakshan, D. Saha and S. Mukhopadhyay, "Optical Scheme of Obtaining Highest Transmission Factor in Case of KDP Based Electro-Optic Crystal by the Adjustment of Suitable Biasing Voltage and Number of Feedback Passing", *Journal of Optical Communication* (2019).

6 Slotted Photonic Crystal Waveguide

An Effective Platform for Efficient Nonlinear Photonic Applications

Mrinal Sen and Tanmoy Datta

CONTENTS

6.1 INTRODUCTION

Optics is experiencing unprecedented advancement in catering the ever-evolving demand of the ultra-high bandwidth in the high-performance processing of information and communication systems. Silicon photonics along with its stupendous potentialities have provided a realizable platform for harnessing its benefits on chip, thereby accelerating the advancement substantially [1–3]. Photonic integrated circuit (PIC) in the silicon photonics platform has been the long-standing goal of the researchers, and untiring nourishment in this field is helping it to come out of its infancy and flourish. Despite being complementary to the electronic integrated circuits, PICs utilize benefits of the existing state-of-the-art experience, technology and scalability of Si-processing techniques and promise Moore's-like scaling of photonic devices. Integrated photonic waveguides play a major role in PICs, since they facilitate light–matter interactions and interconnections underlying to the processing of optical signals. Moreover, performance of photonic devices in terms of the footprint, operating power, etc., are also directly related to optical confinement as offered by waveguides and, hence, their geometries. Although silicon-on-insulator (SOI) wire waveguides have been widely used conventionally [4,5], the nonlinear optical interactions excited therein are not efficient enough. This, in turn, necessitates effective interaction length of the waveguide in the scale of centimetre to millimetre and operating power in the order of few watts which are evidently not suitable for PICs. The limitations have been eradicated to some extend by employing slot waveguides whereby the large index discontinuity offers strong spatial confinement of light in the low-index active material sandwiched in between two high-index slabs [6]. Consequently, the requirements of effective length and operating power have been demonstrated to be reduced drastically [7,8]. On the other hand, slow-light propagation has fascinated the researchers as it has the ability to provide substantial reduction of the footprint and operating power [9]. Slow-light propagation is manifestation of the temporal confinement of light and it adds another dimension to control over light–matter interactions. Photonic crystal waveguides (PCWs) are the most promising dielectric structures offering temporal confinement which originates from the resonant interaction of photons with the periodic dielectric lattice [10,11]. Besides temporal confinement, PCW also offers notable characteristics of light propagation in the sub-wavelength scale making it a competent platform for ultra-compact planar PICs. Now, the temporal confinement enhances the local energy density of the propagating signals inside PCWs through spatially compressing the envelopes in the slow-light regime, which contributes extensive enhancement of the light–matter interactions [10–14]. To characterize this enhancement achieved, a slow-down factor is used which is the ratio between the group index experienced at the operating wavelength in slow-light regime to the refractive index of the medium of propagation [10]. Specifically, the linear and the third-order nonlinear optical effects are, respectively, enhanced proportionally and quadratically with the slow-down factor [10–14]. Now, this phenomenon demonstrates sizeable reduction of the requirement of footprint and operating power for linear and nonlinear optical effects inside PCWs. Stimulated Raman scattering (SRS) is among such potential third-order nonlinear light–matter interactions, which has been demonstrated to get enhanced substantially by a factor as high as four orders of magnitude in slow-light regime inside silicon PCW [15]. The reason behind mentioning SRS particularly in this context is its ability to achieve light amplification and lasing in indirect gap materials such as silicon which is otherwise formidable. SRS, by virtue of the distributed interaction, used to take kilometers of length in silica fiber for amplification of the Stokes wave and, hence, has never been thought for its on-chip applications. Although the requirement of the waveguide length and the operating power has been reduced drastically by the use of SOI wire waveguides [16,17], the values still remain incompetent to PICs. Slow-light propagation in silicon photonic crystal cavities and waveguides, on the other hand, has provided significant miniaturization in both the footprint and operating power of the Raman amplifiers and lasers [18–24]. Now, the miniaturization will continue further if the benefits of strong spatial confinement of light as obtained in slot waveguides can possibly be merged with the temporal confinement in PCWs. Precisely that can be achieved using a slotted photonic crystal waveguide (SPCW) whereby the phenomena of strong spatial and temporal confinements are combined offering 'extreme' optical confinement [25,26]. Recently, silicon-based SPCWs are getting stirring attention of researchers as they offer the most compact solutions to realize nanophotonic devices involving light–matter interactions ranging from nonlinear optics, sensing, electro-optic modulation to quantum optics and optical tweezing [27–29]. The footprint and the operating power of such devices demonstrate phenomenal miniaturization that complies with the requirement in PICs. However, the fact of having optical confinement in the low-index slot region creates challenges of using SPCWs in enhancing SRS interaction, since the refractive index of the material in the slot region needs to be lower than that of the silicon. Silicon nanocrystal embedded in silica host ($SiNC/SiO_2$) is such a promising material in silicon photonics platform, which has the remedy in such a situation [30]. The refractive index of $SiNC/SiO_2$ material lies in between that of Si and SiO_2 depending on the concentration of silicon, and is typically less than 2.2 always [31]. It is also worth mentioning here that $SiNC/SiO_2$ offers an exceptionally high SRS gain coefficient which may be even almost four orders of magnitude higher as compared to the bulk Si, subjected to the specific set of distribution of size and crystallinity of the nanocrystals [32,33]. Now, the inherently high SRS gain of $SiNC/SiO_2$ gets intensified further through the ultra-high optical confinement inside SPCW, and a momentous miniaturization in effective length and threshold power is possible to achieve. Moreover, the SPCWs are of photonic crystal slab-based waveguide geometries which along with the embedding $SiNC/SiO_2$ material in the slot are amenable to convenient fabrication and, hence, incorporation in PICs. Therefore, keeping this theme in mind, this chapter essentially highlights some of the recent theoretical developments of active all-optical functionalities reported by the authors in [34–37], which function based on enhanced SRS interaction inside $SiNC/SiO_2$-embedded SPCWs. In particular, the chapter highlights the design of $SiNC/SiO_2$-embedded SPCW-based Raman amplifier with the smallest footprint and operating power ever reported till date by the authors' best knowledge [34]. Moreover, the substantially miniaturized requirement of operating pump power subsequently opens up a possibility of using a light-emitting diode (LED) as the source of pump [35]. The use of the LED would enhance the possibility of on-chip integration of the amplifier while offering cost-effective optical pumping, which eliminates the intrinsic demerits of having an off-chip pump source. Further, introducing incoherency in pumping, as in the case of LED owing to its wider spectral response, provides several advantages in terms of the faithful amplification of the Stokes wave which, however, is not possible to achieve with a continuous wave (CW) laser pumping. Nevertheless, the miniaturization demonstrated remains ineffective as long as the issues related to the coupling of the pump and the Stokes for their SRS interaction are not being taken into consideration properly. The chapter also emphasizes some potential solutions in this context by highlighting a novel PhC-based structure, as proposed by the author recently, for combining the pump and Stokes waves efficiently inside the $SiNC/SiO_2$-embedded SPCW [36]. Thereafter, developments of the micron-scale Raman amplifier along with the pump combiner structure have been purposefully utilized to design an all-optical pass switch (AOPS) [37]. The proposed switch functions based on the SRS mediated depletion and enhancement of the pump and the Stokes waves, respectively, inside the $SiNC/SiO_2$-embedded SPCW. The proposed switch has the prospect to realize other photonic logic functionalities, e.g. multiplexing and decoding in the same line as demonstrated using a complementary metal-oxide semiconductor transmission gate which is the electronic equivalent of the AOPS.

The rest of the chapter is structured as follows. Section 6.2 describes the model description of the $SiNC/SiO_2$-embedded SPCW and necessary optical characterization as in [34,38]. This section also discusses the unified coupled NLS equations incorporating necessary modifications to encompass most of the possible phenomena in a comprehensive manner particularly emphasizing the SRS interaction [34]. Afterward, in Section 6.3, Raman amplification characteristics have been simulated under CW laser pumping and the respective outcomes are discussed [34]. The characteristics have also been simulated in the LED pumping regime in Section 6.4 showing substantial improvement of the performance [35]. Section 6.5 discusses the PhC-based pump-Stokes combiner structure with the characteristic analysis of the transmittance properties [36]. The above developments are utilized subsequently in Section 6.6 in designing the novel AOPS [37]. The same section also illustrates the operation of the proposed AOPS at both the transient and steady states [37]. Finally, a conclusion is made in Section 6.7.

6.2   MODEL DESCRIPTION AND OPTICAL CHARACTERIZATION

Three-dimensional schematic of $SiNC/SiO_2$-embedded SPCW is shown in Figure 6.1(a) [34] comprising of an air bridge silicon slab embedding two-dimensional photonic crystal (PhC) of air holes arrayed in a triangular lattice. The lattice constant (a) of the bulk PhC is considered as $0.426~\mu m$ with the radius (r) of embedding air holes as $0.3a$. The slab-thickness has been considered as $0.7a$ to maximize the quasi transverse electric (TE) photonic bandgap (PBG) that can be obtained from the PhC, which eventually provides deep in-plane optical confinement. A slot section having a width of $0.2a$ aligned at and along the center of the W1 PCW, i.e. $z = 0$ is considered, which is consisting of the $SiNC/SiO_2$ material of refractive index 1.98 pertaining to its dominant SRS gain coefficient of 438 cm/MW [32]. These design parameters are so chosen to have wavelengths of the operating pump and Stokes waves as $1.427~\mu m$ and $1.5413~\mu m$, respectively, in conforming to the work carried out by Sirleto *et al.* [32], with 15.59 THz frequency detuning corresponding to the dominant peak SRS gain coefficient. The structure as depicted in Figure 6.1(a) is also amenable to convenient fabrication processes as has been evidenced by fabrication of similar or even more complicated photonic structures [39–43]. Nevertheless, a possible procedure for fabricating such structure is also been simulated using a standard process simulation software and is depicted in Figure 6.2 [34]. The steps involved in the simulation are described briefly as follows.

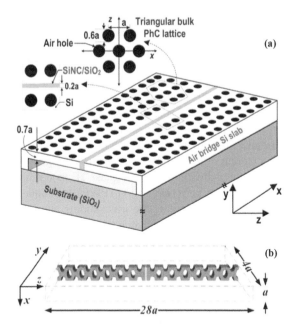

FIGURE 6.1 (a) Schematic of air bridge SiNC/SiO$_2$ embedded SPCW. (b) The supercell considered for the PWE. Reproduced from [34] © IOP Publishing. Reproduced with permission. All rights reserved.

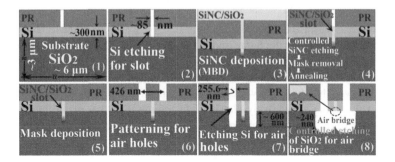

FIGURE 6.2 Possible steps involved for fabricating the air bridge SiNC/SiO$_2$ embedded SPCW using SILVACO TCAD. Reproduced from [34] © IOP Publishing. Reproduced with permission. All rights reserved.

An SOI wafer is considered to begin with, which consists of thickness of the SiO$_2$ and Si layers as 3 μm and 0.3 μm (~0.7a), respectively. In step 1, an 0.085 μm (~0.2a) wide slot section is defined for lithography after depositing photoresist (PR), and the Si layer is etched precisely in the subsequent step using the deep reactive ion etching (DRIE) technique. In step 3, SiO$_x$ material is considered to be deposited inside the slot section using the molecular beam deposition (MBD) technique with the 1:10 volumetric proportion of Si and SiO$_2$. Thereafter, in the same step, annealing of the deposition is performed at high temperature (~1200°C) to form the required SiNC/

SiO_2. Afterward, in step 4, the deposited $SiNC/SiO_2$ lying above the Si layer is etched out. In step 5, PR is deposited again and subsequently patterning is performed for defining air holes of radius 0.128 µm (~0.2a) which are separated by the lattice constant of 0.426 µm (a). In the next step, air holes are etched using the DRIE technique and the etching is allowed to penetrate the SiO_2 layer up to ~0.6 µm. This depth is chosen in such a way that the etching does not hamper the $SiNC/SiO_2$ layer. Nevertheless, an additional depth of ~0.24 µm has been achieved by the wet etching process which eventually provides sufficient air region beneath the Si PhC slab; thereby creating the air bridge $SiNC/SiO_2$-embedded SPCW structure.

Now, dispersion diagram of the SPCW ensuring co-propagation of the pump and the Stokes waves along the x-direction for their quasi-TE modes has been calculated using three-dimensional plane wave expansion (PWE) technique [44]. Dispersion diagram is also useful for characterizing the waveguiding and other related optical characteristics of the SPCW. A suitable supercell of size $a \times 4a \times 28a$ is considered for the purpose, as shown by Figure 6.1(b) [34], which has been discretized into 16, 64, and 1024 levels, respectively, in the x – , y- and z-directions for computing its dielectric distribution. The wavevector range is restricted to its first irreducible Brillouin zone, i.e. $0 \leq k_0 \leq 0.5$, where k_0 is the normalized wavevector. Further, the wavevector and the frequency axis as used in the dispersion diagram are normalized with respect the lattice constant in order to produce the scale invariant values. Now, a portion of the calculated dispersion diagram is shown in Figure 6.3 [34], which shows two propagation bands being in-plane and out-of-plane guided by the PBG and index confinement, respectively. However, the fundamental z-odd band has been considered to be excited in this analysis, since the modes corresponding to this band provide maximum confinement. Moreover, it is also important to consider that the modes are excited in the wavevector range $0.5 \leq k_0 \leq 1$ so as to have forward and co-propagating modes of the pump and the Stokes waves. The variation of the group indices as experienced by the wavelengths is calculated from the inverse of slope of the propagation band and has been illustrated in Figure 6.4 [34]. From the figure, it can be observed that the pump wave is experiencing a group index of ≈ 4.4, while the Stokes is operating at relatively slower light regime than that of the pump with a group index of ≈ 23. Corresponding variation of the dispersion coefficients up the

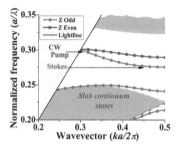

FIGURE 6.3 Dispersion diagram for quasi-TE states depicting the propagation bands of the SPCW in the range $0 \leq k_0 \leq 0.5$. Reproduced from [34] © IOP Publishing. Reproduced with permission. All rights reserved.

third order, as inferred from the variation of the group indices, has also been computed and is shown at the inset of Figure 6.4. Effective modal area is also an important parameter characterizing confinement of modes inside the waveguide and, hence, efficiency of the nonlinear performance. So has been evaluated from the overlap integral which involves the transverse spatial distribution of modes of the operating pump and Stokes waves [45–47]. Now, the transverse distributions of modes have been reproduced here in Figure 6.5 from [45]. Thereafter, the variation of the effective modal area for the entire band of propagation is shown in Figure 6.6. The figure shows that the area increases with the wavelength, which is owing to the fact the higher wavelengths experiences higher group indices implying higher back scattering. This, in turn, results in the higher spatial dispersion of the modal energy into the PhC cladding lowering the effective area.

6.2.1 Coupled NLS Equations for SRS Interaction in Slow-Light Regime

Now, in order to quantify the SRS interaction in terms of energy-transfer between the pump and the Stokes, coupled NLS equations are used which govern evolution of

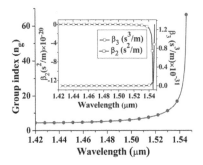

FIGURE 6.4 Wavelength dependence of the group index and the dispersion coefficients (at inset). Reproduced from [34] © IOP Publishing. Reproduced with permission. All rights reserved,

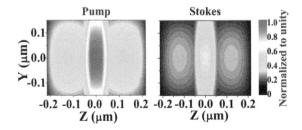

FIGURE 6.5 The transverse modal ($y - z$ plane) distribution of the pump and the Stokes as reproduced from [45]. Republished with permission of [Institution of Engineering and Technology (IET)], from [Tanmoy Datta, and Mrinal Sen, "All-optical logic inverter for large scale integration in silicon photonic circuits," *IET Optoelectronics*, [online] doi: 10.1049/iet-opt.2019.0081]; permission conveyed through Copyright Clearance Center, Inc.

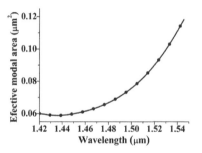

FIGURE 6.6 Wavelength dependence of the effective modal area.

envelopes of the propagating waves inside waveguide. Although the NLS equations have been applied widely in describing nonlinear optical phenomena, the conventional form of the equations need modifications in context to the slow-light propagation as in the case of a PCW/SPCW. The modifications are primarily to incorporate the slow-light induced enhancement of the underlying linear and nonlinear optical effects using a slow-down factor (S) [10]. However, this enhancement comes at the simultaneous expense of temporal dispersion which also gets strengthened at the same time in the slow-light regime reducing the peak intensity of the propagating pulses. This phenomenon, in turn, undermines the local energy density of the pulses thereby diluting the benefits of temporal confinement. A factor ξ, $0 \leq \xi \leq 1$, is used for this purpose to trade off the slow-light induced enhancement of the light–matter interactions with dispersion [10]. Hence, the factor S_j^ξ is used to scale the linear optical effects involving only the jth wave, while the factor $S_j^\xi S_k^\xi$ is used to scale the nonlinear optical effects involving both the jth and kth waves [10,48]. It is worth mentioning that ξ increases with the strength of dispersion. Furthermore, it is also important to mention here that the analysis primarily focuses on the evolution of the Stokes wave sitting at the dominant frequency-shift from the pump frequency. Generation and interaction of the higher order Stokes waves are restricted here by limiting the pump power and the length of the waveguide. Therefore, the coupled NLS equations for the pump and the Stokes waves are reproduced here in terms of their respective envelopes A_P and A_S, respectively,

$$\frac{\partial A_P}{\partial x} - i\sum_{m=1}^{\infty} \frac{i^m \beta_{mP}}{m!}\frac{\partial^m A_P}{\partial t^m} = -\frac{\alpha_{lP}S_P^\xi}{2}A_P - \frac{\sigma_{aP}S_P^\xi}{2}\left(1+i\mu_P\right)NA_P$$

$$+i\left(\gamma_{PP}^e S_P^{2\xi}\left|A_P\right|^2 + 2\gamma_{PS}^e S_P^\xi S_S^\xi\left|A_S\right|^2\right)A_P$$

$$+i\left(\frac{1}{n}\sum_{m=1}^{n}\gamma_{PP}^{R,m}S_P^{2\xi}\int_{-\infty}^{t}h_{R,m}\left(t-\tau\right)\left|A_P\left(\tau\right)\right|^2 d\tau\right.$$

$$+\frac{1}{n}\sum_{m=1}^{n}\gamma_{PS}^{R,m}S_P^\xi S_S^\xi\int_{-\infty}^{t}h_{R,m}\left(t-\tau\right)\left|A_S\left(\tau\right)\right|^2 d\tau\left.\right)A_P$$

$$+iA_S\frac{1}{n}\sum_{m=1}^{n}\gamma_{PS}^{R,m}S_P^\xi S_S^\xi\int_{-\infty}^{t}h_{R,m}\left(t-\tau\right)e^{i\Omega_{PS}\left(t-\tau\right)}A_S^*\left(\tau\right)A_P\left(\tau\right)d\tau, \quad (6.1)$$

$$\frac{\partial A_S}{\partial x} - i\sum_{m=1}^{\infty}\frac{i^m \beta_{mS}}{m!}\frac{\partial^m A_S}{\partial t^m} = -\frac{\alpha_{lS}S_S^{\xi}}{2}A_P - \frac{\sigma_{aS}S_S^{\xi}}{2}(1+i\mu_S)NA_S$$

$$+i\left(\gamma_{SS}^e S_S^{2\xi}|A_S|^2 + 2\gamma_{SP}^e S_S^{\xi}S_P^{\xi}|A_P|^2\right)A_S$$

$$+i\left(\frac{1}{n}\sum_{m=1}^{n}\gamma_{SS}^{R,m}S_S^{2\xi}\int_{-\infty}^{t}h_{R,m}(t-\tau)|A_S(\tau)|^2\,d\tau\right.$$

$$+\frac{1}{n}\sum_{m=1}^{n}\gamma_{SP}^{R,m}S_S^{\xi}S_P^{\xi}\int_{-\infty}^{t}h_{R,m}(t-\tau)|A_P(\tau)|^2\,d\tau\left.\right)A_S$$

$$+iA_P\frac{1}{n}\sum_{m=1}^{n}\gamma_{SP}^{R,m}S_S^{\xi}S_P^{\xi}\int_{-\infty}^{t}h_{R,m}(t-\tau)e^{i\Omega_{SP}(t-\tau)}A_P^*(\tau)A_S(\tau)\,d\tau. \quad (6.2)$$

The parameters β_{mj} and α_{lj} are, respectively, the coefficients of mth order dispersion and linear attenuation. The parameter $\sigma_{aj} = \sigma_{ar}(\lambda_j/\lambda_r)^2$ represents the cross-section of free carrier absorption (FCA) at the wavelength λ_j measured with reference to the value σ_{ar} corresponding to the reference wavelength $\lambda_r = 1.55$ μm. In a similar way $\mu_j = 2k_j\sigma_n/\sigma_{ar}$ incorporates the effects of perturbation of the refractive index induced by free carriers, whereby k_j and σ_n are, respectively, the free space wavevector and free carrier dispersion coefficient [46]. The free carriers are generated through the process of two-photon absorption (TPA) having an average concentration of N inside the waveguide. The dynamics of the free carrier concentration maintains the following well-known continuity equation [47].

$$\frac{\partial N}{\partial t} = \bar{G} - \frac{N}{\tau_0} \quad (6.3)$$

Here $\bar{G}$ is the TPA-dependent average generation rate of free carriers given by the following equation, and τ_0 is the effective lifetime [47]:

$$\bar{G} = \frac{\beta_{PP}^{TPA}S_P^{2\xi}|A_P|^4}{2\hbar\omega_P a_{PP}^{eff\,2}} + \frac{\beta_{SS}^{TPA}S_S^{2\xi}|A_S|^4}{2\hbar\omega_S a_{SS}^{eff\,2}} + 2\frac{\beta_{PS}^{TPA}S_P^{\xi}S_S^{\xi}|A_P|^2|A_S|^2}{\hbar\omega_P a_{PS}^{eff\,2}} \quad (6.4)$$

The parameters β_{jk}^{TPA} and a_{jk}^{eff}, $j, k \in \{P, S\}$, are, respectively, the TPA coefficient of the material of medium, and effective area of the pump and the Stokes modes subtending their SRS interaction. Now, the electronic contribution that is involved in the third-order nonlinear optical effects is described by the parameter γ_{jk}^e which consists of the Kerr and TPA effects as follows [47].

$$\gamma_{jk}^e = \frac{2\pi n_{jk}^{Kerr}}{\lambda_j a_{jk}^{eff}} + i\frac{\beta_{jk}^{TPA}}{2a_{jk}^{eff}} \quad (6.5)$$

Here, n_{jk}^{Kerr} represents the Kerr coefficient of the material of medium. The parameter γ_{jk}^e characterizes the Kerr induced self-phase and cross-phase modulation

(SPM and XPM) phenomena along with the TPA induced nonlinear loss. On the other hand, the parameter γ_{jk}^R characterizes the molecular, i.e. Raman contribution to the third-order optical nonlinearities, which originates from the molecular vibration of the material of the medium. The transience of this phenomenon is incorporated by the Raman response function $h_R(t)$ of the medium which has been modeled here by the composition of possible n numbers of molecular vibrational modes as follows [49]:

$$h_R(t) = \frac{1}{n}\sum_{m=1}^{n} h_{R,m}(t) = \frac{1}{n}\sum_{m=1}^{n} \frac{\tau_{1,m}^2 + \tau_{2,m}^2}{\tau_{1,m}\tau_{2,m}^2} \exp\left(-\frac{t}{\tau_{2,m}}\right)\sin\left(\frac{t}{\tau_{1,m}}\right) \tag{6.6}$$

where $\tau_{1,m}$ and $\tau_{2,m}$ are the two time constants pertaining to the resonance frequency and damping of the mth vibrational modes, respectively. It is important to mention here that the imaginary part of the Fourier Transform of $h_R(t)$ determines the SRS gain spectrum as given by the following expression [47]:

$$\tilde{H}_R(\Omega) = \frac{1}{n}\sum_{m=1}^{n} \tilde{H}_{R,m}(\Omega) = \frac{1}{n}\sum_{m=1}^{n} \frac{\Omega_{R,m}^2}{\Omega_{R,m}^2 - \Omega^2 - 2i\Gamma_{R,m}\Omega} \tag{6.7}$$

Here, Ω denotes the frequency-shift variable. The parameters $\Omega_{R,m}(=\sqrt{\frac{\tau_{1,m}^2 + \tau_{2,m}^2}{\tau_{1,m}^2\tau_{2,m}^2}})$ and $\Gamma_{R,m}$ are the resonance frequency and full-width half maxima (FWHM) of the mth vibrational mode, respectively. Now, the nonlinear Raman parameter γ_{jk}^R has accordingly been modified for the individual resonance as $\gamma_{jk}^{R,m} = \frac{g_{R,m}\Gamma_{R,m}}{a_{jk}^{eff}\Omega_{R,m}}$ where $g_{R,m}$ is the peak SRS gain of that resonance [47]. Thus, the fourth terms in the right-hand side (RHS) of Equations (6.1) and (6.2) model the Raman induced SPM and XPM phenomena as well as the self- and cross-frequency shifts depending on the relative width of the operating pulses as compared to the SRS gain spectrum. Moreover, the SRS induced depletion and enhancement of power of the pump and the Stokes waves are incorporated, respectively, by the last terms in RHS of Equations (6.1, 6.2).

Now, the SRS gain spectrum of the SiNC/SiO$_2$ material, as adopted in this work, has been modeled considering two damped molecular vibrational modes which are centered at the resonance frequencies of 15 THz and 15.59 THz. The different values of time constants [49] and other associated parameters are enlisted here in Table 6.1.

The calculated $\tilde{H}_R(\Omega)$ with these typical values corroborates well with that of the experimentally measured SRS gain spectrum as shown in Figure 6.7 [34]. Corresponding shape of the $h_R(t)$ is shown at the inset. The values of the other rest parameters used in the above equations pertaining to the SiNC/SiO$_2$-embedded SPCW are presented in Table 6.2.

Now, Equations (6.1–6.4) have been solved for the A_P and A_S using the split-step Fourier method (SSFM) in order to get the Raman amplification characteristics inside the SiNC/SiO$_2$-embedded SPCW. Split-step Fourier method is a widely used and quite accurate numerical technique for solving coupled NLS equations [51]. 'Split-step'

TABLE 6.1

Values of the Time Constants and Related Parameters

Description	Resonance	Symbol	Value	Reference
Time constant	15 THz	$\tau_{1,1}$	10.6 fs	[49]
		$\tau_{2,1}$	0.27 ps	[49]
	15.59 THz	$\tau_{1,2}$	10.2 fs	[49]
		$\tau_{2,2}$	1.4 ps	[49]
Raman gain	15 THz	$g_{R,1}$	149 cm/MW	Calculated [34]
coefficient	15.59 THz	$g_{R,2}$	802.78 cm/MW	
FWHM of Raman	15 THz	$\Gamma_{R,1}$	1.178 THz	
gain spectrum	15.59 THz	$\Gamma_{R,2}$	0.23 THz	

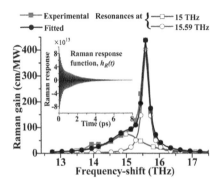

FIGURE 6.7 The SRS gain spectrum and the corresponding the response function, $h_R(t)$, of the SiNC/SiO$_2$. The spectrum has been calculated considering two vibrational resonances at 15 THz and 15.59 THz. Reproduced from [34] © IOP Publishing. Reproduced with permission. All rights reserved.

essentially stands for splitting of length of the operating waveguide into large number of segments whereby the effects of dispersion and nonlinearities are assumed to act independently in each of the segment to evaluate evolution of the propagating envelopes. On the other hand, the terms involving partial derivatives with respect to time are computed in terms of their Fourier transformation (after which the phrase 'Fourier' is included in the name) to reduce them into their equivalent simplified algebraic form. The SSFM has been implemented using coding developed in the MATLAB platform. In this work, the segments of the length and time are taken as 1 nm and 0.5 fs, respectively. The results obtained are depicted and analyzed in the subsequent section.

6.3 RAMAN AMPLIFICATION CHARACTERISTICS UNDER CW LASER PUMPING

At first, simulations are performed for both the CW pump and Stokes waves to evaluate the steady-state performance of the amplifier. Threshold power and effective

TABLE 6.2

Values of Different Parameters Used in Solving the Coupled Equations

Description		Symbol	Value	Reference
Dispersion coefficients	Pump	β_P	[1.5×10^{-8} s/m, -1.0×10^{-23} s²/m, 2.3×10^{-36} s³/m]	Calculated [34]
	Stokes	β_S	[7.6×10^{-8} s/m $-$ 1.2×10^{-20} s²/m 6.1×10^{-33} s³/m]	Calculated [34]
Slow-down factor	Pump	$S_P = \dfrac{n_{gP}}{n_{wg}}$	2.27	Calculated [34]
	Stokes	$S_S = \dfrac{n_{gS}}{n_{wg}}$	11.6	Calculated [34]
Linear loss coefficients	Pump	$\alpha_{lP} S_P^\xi$	127.3 dB/cm	Considered [34]
	Stokes	$\alpha_{lS} S_S^\xi$	242.4 dB/cm	Considered [34]
Kerr coefficient	–	$n_{PS}^{Kerr} = n_{SP}^{Kerr}$	2×10^{-18} m²/W	[7]
TPA coefficient	–	$\beta_{PS}^{TPA} = \beta_{SP}^{TPA}$	50×10^{-11} m/W	[7]
Absorption cross-section of free carriers	–	σ_{ar}	1.45×10^{-22} m²	[50]
Dispersion coefficient of free carriers	–	σ_n	4×10^{-27} m³	[50]
Average lifetime of free carriers	–	τ_0	0.1 ns	[50]

waveguide length are the two important parameters characterizing the steady-state performance [51]. Threshold power of a CW Raman amplifier for a fixed waveguide length is defined as the required input pump power for which the output power of the amplified Stokes and depleted pump become equal at the end of the waveguide. At the other end, the effective waveguide length is defined as the length of the waveguide at which the SRS gain as experienced by the Stokes is exactly balanced by the attenuations originating from the linear and nonlinear losses. In other words, after the effective length the SRS would no longer be able to provide net gain to the Stokes power and it suffers from net attenuation. Now, to calculate variation of the threshold power and effective length of the amplifier as functions of different input parameters, spatial evolution of the Stokes needs to be studied for different input pump power that is depicted in Figure 6.8 [34] in the same line as demonstrated in [38]. In this study, the input power of the Stokes is kept at 50 μW. The value of the factor ξ has been considered unity owing to the absence of the temporal dispersion. From Figure 6.8, it can be seen that the Stokes is getting amplified initially with the propagation distance depleting the pump and eventually reaches to a point of inflection. The point of inflection corresponds to the effective length after which the Stokes suffers from the net attenuation. By the time the Stokes reaches to the effective length

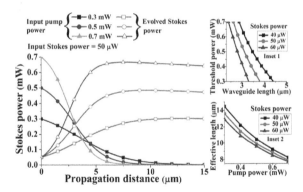

FIGURE 6.8 Spatial evolution of the CW Stokes exploiting the CW pump at the 15.59 THz frequency-shift. Variations of the threshold power and effective length as obtained with respect to the waveguide length and input pump power respectively are depicted at the two insets for different input Stokes power. Reproduced from [34] © IOP Publishing. Reproduced with permission. All rights reserved.

of ~ 8 μm corresponding to the input pump power of 0.7 mW, a significant gain as high as 11 dB has been demonstrated to achieve. This is worthy to note that the gain achieved is the highest and first ever reported to the best of the authors' knowledge for an Si Raman amplifier with such a miniaturized waveguide length and operating power. The variation of the threshold power with the length of the waveguide has been extracted from Figure 6.8 and has been depicted at inset 1. The characteristics have been repeated for different input Stokes power. From the inset, it can be seen that requirement of the threshold power reduces with the increase in the waveguide length, and so is also with the increase in the input Stokes power. This is owing to the fact that the spatial rate of enhancement/depletion of the Stokes and pump, respectively, increases exponentially with the increase in the waveguide length and the individual power of the signals during their evolution. In the same line, variation of the effective length has been depicted at inset 2. The effective length also reduces with the increase in the input pump and the Stokes power.

Thereafter, simulations are performed using Stokes pulses in presence of the CW pump to evaluate the operating repetition rate which is also an important parameter characterizing performance of the amplifier. Results of such simulations need to be demonstrated temporally for understanding the functional characteristics of the output pulses. Meanwhile, this is worth mentioning here that since the pulse-envelopes are traveling with their corresponding group velocities inside the waveguide, it is difficult to demonstrate the output pulses with respect to the primitive time axis (t). Thus, a new retarded time axis (T) is introduced substituting the t in the following manner [51] such that the new time frame moves along the x direction with the group velocity of the Stokes pulses, i.e. v_{gS}.

$$T = t - \frac{x}{v_{gS}} = t - \beta_{1S}x \qquad (6.8)$$

Now, in this new retarded time frame the term $\beta_{1S} \dfrac{\partial A_S}{\partial t}$ is identically vanished, while the corresponding term $\beta_{1P} \dfrac{\partial A_P}{\partial t}$ simplifies to $d_{PS} \dfrac{\partial A_P}{\partial T}$ where d_{PS} is the pulse walk-off parameter. Here, d_{PS} is defined as $\beta_{1P} - \beta_{1S}$. The corresponding walk-off distance (L_W) for a width (T_0) of a pulse is estimated to be $L_W = \dfrac{T_0}{|d_{PS}|}$ [51]. The walk-off phenomenon is important to consider in dispersive mediums as in the cases of PCWs/SPCWs whereby the group velocities that the different frequencies move with are substantially different. Nonlinear optical interactions may get ceased in cases when the faster moving pulse completely walks through the slower moving one. However, the walk-off effect in the present case being investigated is comparatively weaker to really impede the nonlinear interactions owing to the CW pumping.

Now, Gaussian return-to-zero (RZ) optical pulses are considered in this work for representing the Stokes waves. The width T_0 of the pulses needs to be chosen to have minimal temporal dispersion which effectively provides the full capacities of slow-light induced enhancement of the linear and nonlinear light–matter interactions. Hence, T_0 has been considered as $\geq |\beta_{2S}/\beta_{3S}|^{-1}$, which is here ≈ 0.5 ps, to diminish the third-order dispersion induced impairments within the considered length of propagation. Moreover, the second-order dispersion length, i.e. $\dfrac{T_0^2}{|\beta_{2S}|}$ [51] is calculated to be approximately 22 µm for the considered T_0, which is much higher than the length of propagation considered in the work. Hence, the effect of the second order or the group velocity dispersion is also negligibly small within the considered length of propagation. Consequently, the factor ξ has been taken unity throughout the simulations. Results of the simulation have been demonstrated in Figure 6.9 [34]

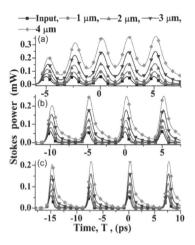

FIGURE 6.9 Temporal evolution of 0.5 ps Stokes pulses with the repetition times of (a) 2.5 ps (b) 5 ps and (c) 7.5 ps, exploiting the CW pump operating at 15.59 THz frequency-shift. Reproduced from [34] © IOP Publishing. Reproduced with permission. All rights reserved.

considering length of the SPCW as 4 μm. The peak power of the Stokes pulses and the CW pump are taken as 60 μW and 600 μW, respectively, and the simulations are repeated for different rates of repetition of the input Stokes pulses, e.g. 400 Gbps, 200 Gbps and 133 Gbps. Moreover, characteristics of the output pulses at every 1 μm interval of propagation length have also been depicted for understanding spatial evolution of the temporal characteristics of the pulses. Figure 6.9 shows that the output pulses suffer from temporal impairments in terms of broadening and intrusion of the trailing edge to the leading edge of the successive pulse. It is worth mentioning here in this connection that the temporal broadening of the output pulses never reduces the peak intensity of the pulses and, thus, the slow-light induced enhancement of the light–matter interactions remain unaffected. The impairments are actually consequences of the spectrum impairments of the operating pulses originating from the non-instantaneous and causal SRS interaction. To understand the phenomena clearly, the SRS gain spectrum of $SiNC/SiO_2$ material has been co-plotted with the spectrum of the input Stokes pulse at the 15.59 THz frequency detuning in Figure 6.10 [34]. The spectrum of the output pulse is also depicted in the same figure after taking the product of these two spectra, since so has been adopted as the mathematical technique to treat the underlying convolution integral in the fifth term at the RHS of Equation (6.2) in presence of the CW pump. The same calculation is also repeated for a train of input pulses to show the corresponding picture in spectral domain and its inferences on the repetition rate of the pulses. It can be seen from Figure 6.10 that the high-frequency components of the spectrum of the input pulse/pulses get restrained at the output as the SRS gain spectrum is substantially narrower as compared to the input pulse spectra. This is also evident from the fact that the time constant ($\tau_{2,2} = 1.4$ ps) determining the FWHM of the resonance at 15.59 THz is more than the pulse width considered by approximately a factor of 3. Consequently, the temporal shape of the output pulse which is obtained after taking inverse Fourier transformation of the output spectrum gets broadened as compared to the input pulse. Moreover, the broadening is appearing at the trailing edges of the output pulses because of the causal nature of the $h_R(t)$ as depicted in Figure 6.7. The broadened

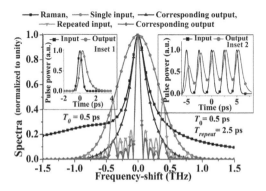

FIGURE 6.10 Spectral domain analyses of the Stokes pulses at the 15.59 THz frequency-shift. Corresponding temporal shapes of the output pulses are shown at the two insets. Reproduced from [34] © IOP Publishing. Reproduced with permission. All rights reserved.

trailing edge intrudes to its immediate successive pulse and increases power corresponding to the minimum level of the pulse train. This residual power, in turn, is
amplified while travelling inside the waveguide and the shape of the pulse loses its
logical identity. Although the trailing edge is diminished considerably by increasing
the repetition time of the operating pulses before reaching the successive pulse as can
be seen from Figure 6.9, this would eventually restrict the operating rate of the amplifier. Now, maintaining the logical identity as far as possible, overall gains of 3 dB,
4.7 dB and 5.6 dB have been achieved at waveguide lengths of 1.5 μm, 3 μm and 4
μm corresponding to the operating repetition rates of 400 Gbps, 200 Gbps and 133
Gbps, respectively.

Now, the only possible way out to get rid of the temporal impairments of the output pulses is to enhance the bandwidth of the SRS gain spectrum effectively so as to
provide near consistent gain over the entire input pulse spectrum. Operating the
Stokes at 15 THz frequency detuning instead of the 15.59 THz is one of such potential techniques to exterminate the impairments to some extent. Although the SRS
gain coefficient as obtained at this frequency-shift is almost five times smaller than
that at the peak, the bandwidth of the resonance at this frequency-shift is simultaneously greater by the same factor as compared to the resonance at 15.59 THz.
Operating the Stokes at a frequency-shift other than the dominant has never been
thought for having amplification in fiber or SOI waveguides, since the gain coefficient corresponding to such a shift is reduced significantly. However, the SRS gain
coefficient in the case of the SiNC/SiO$_2$ material is high enough to persist substantially at a frequency-shift even other than the dominant one. The spectral characteristics, as demonstrated for the 15.59 THz frequency-shift, have also been calculated
at 15 THz as depicted in Figure 6.11 [34] to show and explain the corresponding
effects. The FWHM of the resonance at 15 THz is ≈ 1.18 THz with the associated
time constant $\tau_{2,1} = 0.27$ ps which is now comparable and even lower than the operating pulse width. Thus, the spectral impairments are less in this case and, hence, the

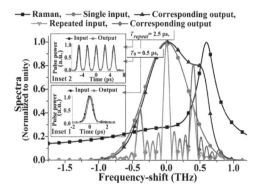

FIGURE 6.11   Spectral domain analyses of the Stokes pulses at 15 THz frequency-shift.
Corresponding temporal shapes of the output pulses are shown at the two insets. Reproduced
from [34] © IOP Publishing. Reproduced with permission. All rights reserved.

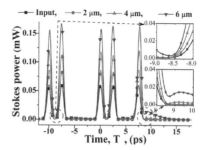

FIGURE 6.12 Temporal evolution of the sequence of the 0.5 ps RZ Stokes pulses ("11,001,101") with the repetition time of 2.5 ps, exploiting the CW pump at the 15.59 THz frequency detuning. Reproduced from [34] © IOP Publishing. Reproduced with permission. All rights reserved.

corresponding temporal shapes of the output pulses are preserved as can be seen from the two insets. However, in contrast to the 15.59 THz frequency-shift, the high-frequency components of the spectrum of the input pulse are strengthen a bit owing to its proximity to the dominant peak SRS gain coefficient at 15.59 THz. The consequences of this phenomenon are observed at the trailing edge of the pulse in terms of the little oscillation as can be seen from inset 1. Moreover, this asymmetrical augmentation of the high-frequency components also leads to the non-uniform peaks of the output pulses which will get further amplified to be noticeable clearly. Now, to confirm these effects in temporal domain, simulations are also performed with some arbitrary chosen pulse trains of width 0.5 ps. In particular, a sequence of RZ pulses, viz. '11001101', has been chosen here for demonstration of the performance of the amplifier as shown in Figure 6.12 [34] with the length of the SPCW as 6 μm. The wavelength of the Stokes has been kept intact at 1.5413 μm, while the pump has been chosen to operate at 1.431 μm maintaining 15 THz frequency-shift. Figure 6.12 shows an overall average gain of ≈ 4.2 dB sustaining the logical identity of the propagating pulses.

6.4   RAMAN AMPLIFICATION CHARACTERISTICS UNDER LED PUMPING

The other and the most promising way out to eradicate completely the temporal impairments of the output pulses is to employ LED pumping instead of the CW laser pumping. Although few works have demonstrated the use of LED in optical pumping inside Er-doped silica waveguide-based amplifiers, noticeable gain can only be achieved in such cases at the expense of very high LED power [52,53]. Moreover, the use of LED pumping in Raman amplifiers has never been explored earlier to the best of the authors' knowledge. Now, the substantially miniaturized threshold power and effective length as investigated here in the SiNC/SiO$_2$-embedded SPCW-based Raman amplifier open up a possibility to use a cost-effective low power LED for

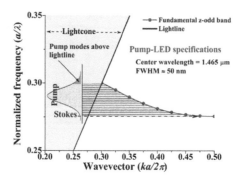

FIGURE 6.13 Fundamental z-odd propagation band and position of the excited pump and the Stokes modes.

optical pumping. This, in turn, widens the scope of its integration on the same platform as of the SiNC/SiO$_2$-embedded SPCW which is otherwise formidable to achieve using CW laser pumping. Most importantly, the LED pumping, owing to its incoherent spectral response, effectively enhances the bandwidth of the SRS gain spectrum considerably which effectively provides consistent gain over the entire spectrum of the operating pulse.

As already mentioned, the fundamental z-odd propagation band has been considered for exciting the operating pump and the Stokes waves inside the waveguide for their SRS interaction. The z-odd band supports propagation of the Stokes at 1.5413 μm along with a band of lower wavelengths considered as pump pertaining to the spectral emission an LED as shown in Figure 6.13. The spectral FWHM of the LED is considered as ≈ 50 nm having its center at 1.465 μm in conforming to the typical InGaAsP LED. These specifications are so chosen to maximize the energy transfer from the pump wavelengths to that of the Stokes based on the constraints like availability of modes on the z-odd propagation band, position of the light line and the frequency-shifts that the pump modes are undergoing. It is important to mention in this connection that the Stokes is experiencing a group index of ≈ 23, while the group index corresponding to the center of the pump wavelengths is calculated as ≈ 4.84 as can be seen from Figure 6.4. Thereafter, the coupled NLS equations presented in Equations (6.1–6.2) need further modification to encounter the incoherent pumping as in LED. In this analysis, a set of discrete spectrum of CW pump scaled with suitable magnitude is considered to reproduce the spectral distribution of the LED. This implies the presence of, say, M number of CW pump signals $\{P_u | 1 \le u \le M\}$ contributing individually to the Stokes, and the overall response can be calculated by taking summation of all the responses obtained individually. Although the accuracy of the response improves with increasing the value of M, an optimally chosen M is always able to provide sufficient precision while keeping time and space complexities of the underlying intensive computation manageable. Thus, the coupled NLS equations for the pump and Stokes waves have been written here in terms of their respective envelopes:

$$\frac{\partial A_{Pu}}{\partial x} - i\sum_{m=1}^{\infty}\frac{i^m \beta_{mPu}}{m!}\frac{\partial^m A_{Pu}}{\partial t^m} = -\frac{\alpha_{lPu}S_{Pu}^{\xi}}{2}A_{Pu} - \frac{\sigma_{aPu}S_{Pu}^{\xi}}{2}\left(1 + i\mu_{Pu}\right)NA_{Pu}$$

$$+ i\left[\gamma_{Puu}^{e}S_{Pu}^{2\xi}\left|A_{Pu}\right|^2 + 2\sum_{v=1(v\neq u)}^{M}\gamma_{Puv}^{e}S_{Pu}^{\xi}S_{Pv}^{\xi}\left|A_{Pv}\right|^2 + 2\gamma_{PuS}^{e}S_{Pu}^{\xi}S_{S}^{\xi}\left|A_{S}\right|^2\right]A_{Pu}$$

$$+ i\left[\frac{1}{n}\sum_{m=1}^{n}\int_{-\infty}^{t}\left(h_{R,m}\left(t-\tau\right)\sum_{v=1}^{M}\gamma_{Puv}^{R,m}S_{Pu}^{\xi}S_{Pv}^{\xi}\left|A_{Pv}\left(\tau\right)\right|^2\right)d\tau\right.$$

$$+ \frac{1}{n}\sum_{m=1}^{n}\int_{-\infty}^{t}\left(h_{R,m}\left(t-\tau\right)\gamma_{PuS}^{R,m}S_{Pu}^{\xi}S_{S}^{\xi}\left|A_{S}\left(\tau\right)\right|^2\right)d\tau\Bigg]A_{Pu}$$

$$+ i\Bigg[A_{S}\left\{\frac{1}{n}\sum_{m=1}^{n}\gamma_{PuS}^{R,m}S_{Pu}^{\xi}S_{S}^{\xi}\int_{-\infty}^{t}\left(h_{R,m}\left(t-\tau\right)e^{i\Omega_{PuS}\left(t-\tau\right)}A_{S}^{*}\left(\tau\right)A_{Pu}\left(\tau\right)\right)d\tau\right\}$$

$$+ \sum_{\substack{v=1\\(v\neq u)}}^{M}A_{Pv}\left\{\frac{1}{n}\sum_{m=1}^{n}\gamma_{Puv}^{R,m}S_{Pu}^{\xi}S_{Pv}^{\xi}\int_{-\infty}^{t}\left(h_{R,m}\left(t-t\right)e^{i\Omega_{Puv}\left(t-\tau\right)}A_{Pv}^{*}\left(\tau\right)A_{Pu}\left(\tau\right)\right)\right\}\Bigg] \qquad (6.9)$$

$$\frac{\partial A_{S}}{\partial x} - i\sum_{m=1}^{\infty}\frac{i^m \beta_{mS}}{m!}\frac{\partial^m A_{S}}{\partial t^m} = -\frac{\alpha_{lS}S_{S}^{\xi}}{2}A_{S} - \frac{\sigma_{aS}S_{S}^{\xi}}{2}\left(1+i\mu_{S}\right)NA_{S} + i\left[\gamma_{SS}^{e}S_{S}^{2\xi}\left|A_{S}\right|^2\right.$$

$$+ 2\sum_{u=1}^{M}\gamma_{SPu}^{e}S_{S}^{\xi}S_{Pu}^{\xi}\left|A_{Pu}\right|^2\Bigg]A_{S} + i\left[\frac{1}{n}\sum_{m=1}^{n}\gamma_{SS}^{R,m}S_{S}^{2\xi}\int_{-\infty}^{t}h_{R,m}\left(t-\tau\right)\left|A_{S}\left(\tau\right)\right|^2 d\tau\right.$$

$$+ \frac{1}{n}\left\{\sum_{m=1}^{n}\int_{-\infty}^{t}h_{R,m}\left(t-\tau\right)\sum_{u=1}^{M}\gamma_{SPu}^{R,m}S_{S}^{\xi}S_{Pu}^{\xi}\left|A_{Pu}\left(\tau\right)\right|^2 d\tau\right\}\Bigg]A_{S}$$

$$+ i\left[\sum_{u=1}^{M}A_{Pu}\left\{\frac{1}{n}\sum_{m=1}^{n}\gamma_{SPu}^{R,m}S_{S}^{\xi}S_{Pu}^{\xi}\int_{-\infty}^{t}h_{R,m}\left(t-\tau\right)e^{i\Omega_{SPu}\left(t-\tau\right)}A_{Pu}^{*}\left(\tau\right)A_{S}\left(\tau\right)d\tau\right\}\right] \qquad (6.10)$$

The $\overline{G}$ becomes for the present situation as.

$$\overline{G} = \sum_{u=1}^{M}\frac{\beta_{Puu}^{TPA}S_{Pu}^{2\xi}\left|A_{Pu}\right|^4}{2\hbar\omega_{Pu}a_{Puu}^{eff\,2}} + \frac{\beta_{SS}^{TPA}S_{S}^{2\xi}\left|A_{S}\right|^4}{2\hbar\omega_{S}a_{SS}^{eff\,2}} + 2\sum_{u=1}^{M}\frac{\beta_{PuS}^{TPA}S_{Pu}^{\xi}S_{S}^{\xi}\left|A_{Pu}\right|^2\left|A_{S}\right|^2}{\hbar\omega_{Pu}a_{PuS}^{eff\,2}}. \qquad (6.11)$$

In this study, the values of parameters corresponding to the Stokes wavelength remain same as mentioned in Table 6.2. On the other hand, the values of parameters pertaining to each of the exciting pump wavelength within the LED spectrum need to be incorporated in order to solve the coupled equations. In this context, it is reasonable to assume that values of some parameters, e.g. the Kerr, TPA, absorption cross-section and dispersion coefficient of free carriers corresponding to each of the pump wavelength do not incur significant changes within the spectral range as encompassed by the LED. Thus, the parameters have realistically been considered to remain constant at the values which are used for CW pump as mentioned in Table 6.2, without affecting much of the accuracy of the calculations. Nevertheless, values of the dispersion coefficients, group indices and effective modal area corresponding to each of the pump wavelength have been calculated from the propagation band. Nevertheless, the linear loss coefficients, α_{lPu}, is assumed to be exponentially increasing with increasing proximity to the light cone as depicted in Figure 6.13. Corresponding value of the α_{lPu} at the centre wavelength is considered as 28 dB/cm that is further scaled to the overall value of $\alpha_{lPu}S_{Pu}^{\xi}$, i.e. 68.4 dB/cm corresponding to the $n_{gPu} = 4.84$ and ξ being considered unity.

Now, each CW pump wavelength assumed within the spectral range of the LED shifts the SRS gain spectrum in the frequency-shift axis, and the same has been demonstrated pictorially in Figure 6.14 [35] for some of the pump wavelengths. The effective overall SRS gain spectrum after putting an envelope joining peaks of the shifted SRS gain spectra is considerably wider than that of the individual spectrum. Moreover, the overlap between the overall gain spectrum with the LED spectrum is not significant. This essentially implies that the SRS interaction within the pump wavelengths is negligible. Thereafter, the spectral characteristics of the operating train of pulses in LED pumping have been calculated and depicted in Figure 6.15 in the same way as demonstrated in Figures 6.10 and 6.11. In this calculation, the overall SRS gain spectrum, as depicted in Figure 6.14, has been utilized which is wide enough to accommodate the entire spectrum of the input pulse. The output pulse

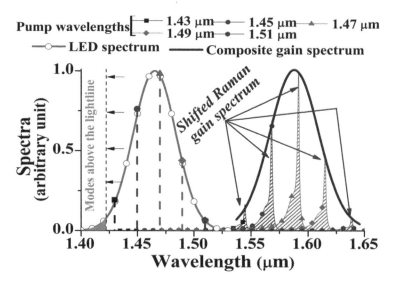

FIGURE 6.14 Overall SRS gain spectrum of the SiNC/SiO$_2$ subjected to the considered LED pump [35]. Reprinted from *Superlattices and Microstructures*, vol. no. 110, Tanmoy Datta, Mrinal Sen, LED pumped micron-scale all-silicon Raman amplifier, Pages no. 273–280, Copyright (2017), with permission from Elsevier.

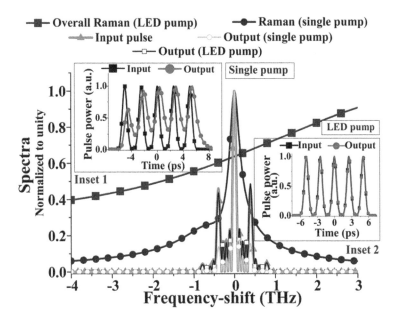

FIGURE 6.15 Spectral domain analyses of the Stokes pulses both in presence of the CW laser pump, at the 15.59 THz frequency-shift, and LED pump with its center wavelength of emission at 1.465 μm. Corresponding temporal shapes of the output pulses are shown at the two insets respectively [35]. Reprinted from *Superlattices and Microstructures*, vol. no. 110, Tanmoy Datta, Mrinal Sen, LED pumped micron-scale all-silicon Raman amplifier, pages no. 273–280, Copyright (2017), with permission from Elsevier.

spectrum as obtained after taking the product shows negligible impairments as compared to its input counterpart. Correspondingly, shape of the output pulses obtained after inverse transformation remains intact as of the input as can be seen from inset 2 of Figure 6.15 [35]. Moreover, the figure also shows the corresponding picture while operating at CW laser pumping at 15.59 THz frequency-shift. The corresponding temporal shape of the output pulses is shown in inset 1 of the figure. It can be clearly seen from inset 2 that the temporal impairments have been eradicated completely which has never been possible to achieve using CW laser pumping.

To confirm these temporal characteristics, simulations are performed using Stokes pulses having peak power of 60 μW in presence of LED pumping with an average optical power of 6 mW. Corresponding results are depicted in Figure 6.16 [35]. From Figure 6.16(a) shows that an overall gain of around 3.22 dB has been obtained at the output of a 4 μm long SPCW corresponding to the repetition rate of 400 Gbps. It is important to note here that, the gain achieved is significantly lower than that in the case of the CW laser pumping. The reason behind this is the spectral distribution of the LED power which no longer remains confined at the dominant frequency-shift. Nevertheless, the gain can be enhanced substantially by increasing length of the waveguide. In such a case the width of the operating pulses needs to be increased to combat the dispersion-induced broadening. Consequently, the repetition rate that can be achieved from the amplifier has to be compromised simultaneously. An overall

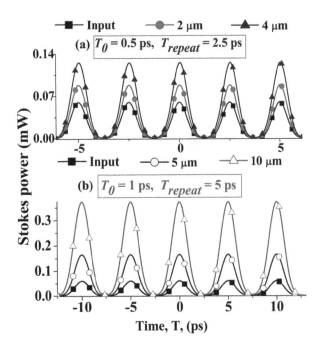

FIGURE 6.16   Temporal evolution of the Stokes pulses with (a) 0.5 ps width and repetition time of 2.5 ps, (b) 1 ps width and repetition time of 5 ps under LED pumping [35]. Reprinted from *Superlattices and Microstructures*, vol. no. 110, Tanmoy Datta, Mrinal Sen, LED pumped micron-scale all-silicon Raman amplifier, pages no. 273–280, Copyright (2017), with permission from Elsevier.

gain of 7.93 dB is demonstrated from an SPCW of length 10 μm corresponding to the repetition rate of 200 Gbps employing pulses of width 1 ps.

6.5   INTEGRABLE PUMP-STOKES COMBINER

The miniaturization, as demonstrated in the footprint and operating power, remains inadequate as long as the issues related to on-chip coupling/combining of the pump and the Stokes power have properly not been taken into consideration. The miniaturization is highly dependent on the coupling efficiencies at the wavelengths of the operating pump and the Stokes waves which are well separated in the spectrum. A weak coupling efficiency even at either of these wavelengths is sufficient enough to worsen the performance of the proposed amplifier considerably. It is worthwhile to note that the coupling of two different wavelengths for their co-propagation in Raman amplifiers as well as other such nonlinear optical devices is accomplished conventionally through splicing of fibers. The splicing is incompetent to be integrated on PICs, and the optical loss incurred at the splicing also inevitably limits the efficiency of coupling [54]. Efficient coupling of optical modes inside SPCWs is quite challenging owing to the mismatch in impedance experienced at the interface between the fast- and slow-light regimes [55]. The picture becomes even more complicated when the modes correspond to the two well-separated wavelengths having substantially different group indices as in the case of Raman amplifier being investigated. Although extensive efforts have been made in designing splitters, multiplexers in PhC platform, efficient on-chip pump-Stokes combiner has never been investigated; particularly in the case of SiNC/SiO$_2$-embedded SPCW-based Raman amplifier. Thus, the chapter also demonstrates a PhC-based novel pump-Stokes combiner structure designed to combine power of the pump and the Stokes inside SiNC/SiO$_2$-embedded SPCW for their efficient SRS interaction. In this work, the coupling has been analyzed only from the PCW to the SiNC/SiO$_2$-embedded SPCW, since the coupling from wire waveguides to the PCWs is well established [56]. The performance of the coupler has been optimized using simulations through the three-dimensional full-vector finite difference time domain (FDTD) technique.

Now, the schematic design of the proposed combiner structure is shown in Figure 6.17 [36], which comprises of an air-bridge silicon PhC slab. The geometrical parameters of the structure, e.g. the lattice constant of the PhC, radius of the air holes and thickness of the slab, have been chosen identical to the design as adopted for SiNC/SiO$_2$-embedded SPCW shown in Figure 6.1. The design of the combiner structure employs a bent (A) and a straight (B) PCW arms to couple the Stokes and pump waves which are operating at 1.5413 μm and 1.427 μm, respectively, corresponding to the 15.59 THz frequency-shift [32]. It is important to note that bending is indispensible in combining two or more optical signals, whereby the deviations to the

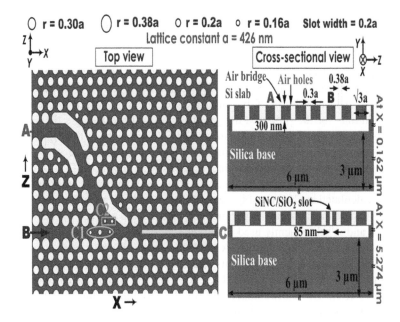

FIGURE 6.17 Design of the proposed pump combiner for SiNC/SiO$_2$ embedded SPCW based Raman amplifier [36]. Reprinted from *Optical Engineering*, vol. 57, no. 7, Tanmoy Datta *et al.*, Efficient pump–signal combiner for stimulated Raman scattering in photonic crystal waveguide, page no. 075103, Copyright (2018), with permission from SPIE.

modal distribution at interface of the bending cause substantial propagation losses. Thus, the combiner is so designed to minimize the number of bending as much as possible, thereby reducing the losses considerably. Photonic crystal structures are well capable to tackle such inconveniences for their notable characteristics of propagation even at sharp bends [57]. Literature has explored several strategies, e.g. adjustment of size [58], position and radii of air holes [59], topography optimization of defects near the interface of the bend [60], for improving the transmission characteristics. Nevertheless, the work presented combines many such strategies exploiting their potential merits. Thereafter, optimization of performance of the design has been performed heuristically and so is primarily relied on inserting, deleting, and/or displacing air holes with engineered radii to provide smooth steering of the signal at the bending. In particular, the design utilizes capsule-shaped topography [61,62] at the bending interfaces by adjoining some adjacent air holes in that region to smoothen the curvature of the bend. This capsule-shaped topography is also amenable to convenient lithography for fabrication. Nevertheless, the challenges are yet to be mitigated completely. The junction of the two coupling arms is accessible to both the arms through which a signal propagating in one arm is prone to intrusion to the other undesired arm. This intrusion weakens the transmittance of both the arms unnecessarily and, hence, needs to be restricted. For this purpose, some additional air holes of optimized radii grouped as C1 and C2 are inserted at the proximity of the junction corresponding to the arms B and A, respectively, to restrict this intrusion. In doing so, these additional holes impede the intrusion only; propagation in the desired arm remains unobstructed.

6.5.1 TRANSMITTANCE OF THE COMBINER

As already mentioned, the three-dimensional full-vector FDTD technique has been adopted in analyzing the performance of the combiner in terms of the transmittance at the respective wavelengths of the pump and the Stokes. However, before taking up any simulations for the proposed design, it is necessary to validate the simulation methodology adopted in this work. An air slot SPCW has been redesigned for this purpose, as reported in [25], and simulated for its transmittance spectrum after impulse excitation with the adopted three-dimensional full-vector FDTD technique. The transmittance spectrum for the quasi TE states ranging from 1.50 μm to 1.64 μm has been calculated as shown in Figure 6.18 [36], which shows sufficient corroboration with that of the measured experimentally. Hence, the adopted FDTD technique is verified and is subsequently used for simulating the proposed combiner structure.

Transmittance spectra of the arms A and B have been obtained independently after an impulse excitation at both the inputs and monitoring the corresponding transmittance at the junction of the two arms. The corresponding spectra have been plotted in Figure 6.19 [36]. The transmittance spectra show ripples owing to the reflection, scattering and out-of-plane losses incurred by the irregularities in the design at the bending and junction interface which cause substantial reduction in transmittance. Nevertheless, the figure shows transmittance of $\approx$ −3.5 dB and −3 dB for the arms A and B at 1.5413 μm and 1.427 μm which act as wavelengths of the Stokes and the pump, respectively. Thereafter, the electric field distribution in the $x - z$ plane during

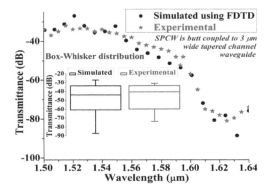

FIGURE 6.18 Comparison of the transmittance spectra (with Box-Whisker plot) that are obtained after calculation using the adopted three-dimensional FDTD method and experimental measurement as reported in [25]. Reprinted from *Optical Engineering*, vol. 57, no. 7, Tanmoy Datta *et al.*, Efficient pump–signal combiner for stimulated Raman scattering in photonic crystal waveguide, page no. 075103, Copyright (2018), with permission from SPIE. Experimental data have been reprinted from *Photonics and Nanostructures – Fundamentals and Applications*, vol no. 6, no. 1, A. Di Falco, L. O'Faolain, and T. F. Krauss, Photonic crystal slotted slab waveguides, pages no. 38–41, Copyright (2008), with permission from Elsevier.

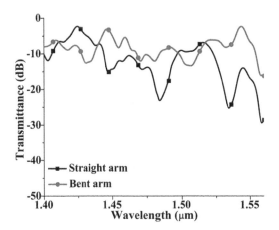

FIGURE 6.19 Transmittance spectrum of both the straight and bent arms chosen respectively for the pump and Stokes waves [36]. Reprinted from *Optical Engineering*, vol. 57, no. 7, Tanmoy Datta *et al.*, Efficient pump–signal combiner for stimulated Raman scattering in photonic crystal waveguide, page no. 075103, Copyright (2018), with permission from SPIE.

the steady-state propagation for the pump and the Stokes waves, respectively, has been shown in Figure 6.20 [36]. The figure shows that the bending and the junction interface alter the waveguiding characteristics by locally exciting higher order modes from the one which has been set for propagation. These higher order modes behave differently with wavelengths, and suffer from both in-plane and out-of-plane losses while propagation causing substantial reduction in transmittance. Henceforth,

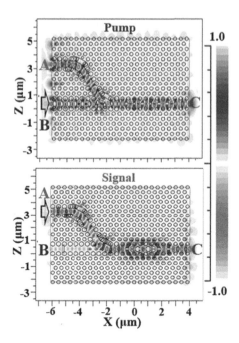

FIGURE 6.20 Electric field distribution in the $x - z$ plane during the steady-state propagation for the pump and the Stokes waves respectively [36]. Reprinted from *Optical Engineering*, vol. 57, no. 7, Tanmoy Datta *et al.*, Efficient pump–signal combiner for stimulated Raman scattering in photonic crystal waveguide, page no. 075103, Copyright (2018), with permission from SPIE.

simulations are also performed to justify the insertion of C1 and C2 to prevent intrusion of the propagation in the undesired arms. Transmittance spectra of both the arms in absence of C1 have been obtained and are shown in Figure 6.21 [36]. The steady-state distribution of the electric field in the $x - z$ plane is also depicted at the bottom insets. From the spectra, it can be seen that the transmittance at the pump wavelength remains almost unaffected, while that at the Stokes wavelength increases substantially to ~ 18 dB. The same can also be inferred from the corresponding distribution of the electric field in the $x - z$ plane. Moreover, simulations are also performed in absence of the C2 and the corresponding transmittance spectra for both the arms along with the corresponding steady-state distribution of the electric field are shown in Figure 6.22 [36]. In such a case, transmittances at both the pump and the Stokes wavelengths are increased to ~ 10 dB. Thus, the presence of the C1 and C2 improvises the performance of the combiner considerably. Nevertheless, there still exist opportunities to improve the performance of the combiner further. Evolutionary optimization techniques may be employed for this purpose.

6.6   DESIGN OF ALL-OPTICAL PASS SWITCH

In this section, the potentialities of the developments as demonstrated above will be explored in designing an all-optical switch. All-optical switching or, in other words,

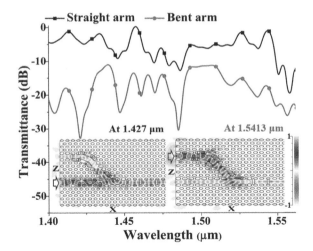

FIGURE 6.21 Transmittance spectra of the straight and bent arms in the absence of air holes C1 [36]. Reprinted from *Optical Engineering*, vol. 57, no. 7, Tanmoy Datta *et al.*, Efficient pump–signal combiner for stimulated Raman scattering in photonic crystal waveguide, page no. 075103, Copyright (2018), with permission from SPIE.

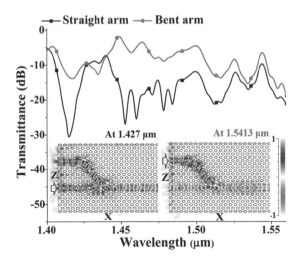

FIGURE 6.22 Transmittance spectra of the straight and bent arms in the absence of air holes C2 [36]. Reprinted from *Optical Engineering*, vol. 57, no. 7, Tanmoy Datta *et al.*, Efficient pump–signal combiner for stimulated Raman scattering in photonic crystal waveguide, page no. 075103, Copyright (2018), with permission from SPIE.

switching a beam of light using another has been the long-goal of researchers in catering the ever-evolving demand of the high-performance processing of information [63,64]. Different strategies have been investigated tirelessly over the past several years in realizing this goal. However, it is worth mentioning here that the competences of Raman amplifier have not been nourished properly designing in all-optical switches, despite being efficient enough for such purposes. Having said that, functionality of a novel AOPS is proposed based on the above developments, which is an electronic analogue to the semiconductor transmission gates. Pass switch, in general, refers to the switching operation whereby the input signal is allowed to pass to the output preserving its logic state as in the input. However, a control signal functions as the administrator and thereby enables this transfer at its particular logic state, which in otherwise prohibits this transfer. Hence, the general topology of the switch consists of three ports namely input, control and output.

6.6.1 DEVICE ARCHITECTURE OF THE ALL-OPTICAL PASS SWITCH

The complete architecture of the proposed AOPS is shown in Figure 6.23 [37]. The architecture can be divided into three segments based on the architectural parameters and, hence, their functionalities. Segment 1 is the combiner structure as just discussed with ~ 3 dB coupling losses at both the pump and the Stokes wavelengths. Segment 2 is the SiNC/SiO$_2$ embedded SPCW-based Raman amplifier as illustrated in Section 2 which is considered to be operated under CW pumping. This section plays the central role behind the successful switching operation. The operating principle of the proposed switch relies on the SRS mediated depletion of the pump and enhancement of the Stokes waves which are, respectively, employed as the input and control signal of the switch. For the sake of clarity, it is important to note that the wavelengths of the operating pump and the Stokes waves have been considered as 1.427 μm and 1.5413 μm which are at the 15.59 THz frequency detuning conforming to the peak SRS gain in SiNC/SiO$_2$ [32]. To elucidate, the complete absence of the Stokes or its presence below a specified limit, considered as logic '0' of the control

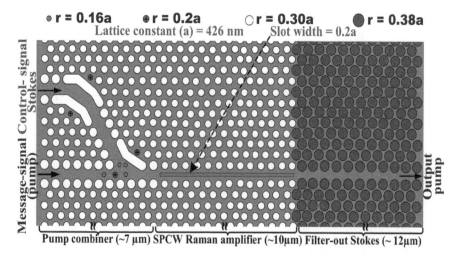

FIGURE 6.23 Schematic of the proposed design of the proposed AOPS [37]. Republished with permission of [Institution of Engineering and Technology (IET)], from [Integrable all-optical pass switch, Tanmoy Datta, and Mrinal Sen, vol. 54, issue 25, 2018]; permission conveyed through Copyright Clearance Center, Inc.

signal, is not able to incite the SRS interaction inside Segment 2. Consequently, the pump power remains effectively undepleted throughout the length of Segment 2 except the linear and nonlinear losses which may be kept negligible by appropriately choosing the length of the waveguide. Hence, the logic state of the pump or rather the input signal is maintained at the output for the logic '0' of the control signal. On the other hand, the presence of the Stokes above a specified limit, considered as logic '1' of the control signal, is well able to incite the SRS interaction; and thereby depleting the pump power significantly while propagating inside Segment 2. Hence, for an appropriately chosen length of the underlying SPCW, the pump power gets summarily attenuated at the output of Segment 2 leaving it to logic '0' at the output irrespective of its input logic state. Finally, the undesired control signal needs to be eliminated to maintain cascadability [64] of the proposed switch in terms of the operating wavelength and power. This is accomplished by the filter section in Segment 3 of the proposed AOPS which filters out the undesired Stokes power substantially and delivers the pump power at the output port. This segment has been realized by increasing the radii of the air holes that are lying on the crystal cladding of the W1 PCW to $0.4a$. The chosen hole radius consequences the Stokes mode to be operating at the extended state corresponding to the slab continuum of the PhC cladding. Nevertheless, the pump mode still remains within the bandgap of the PhC. Transmittance characteristics of the filter have been investigated using full-vector three-dimensional FDTD simulations. The characteristics show transmittances of ~ −0.4 dB and −26 dB at wavelengths of the pump and the Stokes, respectively, for the 12 μm long Segment 3, which are sufficient enough to deliver the pump power only at the output port.

Thereafter, the coupled NLS equations as applicable in this context, i.e. Equations (6.1–6.4), have been solved with the parameters as presented in Table 6.2 for analyzing performance of the proposed switch. The results and corresponding discussions are presented in the subsequent section.

Republished with permission of [Institution of Engineering and Technology (IET)], from [Integrable all-optical pass switch, Tanmoy Datta, and Mrinal Sen, vol. 54, issue 25, 2018]; permission conveyed through Copyright Clearance Center, Inc.

6.6.2 PERFORMANCE OF THE AOPS

Steady-state and transient analyses of the AOPS are performed in this work to explain its complete phenomenal characteristics. For steady-state analysis, spatial evolution of CW pump with 500 μW input power, considered as logic '1', has been studied inside Segment 2 for both (i) in presence of the CW Stokes with 500 μW input power, considered as logic '1', and (ii) in complete absence of the Stokes considered as logic '0'. The outcomes of this study have been depicted in Figure 6.24 [37]. In the first case of study both the pump and the Stokes undergo ~3 dB coupling losses at Segment 1 and reduce to ~ 250 μW which is high enough to incite the SRS interaction substantially at Segment 2. Owing to this SRS interaction, the pump power is getting depleted exponentially while propagating through the SPCW. The depletion eventually reaches to ~0.45 μW, which can reasonably be considered as logic '0', at a length of 10 μm of Segment 2 as can be seen from Figure 6.24. The undesired Stokes power in subsequence of the interaction is filtered out by Segment 3. Thus, the logic '1' of the

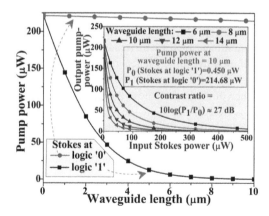

FIGURE 6.24 Spatial evolution of the CW pump in steady state analysis of the proposed switch. Inset shows a set of transfer characteristics curves subjected to the variation of the waveguide length [37].

control signal rejects the logic '1' of the input signal and thereby delivers logic '0' at the output.

In the second case of study, the absence of the Stokes or the logic '0' of the control is unable to incite the SRS interaction in Segment 2. Thus, the pump power remains essentially undepleted during propagation except for the losses incurred that are originated from linear and nonlinear attenuations. Consequently, the pump power is attenuated to ~ 215 μW at a length of 10 μm of Segment 2 from its ~ 250 μW at input of Segment 2, which can reasonably be considered as logic '1'. Thus, the logic '0' of the control signal maintains the logic of the input signal at the output.

Moreover, it is also to note for the sake of generalization that the logic '0' of the input signal remains at the same logic state at the output irrespective of any logic of the control signal.

Therefore, to summarize, the logic '1' of the control signal blocks the transfer of input signal and delivers logic '0' at the output irrespective of logic states of the input. On the contrary, the logic '0' of the control signal allows the transfer of input signal through maintaining logic state of the input at the output. The above observations can be represented in the following Boolean truth table in Table 6.3. Further, it

TABLE 6.3

Truth Table Depicting the Operation of the Proposed AOPS

Input	Control	Output	Operation
0	0	0	Passed
0	1	0	Blocked
1	0	1	Passed
1	1	0	Blocked

is also worth mentioning here that the extinction ratio in terms of the output power corresponding to its logic '0' and logic '1' has been calculated to be ~ 27 dB for the considered length of 10 μm of Segment 2, which is sufficient enough to combat the noise-induced perturbations.

Nevertheless, the output characteristics also substantially depend on the length of the SPCW. Simulations are performed to obtain power transfer characteristics of the switching operation for different sets of length of Segment 2 and the same have been shown at the inset of Figure 6.24. Increasing the length of Segment 2 increases slopes of the characteristic curves, thereby making the switching operation more prominent. However, in doing so consequence the simultaneous reduction of integration density of the proposed switch in PICs. Now, identification of the threshold of logic in both the input and the output signals is important to establish the cascadability of any logic device. The transfer characteristic curve corresponding to 10 μm length of Segment 2 infers the power level of 50 μW as the logic threshold. This implies that the logic '1' corresponds to the power level lying above or equal to this threshold which in otherwise corresponds to the logic '0'.

For transient analysis, output of the AOPS has been studied temporarily for both the cases of (i) switching of a CW input signal by a pulsating control signal, and (ii) switching of a pulsating input signal by another pulsating control. The study has also been repeated for different repetition rates of the operating pulses.

In the first case of study, width of the control pulses is chosen as 1 ps for which the considered length of the waveguide, i.e. 10 μm becomes substantially smaller than the calculated dispersion length of 85 μm. This is to ensure negligible temporal broadening of pulses as induced by the GVD and, hence, $\xi = 1$ in Equations (6.1–6.4). Nevertheless, the SRS-induced temporal impairments, despite being unavoidable owing to the non-instantaneous and causal SRS interaction, never hamper the slow-light-induced enhancement as already argued in the previous section. The results of the simulation have been depicted in Figure 6.25(a) [37] for repetition rates of 75 Gbps, 100 Gbps and 125 Gbps. The successful switching operation can be understood by noting the simultaneous decrement in the output power level to almost logic '0' whenever a control pulse is appearing as logic '1'. However, the logic levels are getting deteriorated by increasing the repetition rates of the pulses.

Moreover, in the second case of study, the input signal is taken in the identical fashion as investigated in the first case. The control signal is also considered as pulsating in this case having a width of 6 ps with a repetition rate of 30 Gbps. Although width and repetition rate of the control pulses can be chosen arbitrarily, the values so chosen here to accommodate multiple pulses of the input signal under a single control pulse. Results depicting successful switching operation are shown in Figure 6.25(b) with the varying repetition rates of the pulses of the input signal. However, it is important to note from the figure that a considerable residual power exists at the output corresponding to the input pulses which are appearing in align to the leading and trailing edges of the control pulses. This phenomenon is because the optical power of the control pulses is mostly concentrated within their FWHM. Hence, the power available corresponding to the leading and trailing edges of the pulses is not sufficient enough to attenuate the input pulses completely which, in turn, may result in the substantial loss of the logical identity of the output at those instants.

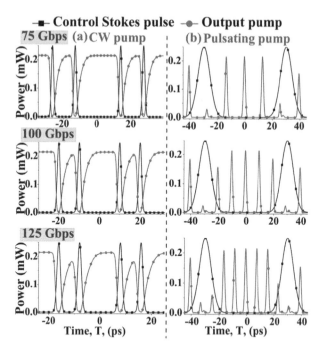

FIGURE 6.25   Switching characteristics for both (a) CW input, and (b) pulsating input with different repetition rates [37]. Republished with permission of [Institution of Engineering and Technology (IET)], from [Integrable all-optical pass switch, Tanmoy Datta, and Mrinal Sen, vol. 54, issue 25, 2018]; permission conveyed through Copyright Clearance Center, Inc.

This limitation can be overcome by increasing the peak power of the control pulses so as to strengthen the power level corresponding to the leading and trailing edges.

6.7   CONCLUSION

SPCWs have been an effective platform for efficient light–matter interactions owing to the ultra-high optical confinement as attained after merging the strong spatial and temporal confinement of the slotted and the photonic crystal waveguides, respectively. Light–matter interactions, especially the nonlinear optical phenomena typically require high operating power and/or large footprint of the underlying waveguides which are the fundamental constraints limiting their applicability in integrated photonics. SRS has been such a potential nonlinear distributed light–matter interaction; that conventionally takes kilometers of length of the optical fiber along with the pump power lying in the order of few hundreds of Watt for observable gain. Despite the untiring research being carried out over the past several years in miniaturizing the effective interaction length and the threshold power of Raman amplifiers, the miniaturization is yet to be competent enough to PICs for exploring on-chip active optical functionalities. The prospects and possibilities of SRS in $SiNC/SiO_2$-embedded SPCW have been investigated in this chapter towards circumventing the limitations. The ultra-high SRS gain in $SiNC/SiO_2$ material is fortified further by the intense spatio-temporal confinement of SPCWs, which, thereby, is able to provide unprecedented miniaturization of the effective length and threshold power. So has been demonstrated in this work by designing micron-scale all-silicon Raman amplifier based on $SiNC/SiO_2$-embedded SPCW. It is important to mention in this context that the coupled NLS equations, which quantify the energy transfer between the pump and the Stokes waves and their envelope evolution, have suitably been modified to incorporate enhancement of the linear and nonlinear interactions in slow-light regime. The Raman response function of the $SiNC/SiO_2$, characterizing transience of the SRS interaction, has been modeled considering two damped molecular vibrational modes that are centered at the resonance frequencies of 15 THz and 15.59 THz.

At first, simulations are performed after obtaining solutions of the coupled NLS equations for the proposed model of the Raman amplifier under CW pumping to obtain the phenomenal characteristics at both the CW and pulsating input operating Stokes. A significantly high overall gain of almost 11 dB for the CW Stokes has been achieved within a length of merely 8 μm of the SPCW using a CW pump power of 0.7 mW operating at the 15.59 THz frequency detuning from the Stokes. Nevertheless, the pulsating Stokes suffer from temporal broadening and impairments under CW pumping while propagation, which inevitably limits the repetition rate of the operating pulses for their faithful amplification. The broadening is originating from the non-instantaneous and causal nature of the SRS interaction since the effect of GVD is negligibly small for both the considered pulse width and length of the propagation inside the SPCW. Operating the pump at 15.59 THz frequency detuning; the amplifier provides overall gains of 3 dB, 4.7 dB and 5.6 dB, respectively, at the waveguide lengths of 1.5 μm, 3 μm and 4 μm maintaining logical identities of the output Stokes pulses corresponding to the repetition rates of 400 Gbps, 200 Gbps and 133 Gbps. The broadening of the pulses has been exterminated to some extend after operating the pump at 15 THz frequency detuning; however, at the cost of achievable gain. In this case, an average gain of around 4 dB has been obtained at the waveguide length of 6 μm and for a repetition rate of 400 Gbps.

On the other hand, broadenings and impairments of the output pulses are demonstrated to be eradicated completely by use of LED as the pumping source owing to its incoherent spectral response. The amplifier in this case offers an overall gain of 3.22 dB at an operating repetition rate of 400 Gbps with a 4 μm long SPCW and 6 mW average LED pump power. The gain has further been heightened to 7.93 dB while taking 10 μm long SPCW; however, with a subsequent reduction of the operating speed to 200 Gbps. The LED can be integrated in the same platform as of the $SiNC/SiO_2$-embedded SPCW, thereby providing cost-effective and convenient solution to the optical pumping. Furthermore, the electroluminescence of SiNC may also purposefully be utilized as the integrated pumping source which essentially eliminates the need of the external laser source in realizing the all-silicon Raman.

Nevertheless, the miniaturization achieved remains ineffective as long as the issues related to the coupling of the pump and the Stokes for their SRS interaction are not being taken into consideration. Thus, the work proposes a novel PhC-based structure for combining of the pump and Stokes efficiently for SRS interaction inside the $SiNC/SiO_2$-embedded SPCW. The structure has been designed by adjoining two asymmetrical PCWs consisting of a bent and a straight PCW, respectively, to couple the Stokes and the pump. Optimization of performance of the proposed design has been carried out heuristically using three-dimensional full vector FDTD method following its validation by comparing the simulation results with that of some experimentally measured data. Coupling losses of ~3 dB have been obtained from the transmittance spectra at both the exciting pump and the Stokes wavelengths within an overall footprint as small as ~54 μm^2.

Finally, the potentialities of the developments demonstrated are explored in designing an all-optical switch. In this work, the functionality of a novel AOPS is proposed based on the SRS mediated depletion and enhancement of the pump and the Stokes waves inside the $SiNC/SiO_2$-embedded SPCW, which are, respectively, employed as the input signal and the control wave of the switch. Logic '0' at the control signal maintains the logic state of the input signal throughout the device. On the other hand, logic '1' of the control signal summarily rejects the input signal power during propagation through the device, leaving it to logic '0' at the output irrespective of its input logic state. The overall footprint and the input operating power of the proposed switch have been miniaturized significantly to $\sim 7.5 \times 29~\mu m^2$ and 0.5 mW, respectively. The switch offers successful operation at a high repetition rate of the pulses, i.e. 125 Gbps, with an extinction ratio of ~27 dB between the output logic levels; beyond which the performance gets limited by the non-instantaneous and causal SRS interaction. Operation of the switch also confirms cascadability both in terms of the input–output operating wavelength and power level which ensure its application in a long-chain connection of similar devices. Moreover, the proposed switch, by virtue of its operation, has the ability to realize other photonic logic functionalities, e.g. multiplexing and decoding in the same line as demonstrated using complementary metal-oxide semiconductor transmission gate which is the electronic equivalent of the AOPS.

Therefore, with the potential merits in terms of the requirement of ultra-compact footprint, miniaturized operating power, cascadability and ultra-high speed of operation, the proposed $SiNC/SiO_2$-embedded SPCW-based developments may serve as basic building blocks all-optical signal processing in PICs.

REFERENCES

1. L. Pavesi, and D. J. Lockwood, Eds., *Silicon Photonics*, Springer, New York, 2004.
2. G. T. Reed and A. P. Knights, *Silicon Photonics: An Introduction*, Wiley, Hoboken, NJ, 2004.
3. B. Jalali, and S. Fathpour, "Silicon Photonics," *Journal of Lightwave Technology*, vol. 24, no. 12, pp. 4600–4615, 2006.
4. Ö. Boyraz et al., "All optical switching and continuum generation in silicon waveguides," *Optics Express*, vol. 12, no. 17, pp. 4094–4102, 2004.
5. C. Koos et al., "Nonlinear silicon-on-insulator waveguides for all-optical signal processing," *Optics Express*, vol. 15, no. 10, pp. 5976–5990, 2007.
6. V. R. Almeida et al., "Guiding and confining light in void nanostructure," *Optics Letters*, vol. 29, no. 11, pp. 1209–1211, 2004.
7. A. Martínez et al., "Ultrafast all-optical switching in a silicon-nanocrystal-based silicon slot waveguide at telecom wavelengths," *Nano Letters*, vol. 10, no. 4, pp. 1506–1511, 2010.

8. I. D. Rukhlenko, and V. Kalavally, "Raman amplification in silicon-nanocrystal wave-guides," *Journal of Lightwave Technology*, vol. 32, no. 1, pp. 130–134, 2014.

9. T. F. Krauss, "Why do we need slow light?," *Nature Photonics*, vol. 2, no. 8, pp. 448–450, 2008.

10. T. F. Krauss, "Slow light in photonic Crystal waveguides," *Journal of Physics D: Applied Physics*, vol. 40, no. 9, pp. 2666–2670, 2007.

11. T. Baba, "Slow light in photonic crystals," *Nature Photonics*, vol. 2, pp. 465–473, 2008.

12. Y. A. Vlasov et al., "Active control of slow light on a chip with photonic crystal wave-guides," *Nature Letters*, vol. 438, pp. 65–69, 2005.

13. M. Soljačić, and J. D. Joannopoulos, "Enhancement of nonlinear effects using photonic crystals," *Nature Materials*, vol. 3, pp. 211–219, 2004.

14. C. Monat et al., "Slow light enhancement of nonlinear effects in silicon engineered photonic crystal waveguides," *Optics Express*, vol. 17, pp. 2944–2953, 2009.

15. J. F. McMillan et al., "Enhanced stimulated Raman scattering in slow-light photonic crystal waveguides," *Optics Letters*, vol. 31, no. 9, pp. 1235–1237, 2006.

16. R. Claps et al., "Observation of stimulated Raman amplification in silicon waveguides," *Optics Express*, vol. 11, no. 15, pp. 1731–1739, 2003.

17. R. L. Espinola et al., "Raman amplification in ultrasmall silicon-on-insulator wire wave-guides," *Optics Express*, vol. 12, no. 16, pp. 3713–3718, 2004.

18. M. Woldeyohannes, S. John, and V. I. Rupasov, "Resonance Raman scattering in pho-tonic bandgap materials," *Physical Review A*, vol. 63, no. 1, pp. 013814–013827, 2000.

19. R. G. Zaporozhchenko, S. Y. Kilin, and A. G. Smirnov, "Stimulated Raman scattering of light in a photonic crystal," *Quantum Electronics*, vol. 30, no. 11, pp. 997–1001, 2000.

20. J. F. McMillan et al., "Observation of spontaneous Raman scattering in silicon slow-light photonic crystal waveguides," *Applied Physics Letters*, vol. 93, p. 251105, 2008.

21. H. Oda et al., "Light amplification by stimulated Raman scattering in AlGaAs-based photonic-crystal line-defect waveguides," *Applied Physics Letters*, vol. 93, p. 051114, 2008.

22. X. Checoury, Z. Han, and P. Boucaud, "Stimulated Raman scattering in silicon photonic crystal waveguides under continuous excitation," *Physical Review B*, vol. 82, p. 041308, 2010.

23. I. H. Rey, "Active slow light in silicon photonic crystals: tunable delay and Raman gain," PhD Dissertation, University of St. Andrews, 2012.

24. Y. Takahashi et al., "A micrometre-scale Raman silicon laser with a microwatt thresh-old," *Nature Letters*, vol. 498, pp. 470–474, 2013.

25. A. D. Falco, L. O'Faolain, and T. F. Krauss, "Photonic crystal slotted slab waveguides," *Photonics and Nanostructures-Fundamentals and Applications*, vol. 6, no. 1, pp. 38–41, 2008.

26. C. Caër et al., "Extreme optical confinement in a slotted photonic crystal waveguide," *Applied Physics Letters*, vol. 105, no. 12, p. 121111, 2014.

27. M. G. Scullion, T. F. Krauss, A. Di Falco, "Slotted Photonic Crystal Sensors," *Sensors*, vol. 13, pp. 3675–3710, 2013.

28. C. Y. Lin et al., "Electro-optic polymer infiltrated silicon photonic crystal slot waveguide modulator with 23 dB slow light enhancement," *Applied Physics Letters*, vol. 97, p. 093304, 2010.

29. S. Lin et al., "Design of nanoslotted photonic crystal waveguide cavities for single nanoparticle trapping and detection," *Optics Letters*, vol. 34, no. 21, pp. 345–3453, 2009.

30. *Silicon Nanophotonic*: Basic Principles, *Present Status and Perspectives* (Edited by L. Khriachtchev), Pan Stanford Publishing, Singapore, 2009.

31. S. Hernández et al., "Linear and nonlinear optical properties of Si nanocrystals in SiO_2 deposited by plasma-enhanced chemical-vapor deposition," *Journal of Applied Physics*, vol. 103, no. 6, p. 64309, 2008.

32. L. Sirleto et al., "Giant Raman gain in silicon nanocrystals," *Nature Communications*, vol. 3, p. 1220, 2012.

33. T. Nikitin, S. Novikov, and L. Khriachtchev, "Giant Raman gain in annealed silicon-rich silicon oxide films: Measurements at 785 nm," *Applied Physics Letters*, vol. 103, p. 151110, 2013.

34. T. Datta, and M. Sen, "Micron-scale Raman amplifier: another progressive step towards the maturity of silicon photonics," accepted for publication in Journal of Optics (IOP science), [online] doi:10.1088/2040-8986/aa74d3.

35. T. Datta, and M. Sen, "LED pumped micron-scale all-silicon Raman amplifier," *Superlattices Microstructures*, Elsevier Publishing, vol. 110, pp. 273–280, 2017, doi:10.1016/j.spmi.2017.08.033.

36. T. Datta, M. Sen, Shatrughna Kumar, Akash Kumar Pradhan, "Efficient pump–signal combiner for stimulated Raman scattering in photonic crystal waveguide," *Optical Engineering*, vol. 57, no. 7, 075103, 2018, doi:10.1117/1.OE.57.7.075103.

37. T. Datta, M. Sen, "Integrable all-optical pass switch," *Electronics Letters*, vol. 54, no. 25, pp. 1446–1448, 2018, doi:10.1049/el.2018.6350.

38. T. Datta, and M. Sen, "Characterization of slotted photonic crystal waveguide and its application in nonlinear optics," *Superlattices Microstructures*, Elsevier Publishing, vol. 109, pp. 107–116, 2017, doi:10.1016/j.spmi.2017.04.039.

39. S. Matsuo et al., "High-speed ultracompact buried heterostructure photonic-crystal laser with 13 fJ of energy consumed per bit transmitted," *Nature Photonics*, vol. 4, pp. 648–654, 2010.

40. S. Matsuo et al., "20-Gbit/s directly modulated photonic crystal nanocavity laser with ultra-low power consumption," *Optics Express*, vol. 19, no. 3, pp. 2242–2250, 2011.

41. S. Matsuo et al., "Room-temperature continuous-wave operation of lateral current injection wavelength-scale embedded active-region photonic-crystal laser," *Optics Express*, vol. 20, no. 4, pp. 3773–3780, 2012.

42. K. Takeda et al., "Few-fJ/bit data transmissions using directly modulated lambda-scale embedded active region photonic-crystal lasers," *Nature Photonics*, vol. 7, pp. 569–575, 2013.

43. K. Nozaki et al., "Photonic-crystal nano-photodetector with ultrasmall capacitance for on-chip light-to-voltage conversion without an amplifier," *Optica*, vol. 3, no. 5, pp. 483–492, 2016.

44. D. W. Prather et al., Photonic crystals, theory, applications and fabrication, Wiley series in pure and applied optics, 2009.

45. T. Datta, and M. Sen, "All-optical logic inverter for large scale integration in silicon photonic circuits," accepted for publication in *IET Optoelectronics*, [online] doi:10.1049/iet-opt.2019.0081.

46. I. D. Rukhlenko, M. Premaratne, and G. P. Agrawal, "Nonlinear silicon photonics: Analytical tools," *IEEE Journal of Selected Topics in Quantum Electronics*, vol. 16, no. 1, pp. 200–215, 2010.

47. Q. Lin, O. J. Painter, and G. P. Agrawal, "Nonlinear optical phenomena in silicon waveguides: Modeling and applications," *Optics Express*, vol. 15, no. 26, pp. 16604–16644, 2007.

48. T. Datta, A. K. Pradhan, S. Kumar, and M. Sen, *"Unified Coupled Equations for Raman Mediated Interaction in Slow-light Regime," Presented at IEEE III International Conference on Microwave and Photonics (ICMAP) 2018*, (Print ISBN: 978-1-5386-0934-7), Indian Institute of Technology (Indian School of Mines) Dhanbad, Jharkhand, India, February 09–11, 2018, doi:10.1109/ICMAP.2018.8354516

49. M. Sen, and M. K. Das, "High-speed all-optical logic inverter based on stimulated Raman scattering in silicon nanocrystal," *Applied Optics*, vol. 54, no. 31, pp. 9136–9142, 2015.

50. I. D. Rukhlenko, "Modeling nonlinear optical phenomena in silicon nanocrystal composites and waveguides," *Journal of Optics*, vol. 16, p. 015207, 2014.

51. G. P. Agrawal, *Nonlinear Fiber Optics*, 3ʳᵈ ed., Academic Press, California, USA, 2001.

52. J. Lee, J. H. Shin, and N. Park, "Optical gain at 1.5 μm in nanocrystal Si-sensitized Er-doped silica waveguide using top-pumping 470 nm LEDs," *Journal of Lightwave Technology*, vol. 23, no. 1, pp. 19–25, 2005.

53. H. Lee, J. H. Shin, and N. Park, "Performance analysis of nanocluster-Si sensitized Er-doped waveguide amplifier using top-pumped 470 nm LED," *Optics Express*, vol. 13, no. 24, pp. 9881–9889, 2005.

54. T. Theeg et al., "Pump and signal combiner for bi-directional pumping of all-fiber lasers and amplifiers," *Optics Express*, vol. 20, no. 27, pp. 28125–28141, 2012.

55. X. Chen et al., "20 dB-enhanced coupling to slot photonic crystal waveguide using multimode interference coupler," *Applied Physics Letters*, vol. 91, no. 9, p. 091111, 2007.

56. H. S. Dutta et al., "Coupling light in photonic crystal waveguides: A review," *Photonics and Nanostructures-Fundamentals and Applications*, vol. 20, pp. 41–58, 2016.

57. R. D. Meade et al., "Novel applications of photonic bandgap materials: Low-loss bends and high Q cavities," *Journal of Applied Physics*, vol. 75, no. 9, pp. 4753–4755, 1994.

58. A. Chutinan, M. Okano, and S. Noda, "Wider bandwidth with high transmission through waveguide bends in two-dimensional photonic crystal slabs," *Applied Physics Letters*, vol. 80, no. 10, pp. 1698–1700, 2002.

59. L. H. Frandsen et al., "Broadband photonic crystal waveguide 60° bend obtained utilizing topology optimization," *Optics Express*, vol. 12, no. 24, pp. 5916–5921, 2004.

60. B. Miao et al., "High-efficiency broad-band transmission through a double-60° bend in a planar photonic crystal single-line defect waveguide." *IEEE Photonics Technology Letters*, vol. 16, no. 11, pp. 2469–2471, 2004.

61. J-H. Chen et al., "Design, fabrication, and characterization of Si-based ARROW photonic crystal bend waveguides and power splitters," *Applied Optics*, vol. 51, no. 24, pp. 5876–5884, 2012.

62. L. H. Frandsen et al., "Broadband photonic crystal waveguide 60° bend obtained utilizing topology optimization," *Optics Express*, vol. 12, no. 24, pp. 5916–5921, 2004.

63. H. J. Caulfield, and S. Dolev, "Why future supercomputing requires optics," *Nature Photonics*, vol. 4, no. 5 pp. 261–263, 2010.

64. D. A. B. Miller, "Are optical transistors the logical next step?," *Nature Photonics*, vol. 4, no. 1 pp. 3–5, 2010.

7 Performance Evaluation of Raman Amplifier- Embedded Optical Fibre Communication System at Both Minimum Dispersion and Minimum Attenuation Windows

Rajarshi Dhar and Arpan Deyasi

CONTENTS

7.1   INTRODUCTION

After the pioneering inventory of Sir Alexander Graham Bell for successfully conversion of voice signal into electrical signal, and thereafter its transmission through copper wire, the revolution in the field of telecommunication sector started. Sir Bell completed the circular path when the same electrical signal returned to the receiver and converted back to voice signals. Subsequently, a continuous need for increasing bandwidth is generated as per the demand of customers owing to potential increase of number of users consequently the need of bandwidth, which was failing to support such huge loads. Customer requirement drives the scientists and research communities to find new ways of communication which will be meeting the aforesaid demands. Thereafter technologies of RF communication, microwave communication andmillimetre-wave communication are originated chronologically (Huurdeman2003) and when even those fail to meet the demands, optical communication showed some promises (Agrawal 2010) and hence started the age of light. Twenty-first century is rightly called as the age of photonics, where the most secured, advanced and faster ways of communication are the optical fibre communication (Hecht1999) owing to several propagation advantages in THz and beyond THz ranges. Communication of data and information using light as a carrier has a lot of advantages such as huge bandwidth, extremely high data rate, higher speed, high prospects of security, less prone to noise and lots more. Research is being conducted across the world to explore this highly promising field of optical communication.

7.1.1   THE BEGINNING AND NEED OF OPTICAL COMMUNICATION

Researches on communication have always been focused on a few basic things, namely, higher data transfer speeds, improved security of data transfer, higher bandwidth or wider-band data transmission (Nouchi *et al.,* 2003) and low-loss medium for data transfers. The first stages of communications started at frequencies of about 30 MHz as at those times the bandwidth requirement was low.But with time, the number of users increased which demanded a larger bandwidth which led to the increase of operating frequencies. The medium which was used for the transmission of signals with operating frequencies up to about 1 GHz were coaxial cables with a loss figure of about 20 dB/km (Keiser 2000). With increase of operating frequency in search of higher bandwidth to incorporate more channels, coaxial cables are discarded due to their lossy nature. Therefore, people thought about guided wave transmission, gives birth to the new concept, termed as waveguide, which is nothing but a hollow structure for proper guiding of electromagnetic energy propagation. With further moving up in the frequency scale, configuration of single-layer hollow waveguides too proved inadequate due to lack of associated electronic circuits compatible to be worked on such high frequencies (Senior 1992; Miller & Kaminow 1988).The reason for this was that the electronic components, at such frequencies, show variation in their behaviour and stop behaving as lumped elements (Ghatak & Thyagrajan 1989). They start to act as distributed elements, which are described as elements whose circuit characteristics are basically defined by the wavelength, hence the frequency of the signals they are working with (Ghatak & Thyagrajan 1998).Thus, the need for other alternatives became very important, because there appeared a halt in the available technology.

After the novel discovery of Charles Kuo in 1966, researches are being carried out all over the world to explore the optical domain beyond the framework of laboratory experiments. The sole objective was to utilize the well-established relationship between operating frequency (f_0) and bandwidth (BW) at THz domain which may provide a 10^3–10^4 fold increase of potential bandwidth (Smith 2000). Detailed investigations revealed that THz domain has the potentiality for creating a new impact of the existing communication system (Ghatak & Thyagrajan 1998). This is closely related with the making of compatible transmitter and receiver at that wide-band frequencies having loss-less window for effective communication (Keiser 2000).

Now, air as a medium can prove to be very effective for light transmissions, but only up to a few kilometres(Agrawal 2010). After that signal will start to diminish and attenuation prevail. So search of new media to carry light begin and the most common material for that was initially found to be glass. Though glasses with impure fabrications had a loss factor of 1000 dB/km, which was far better than the existing cables and henceforth, optical communication becomes inappropriate over such medium. But with innovative and decontaminated fabrications of glasses, loss factor considerably reduced to a value of 20 dB/km (Nayayama *et al.,* 2002) and therefore, can now become equivalent with coaxial cable (Agrawal 2016). This made the revolutionary transition in optical communication system design, with a large increase in bandwidth (Oh *et al.,* 2005), and researched is progressed worldwide. This paved the way of fabricating optical fibre using glass, and journey of optical communication begins (Hui 2019).

Another critical problem was about the availability of source of electromagnetic radiation. This answer needs to be evaluated correctly though several options are available with a wider difference in properties and cost. But the problem was whether the available light sources were enough to be used as carrier signals. From the basic concept of communication principle, it can be said that carrier signal carries information after modulating it's any one of the fundamental properties such as phase, amplitude, frequency, etc. The problem therefore converges as choice becomes limited. Normal sources of light, e.g. incandescent bulbs, were not suitable for fast switching; thus, they failed to be used as suitable optical sources. In this case, it may be mentioned that the spectral width is primarily responsible for turning ON and OFF of the optical source. Now for high-frequency communication, this width should be small as possible (Fu & Xi 2016). Therefore, research is focused to design sources having narrow spectral width as far as practicable (Al-Taiy *et al.,* 2014).

As the search of narrow spectral width device is in progress, concept of LASER and its fruitful experimental realization comes into picture. It has all such requirements, narrow spectral width, higher directivity, and most important, extreme low loss (Smith 2000). Therefore, compatibility is arrived between optical media and optical sources (Garmire2002). With progress of work, several novel amplifiers are proposed in order to compensate the loss (attenuation and dispersion) inside the system, and complex systems are proposed by design engineers. Along with transmitter, compatible receivers are made, associated with other circuit equipments such as coupler, splitter, splices, etc. A composite optical communication system for the present day is shown in Figure 7.1.

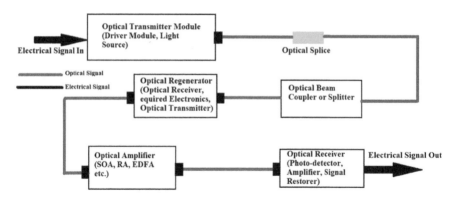

FIGURE 7.1   Basic optical fibre communication link.

7.1.2 NEED AND CHOICE OF DIFFERENT AMPLIFIERS IN OPTICAL COMMUNICATION

While designing an optical fibre link, certain parameters are needed to be maintained such as modulation format, link length (attenuation limitation), system fidelity (BER, SNR), Bit Rate (Dispersion Limitation), upgradeability, cost and commercial availability. The optical length is an intricate parameter for the link design as it helps to closely determine the position for the optical repeaters in the total channel length (between transmitter and receiver). Another essential parameter in the design is its capability to receive signals with acceptable BERs (Bit Error Rate). As the data rate slowly increases, average power also increases (here the power is measured to get the acceptable value of BER). But noise adds a negative effect on it, and therefore, minimum value of detectable power is higher with the increased rate of bit transfer in presence of noise. Any system in reality will have losses and noise disturbances. Thus, it is very critical to design the system in such a way so as to achieve the desired SNR or BER at user (receiver) end. Such estimations and calculations are known as power budget calculations. The total loss of the system is given by (Keiser 2000)

$$\alpha_{fiber} = \alpha_{max} - \left(\alpha_{connectors} + \alpha_{system} + \alpha_{splices}\right) \tag{7.1}$$

The expression of maximum possible length (L_{Pmax}) when BER remains unaffected can be determined as (Keiser 2000)

$$L_{P\,max} = \frac{\alpha_{max}}{Loss} \tag{7.2}$$

where loss is measured in per unit km. This length, obtained from Equation (7.2), can be termed as Power Budget Limited Link Length. This is such a critical value for communication system design in the way that if anyone used length as greater than that, BER of the system degraded, and consequently the system performance.

There is another aspect that determines the link of the optical fibre and that is the rise time estimated calculations. The rise time of an optical system, like any real system, is a non-zero value, be it as small as possible. The rise time is the time which is required by the system to reach a steady state so that any variations in the input signal don't affect the system parameters. The rise time of the system gives rise to another expression for the determination of the length of the optical link.

The system rise time (t_{sys}) can be calculated as (Keiser 2000)

$$t_{sys} = \sqrt{t_{tx}^2 + \left(D\sigma_\lambda\right)^2 + t_{rx}^2} \tag{7.3}$$

Hence, the link length due to the rise time of the system is given by (Keiser 2000)

$$L_{RT\,max} = \frac{1}{D\sigma_\lambda}\left(\sqrt{t_{sys}^2 - \left(t_{tx}^2 + t_{rx}^2\right)}\right) \tag{7.4}$$

The minimum of these two lengths is considered as the final link length value as that will help both the power budget and rise time to be under acceptable values. The link length thus basically gives the position for the placement of the optical repeaters (Ramaswami et al., 2009). But optical repeaters are costly devices and are used for rectification of distortions in the signal. But since the link length is taken under the rise time length; hence, the distortions are not as much that repeaters are needed. The main thing that gets hampered is the SNR of the signal. For improving that a simple optical amplifier is enough (Senior 1992). Thus, amplifiers are needed to be placed at different intervals in the optical link.

The optical amplifiers were developed with the sole purpose of working in the optical domain, which eliminates the need of optical to electrical signal. As mentioned earlier, the repeater systems were used for the regeneration of the signal at different intervals (Ramaswami et al., 2009), but that system was extremely costly and a bit unnecessary. Over the years, the optical amplifiers have thus proved to be much more useful for every purpose regarding the signal amplification, pulse shapers, routing switches and many more.

The amplifiers that are available till date for the optical systems are of two kinds:

(i)   Semiconductor optical amplifiers or SOA
(ii)   Fibre amplifiers or FA

All these have varied uses and are utilized in the optical system according to the application required. It is very evident that all of them very wide spectral width and thus can be easily used wide-band amplification. Making a vis-à-vis analysis, SOA demonstrates the property of lower power consumption and due to the inherent architecture, they are custom-tailored with single-mode fibre (Agarwal & Agrawal 2018); fibre amplifiers, on the other hand, show less compatibility issues with inline interconnection inside the links (Wasfi 2009).

The major advantage of using SOA in optical communication system is their duality window compatibility, i.e. they can be applied in the systems designed for either 1330 nm or 1550 nm (Zoiros 2018). It is nothing but a conventional amplifier using semiconducting material where facet reflectivites at output side vary between 30% and 35% (Zhou et al., 2017). Output power obtained from this type of amplifier is high, and this is obvious as the operating principle is nothing but similar to a LASER.

Their constituent advantage makes its wider integrability with semiconductor-based optolelectronic circuits. However, the main disadvantages with the SOAs are that they have high coupling loss and high non-linearity (Chi 2017).

The fibre amplifiers or FAs on the other hand provide gain by utilizing the concept of Raman scattering or Brioullin scattering or by rare-earth dopants. They provide advantages over the non-linear problems as well as the coupling problems of the SOAs along with being low noise amplifiers, but they are only functional in the 1550 nm window which gives a kind of limitation on the bandwidth of operation (Ivaniga & Ivaniga 2017). It consists of the signal source and a pump which can be co-coupled. So, the signal and pump are flowing in the same direction and power is transferred from the pump to signal from lower wavelengths to higher wavelengths. Practical pump wavelengths mainly include 807 nm, 980 nm and 1480 nm. The materials that are mainly used for the DFAs or doped fibre amplifiers are neodymium or erbium or ytterbium (Karimi 2018). Among these different materials, erbium-doped fibre amplifiers or EDFAs are the most popular and have been studied extensively (Bebawi *et al.*, 2018; Putra *et al.*, 2018). They provide a very good gain of approximately 30–40 dB with pump power having a range from 50 to 100 mW. The main disadvantage of the EDFAs is that it reduces gain performance significantly in the excited state absorption (ESA). Moreover, the gain of EDFAs mainly available on the 1550 nm band while the Raman amplification fibre amplifiers can provide significant amplification over both 1550 nm and the 1310 nm band. This is discussed in the following section.

7.1.3 CHOICE OF DIFFERENT FREQUENCY SPECTRA: PROS AND CONS

Gain-bandwidth curve determines the range of operating wavelength spectra for a particular system before application. Good bandwidth is that spectra, where atmospheric attenuation is lowest, in other words, gain of the system is significantly higher. In general, communication system designed at beyond THz region, 20–40 dB gain is considered as standard. A good gain means the signal can travel a longer distance without much detoriation. In general, systems are designed at 850 nm, 1330 nm and 1550 nm, considering fibre optics as the guiding medium. All of these characteristic wavelengths have specific advantages owing to minimization of specific losses.

One of the major types of attenuation in high-frequency communication is called absorption. This physical phenomena cause's loss of light when it passes through water vapor, or small metal particles present inside the glass. Water band is that frequency spectrum where maximum attenuation is taking place. Starting form microwave range there are specific window spectra where atmospheric attenuation is less, even minimum. Therefore, these small spectra are perfectly chosen by design engineers for making optical transmitters and detectors, connected by optical waveguide.

The major spectra for fibre-optic communication are [i] first optical window, defined from 800 to 900 nm, least value of loss is around 4 dB/km; [ii] second optical window, taken around 1330 nm, loss is further less of approximately 0.5 dB/km; [iii] third optical window, defined around 1550 nm, with a very low loss of 0.2 dB/km. Henceforth, it is the choice of optical engineers to design the system at 1550 nm, and is justified by numerous publications in last decade. However, when the system is embedded with different amplifiers with suitable input power, then other systems also considered worthwhile. Sometimes, their performance exceeds the features offered by the system centered around 1550 nm, in terms of gain and output signal level.

7.2 RAMAN AMPLIFIER

Raman amplifiers work on the principle of non-linear effects in optical domain. The basic principle behind the Raman amplifier is the phenomenon of Raman scattering (Islam 2002) which is also one of the non-linear effects in an optical fibre which is depicted in Figure 7.2. The frequencies depicted by 'f_p', 'f_s' and 'f_a' represent pump frequency, Stoke frequency and anti-Stokes frequency, respectively. The two energy levels or states that are shown here are the intermediate and ground states.

When electrons are energized from the lower energy states to the meta-states (virtual), the electrons then get down from their positions after overcoming life-time to the lower energy states at two different frequencies other than the pumping frequency, namely the Stokes and anti-Stokes frequencies (Bromage 2004). The frequency at which the electrons are energized from ground levels to virtual levels, they release their energies to jump back to vibrational states is called the Stoke frequency or Stokes and denoted by 'f_s'. The other frequency at which the electrons release their energy to get down to the ground levels after being energized to virtual levels from intermediate states is called the anti-Stoke frequency or the anti-Stokes and is denoted by 'f_a'. This is the basic principle of Raman scattering. The spectral representation of the energies is given in Figure 7.3.

When the input signal of fibre is self-phase modulated, high gain can be achieved. This is done by pumping a high power signal into the fibre. As a result of this, wider gain profile can be achieved both at 1310 nm and 1550 nm windows. This phenomenon is very much desirable for communication engineers when implementing WDM or DWDM networks (Emori *et al.*, 2002). Literatures showed that gain of Raman amplifier depends both on pump switching and on fibre length. With silica

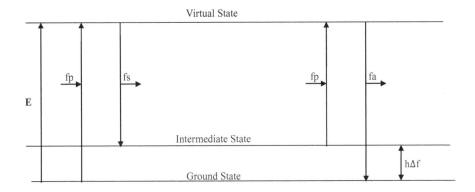

FIGURE 7.2 Energy level diagram for Raman scattering.

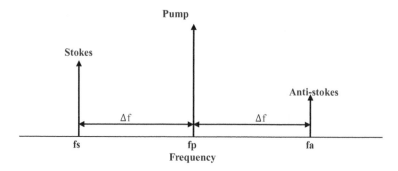

FIGURE 7.3   Spectral representation of the three frequencies.

fibres and proper choosing of pump power the gain of the overall system can be effectively increased. Already moderate gain of ~20 dB is reported for silica glass fibre. Researchers establish its utility in designing communication system around 100 THz.

7.2.1   IMPORTANCE OF RAMAN AMPLIFIER OVER OTHER OPTICAL FIBRE AMPLIFIERS

Raman amplifiers can provide significant gain over both the optical windows as mentioned before. Since RAs can be configured in both co- and counter-propagating architectures, hence simultaneous forward and backward amplification can be achieved which is much more effective than any other method of amplification. Figure 7.4 shows the two architectures in their very basic form. In co-propagating structure, both the signal and pump photons flow in the same direction and thus result

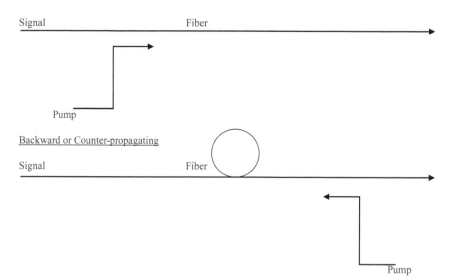

FIGURE 7.4   Co and counter propagating RA architecture.

in forward amplification while in counter-propagating architecture the pump and signal photons flow in opposite directions and hence result in backward amplification. Hence, if both these architectures are used simultaneously, maximum amplification is obtained in the middle of the fibre which is beneficial for long-haul DWDM networks (Seraji & Kiaee 2018). Also these amplifiers can be integrated with EDFAs for further increase in performance parameters as reported in the literature (Hambali & Pamukti 2017).

It has been seen from mathematical analysis that the Stokes Raman scattering is far more less than Rayleigh scattering. But the Rayleigh scattering cannot be used for having too high intensity. Hence, focus is concentrated in the Stokes and the anti-Stokes scattering spectra. According to the Boltzmann's distribution given by $N \propto \exp\left(-\dfrac{E}{k_B t}\right)$, the number of molecules in the higher energy states decrease exponentially, and since these molecules are responsible for the anti-Stokes scattering, the intensity at the anti-Stokes is less than at the Stokes. Thus most of the energy of the pump gets down-converted to the Stokes frequency, hence to achieve amplification the signal frequency must coincide with the Stoke frequency. The difference in frequency that occurs during the up-conversion or down-conversion depends on the value of the energy level. In stimulated Raman scattering (SRA), i.e. in systems where the input signal is given, the anti-Stokes scattering is completely swamped and only Stokes frequency is obtained. Figure7.5 gives the basic process for SRA.

In reality, the discrete nature of frequency bands do not occur, instead there are finite bands of frequencies that surround the Stokes frequency when the material is exposed to the pumping frequency. Therefore, all the frequency bands are closely spaced in the spectral domain; hence, they provide consecutive amplification of consecutive bands. Thus, the Raman Gain Profile is not the same for all frequencies.

The Raman gain G_R is dependent on a number of factors including the fibre length, the fibre attenuation and the fibre core diameter. It is given as a function of pump power by the following expression (Boiyo *et al.*, 2012):

$$G_R = \exp\left(\frac{g_R L_{eff} P_p}{A_{eff} k} - \alpha_s L\right) \qquad (7.5)$$

where g_R is the Raman Gain coefficient, L is the actual length, L_{eff} is the effective length of the fibre, P_p is pump power, A_{eff} is the effective area of interaction between the signal and pump photons and k is the polarization factor.

FIGURE 7.5   Schematic diagram for stimulated Raman scattering.

The effective area is given by the following expression:

$$A_{eff} = \pi r_{eff}^2 \tag{7.6}$$

where r_{eff} is the effective core radius.

Equation (7.7) gives the expression for the effective length of the fibre

$$L_{eff} = \frac{1 - \exp(-\alpha_p L)}{\alpha_p} \tag{7.7}$$

where α_p is the attenuation constant.

Thus, RAs provide a lot of advantages over normal DFAs and SOAs in terms of large bandwidth, flat band gain, easier to implement and flexibility to be used with other components. Hence, recent researches have focus on pure and hybrid RA as the main amplification component in Optical Communications.

7.2.2 CHOICE OF FREQUENCY SPECTRUM

Raman Gain Profile exhibits that the gain of an RA is highest with when there is a frequency shift of 13 THz which in wavelength scale equals nearly 100 nm (Agrawal 2013). Thus for effective RA it is recommended to have a difference of 100 nm between the Pump and Signal wavelengths. However in the 1310 nm window where the Raman Gain is essentially flat band, the practical difference of wavelengths is recommended to be kept nearly at 70 nm, which has been shown in the literature (Czyżak et al., 2014). The other window for the system to achieve efficient gain is the 1550 nm window.

In the present work, the optical properties at the 1310 nm, where the flat band gain occurs, are investigated. The 1550 nm window gives much higher gain but for efficient WDM or DWDM networks the effectiveness of different optical properties at the flat band gain at 1310 nm is studied.

7.2.3 OPTICAL PROPERTIES TO BE INVESTIGATED

The main factors that mainly matter in an amplifier system are the gain of the system, the Q-factor of the system as well as detailed analysis of the Eye Diagram of the system. The gain of the system is studied by individually observing the amplitudes and nature of the input signals before and after Raman Amplification. The Q-factor and the Eye Diagram gives further results on the effective length of the fibre as well as the effective bit rate to be used for the system.

7.2.4 NOVELTY OF THE PRESENT WORK

The proposed work is significant in the aspect that it incorporates a novel design of fibre communication system embedded with RAMAN amplifier such that it can provide significant out even at a very high distance from the source along with considerable gain Very high Q-factor can be obtained when operated at either 1310 nm or 1550 nm. Clear eye diagram can be obtained which speaks in favour of negligible jitter. A vis-à-vis analysis is carried out for BER, sensitivity, Q-factor, signal power before and after amplification and gain. Results are simulated for three different bit rates, which expands the operating region of the proposed system.

7.3 RESULTS

The whole study has been in the Opti-System Software environment. The simultaneous co- and counter-propagating structure has been used for the present work. It has been simulated and the said results have been obtained.

7.3.1 OPTICAL PROPERTIES CALCULATED AT 1330 NM

The following results are obtained at 1310 nm bandwidth. The information signal was kept at 1310 nm with a power of 0 dBm and the pump was kept at 1240 nm with a power of 26.02 dBm or 400 mW. Figures 7.6–7.21 were taken at a constant bit rate of 5 Gbps and distance was varied. Figures 7.38–7.49 are taken with distance kept fixed at 60 km and 130 km and the bit rate was varied between 2.5 Gbps, 5 Gbps and 7.5 Gbps, respectively.

Figures 7.6–7.9 show the change in the nature of Q-factor over an entire bit period. The bit rate is kept constant at a moderate value 5 Gbps. From the plots, we can observe that though the nature of the Q-factor variation remains quite similar but the value of the peak value of Q-factor decreases with fibre length, with value less than 1800 at a fibre length of 160 km. The validation for such a result can be attributed to the fact that to maintain a minimum BER with sustainable gain with a constant pump power, the stored energy of the system decreases in respect to the energy supplied to the system and hence by the definition of Q-factor, its value also decreases.

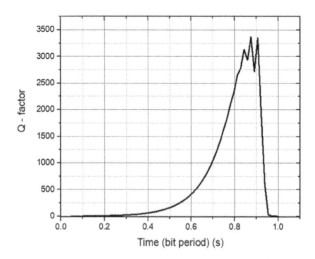

FIGURE 7.6   Q-factor for 20 km at 5 Gbps.

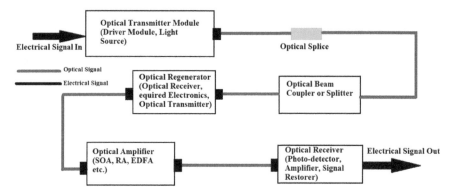

FIGURE 7.7   Q-factor for 80 km at 5 Gbps.

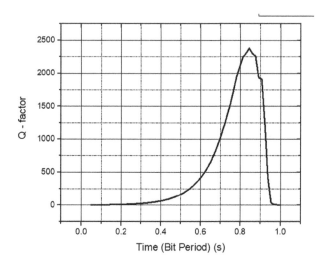

FIGURE 7.8   Q-factor for 120 km at 5 Gbps.

Figures 7.10–7.13 show the eye diagrams (ED) for equal distance where the Q-factors are taken and for the same system parameters. It can be observed that the eye height (EH) decreases for all the distances except getting a rise at 80 km. At a distance of 50 km, the maximum gain occurs, where the EH is maximum, after that it starts to fall again but gradually. Hence at the 80 km length the value does not decrease much and thus attains a maximum value among the taken measurements. The value of the EH gives a knowledge about the purity of the bit and the threshold for the correct detection of information. Hence, larger EH near the maximum gain implies the signal is the purest and easily detectable with a minimum threshold at that length.

Figures 7.14–7.17 show the input signal spectrum at the aforesaid distances. As said before, the input power of the signal was taken to be 0 dBm or 1 mW. Therefore, all the spectra show the same peak power as well as the nature of the signal. The peak

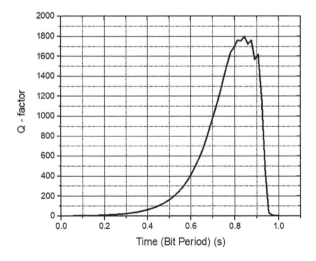

FIGURE 7.9   Q-factor for 160 km at 5 Gbps.

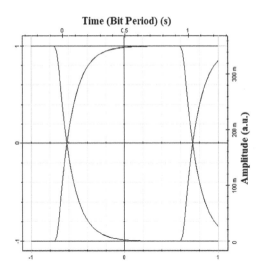

FIGURE 7.10   BER for 20 km at 5 Gbps.

power is a little less than 0 dBm since some of the power is used in the modulation process. The effect of noise can also be seen by the generation of other signals at the side bands of 1310 nm.

Figures 7.18–7.21 show the spectra of the output signal that is the spectrum of the signal after amplification. It can be observed that the output signal peak power is maximum at the 80 km length since the gain is highest at the 80 km length which among the taken distances which again follows the same explanation about the 80 km length being near the 50 km length. The gain is highest at 50 km because of the simultaneous co- and counter-propagating architecture where the forward and

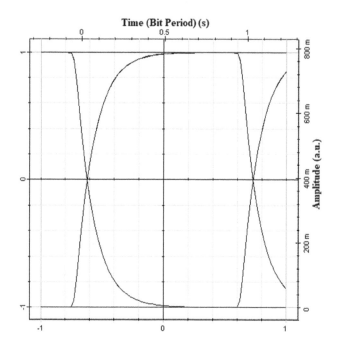

FIGURE 7.11   BER for 80 km at 5 Gbps.

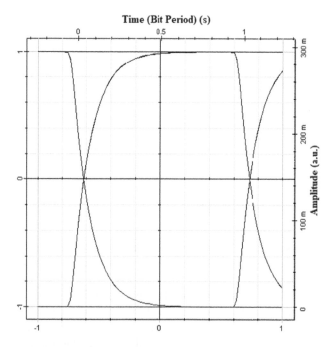

FIGURE 7.12   BER for 120 km at 5 Gbps.

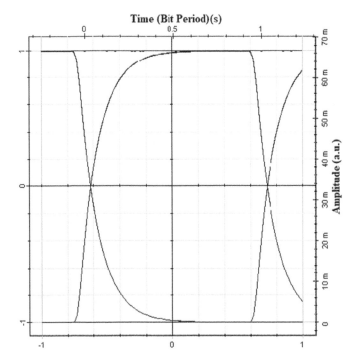

FIGURE 7.13   BER for 160 km at 5 Gbps.

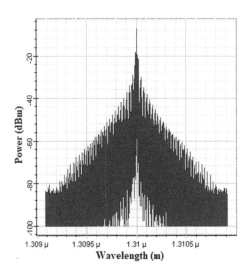

FIGURE 7.14   Signal before amplification for 20 km at 5 Gbps.

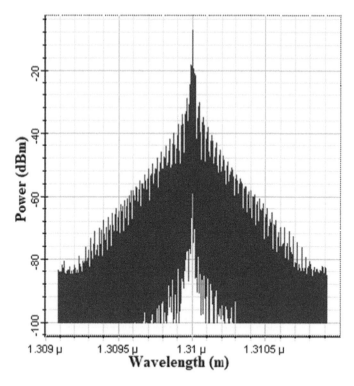

FIGURE 7.15   Signal before amplification for 80 km at 5 Gbps.

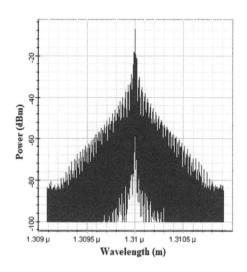

FIGURE 7.16   Signal before amplification for 120 km at 5 Gbps.

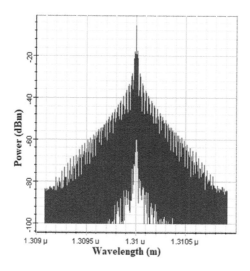

FIGURE 7.17 Signal before amplification for 160 km at 5 Gbps.

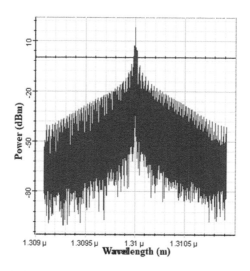

FIGURE 7.18 Signal after amplification for 20 km at 5 Gbps.

backward pumping occurs simultaneously and hence the maximum gain occurs at the mid-length of the fibre. The output signal power at the maximum gain point is around 20 dBm which gives a gain of about more than 20 dB. After that length the gain gradually decreases and hence the signal peak power reduces too and reaches a value of 11 dBm at a fibre length of 160 km. The harmonics also get amplified by the system but by using proper filtering these can be suppressed and pure signal can be received.

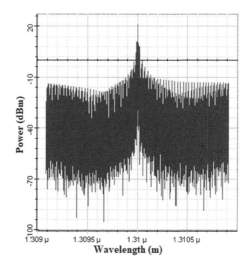

FIGURE 7.19   Signal after amplification for 80 km at 5 Gbps.

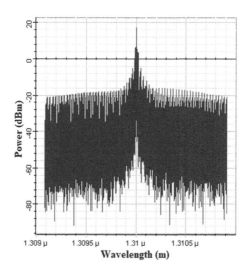

FIGURE 7.20   Signal after amplification for 120 km at 5 Gbps.

Figures 7.22–7.24 show the Q-factor variation with the bit rate at a fixed distance of 60 km. It can be observed that as the bit rate increases the Q-factor significantly decreases. As the bit rate increases the energy required to transmit the bit with as minimum BER as possible also increases, hence the stored energy of the system decreases. Since Q-factor is the ratio of the stored energy to the energy supplied, hence decrease in stored energy also decreases the value of the Q-factor.

Another thing to observe is that the variation of Q-factor is spread in case of 2.5 Gbps and it gets narrowed at 7.5 Gbps. At low data rates the storage of energy starts at a lower bit period than at higher data rates. Hence the Q-factors start to increase

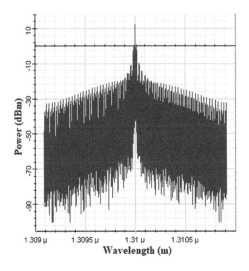

FIGURE 7.21   Signal after amplification for 160 km at 5 Gbps.

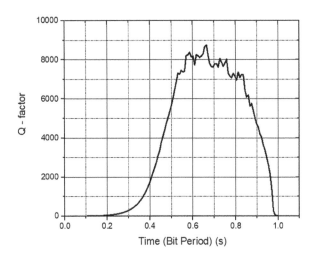

FIGURE 7.22   Q-factor for 60 km at 2.5 Gbps.

earlier in case of low data rates and are able to maintain it for a longer length of the period.

Figures 7.25–7.27 show the ED of the system at the same data rates and the same distance of 60 km. It can be seen that the EH decreases with the increase in data rates which implies that the threshold of detection of signal is lower at low data rates and the threshold increases as the data rate is increased demanding the system to use more power.

Figures 7.28–7.30 show the input signal spectrum at the different data rates for the distance of 60 km. as before the input signal is taken at 0 dBm but due to usage of

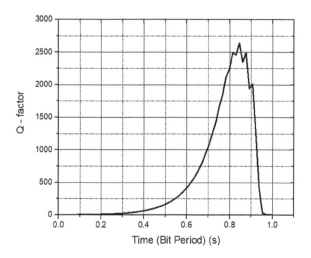

FIGURE 7.23   Q-factor for 60 km at 5 Gbps.

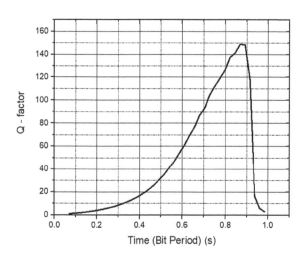

FIGURE 7.24   Q-factor for 60 km at 7.5 Gbps.

power during modulation the measured power is a little less than 0 dBm. As before the harmonics are present here too and will also be present in the successive spectra too. Since the reasons and explanation for all of them are same, they won't be addressed henceforth.

Figures 7.31–7.33 exhibit the spectra of the output signal that is the spectrum of the signal after amplification for the above input signals. It can be observed that the output signal peak powers are nearly around 20 dBm with very little variations. It goes on to show that the gain depends mainly on the distance and very little on the bit rate of the signals.

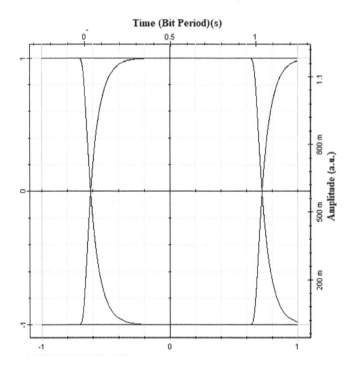

FIGURE 7.25   BER for 60 km at 2.5 Gbps.

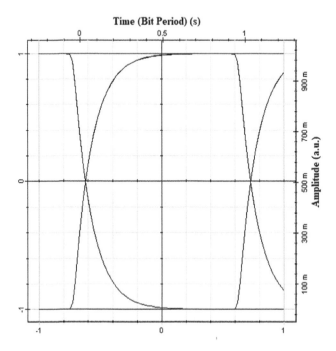

FIGURE 7.26   BER for 60 km at 5 Gbps.

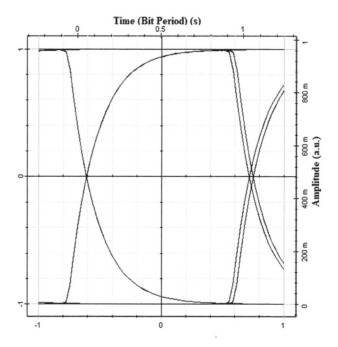

FIGURE 7.27 BER for 60 km at 7.5 Gbps.

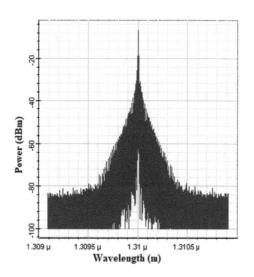

FIGURE 7.28 Signal before amplification for 60 km at 2.5 Gbps.

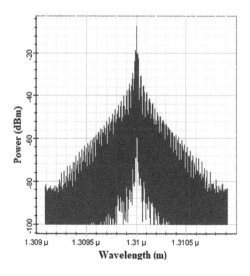

FIGURE 7.29   Signal before amplification for 60 km at 5 Gbps.

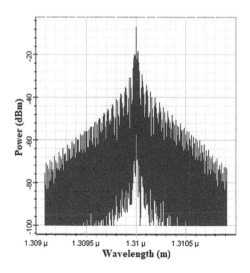

FIGURE 7.30   Signal before amplification for 60 km at 7.5 Gbps.

Figures 7.34–7.36 illustrate the Q-factor variation with the bit rate at a fixed distance of 130 km. It can be observed that as the bit rate increases the Q-factor significantly decreases and also less in value than those taken at 60 km, the reason for which can be explained by the requirement of more energy for transmitting the bits for longer distances. The decrease of the values with the increase in bit can be

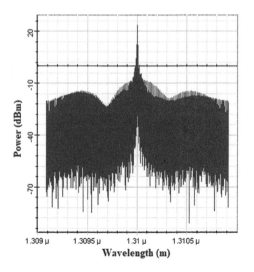

FIGURE 7.31   Signal after amplification for 60 km at 2.5 Gbps.

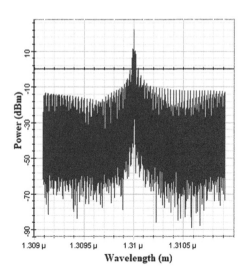

FIGURE 7.32   Signal after amplification for 60 km at 5 Gbps.

explained as before. Similarly, the spreading of the Q-factor curve can also be explained with the same explanation as before.

Figures 7.37–7.39 demonstrate the EDs of the system at the same data rates and the same distance of 130 km. It can be seen that the EH decreases with the increase in data rates, with values less than those at 60 km which implies that the threshold of detection of signal decreases with increasing distance even though the data rate is made higher, thus demanding the system to be more sensitive to the received signal. Thus, it can be said that the dependence of EH and threshold on distance is more than on the data rate.

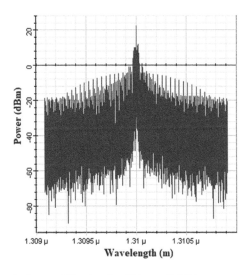

FIGURE 7.33   Signal after amplification for 60 km at 7.5 Gbps.

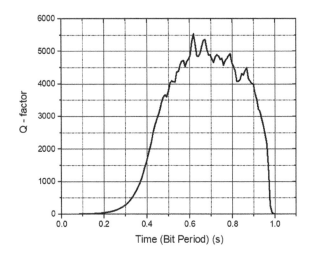

FIGURE 7.34   Q-factor for 130 km at 2.5 Gbps.

Figures 7.40–7.42 show the input signal spectrum at the different data rates for the distance of 130 km. As before the input signal is taken at 0 dBm but due to usage of power during modulation the measured power is a little less than 0 dBm.

Figures 7.43–7.45 show the spectra of the output signal that is the spectrum of the signal after amplification for the above input signals. It can be observed that the output signal peak powers are nearly around 15 dBm with very little variations. It goes on to show that with increasing distance the gain decreases and moreover its variation on data rates is very little.

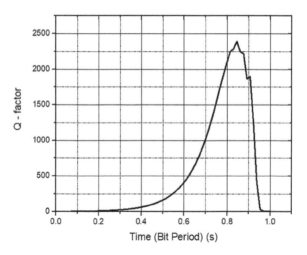

FIGURE 7.35   Q-factor for 130 km at 5 Gbps.

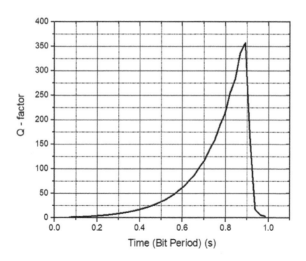

FIGURE 7.36   Q-factor for 130 km at 7.5 Gbps.

7.3.2   COMPARATIVE STUDY WITH PROPERTIES OBTAINED AT 1550 NM

The following figures are taken at the 1550 nm window for the three data rates as used before at two distances of 60 km and 130 km. the results are compared with the same taken at the 1310 nm window.

Figures 7.46–7.51 show the Q-factor variation at 60 km (50–52) and 130 km (53–55) distance for different data rates. It can be seen that the value of Q-factor decreases as per the previous discussion given for 1310 nm window as well as the spreading and narrowing of the curve and the decrease in value of the same with distance has

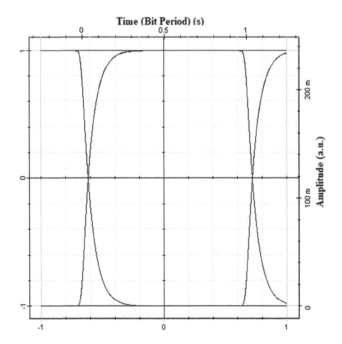

FIGURE 7.37   BER for 130 km at 2.5 Gbps.

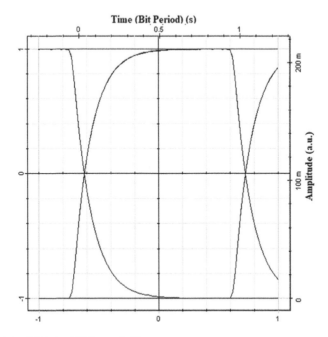

FIGURE 7.38   BER for 130 km at 5 Gbps.

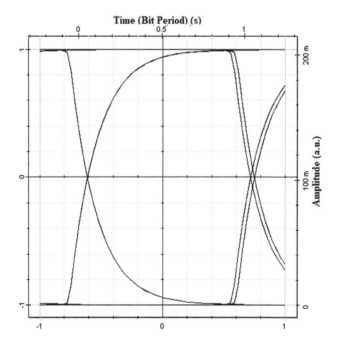

FIGURE 7.39   BER for 130 km at 7.5 Gbps.

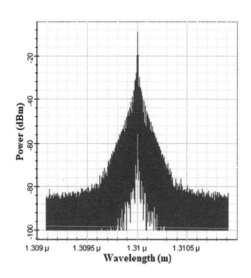

FIGURE 7.40   Signal before amplification for 130 km at 2.5 Gbps.

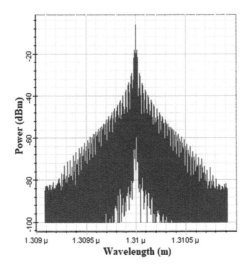

FIGURE 7.41   Signal before amplification for 130 km at 5 Gbps.

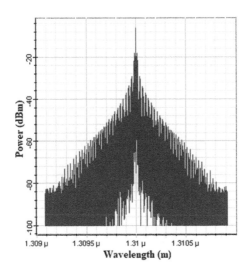

FIGURE 7.42   Signal before amplification for 130 km at 5 Gbps.

also been explained before in Section 3.1, but the individual peak values of Q-factor for both the sets are less than those taken at the 1310 nm window. Since attenuation is lesser at the 1550 nm window than the 1310 nm window, the gain is higher for the 1550 nm window. Therefore, to provide more gain for the same fibre length at the 1550 nm band, the stored energy decreases and thus the Q-factor too measures less at this band than at the 1310 nm band.

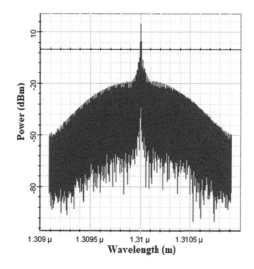

FIGURE 7.43   Signal after amplification for 130 km at 2.5 Gbps.

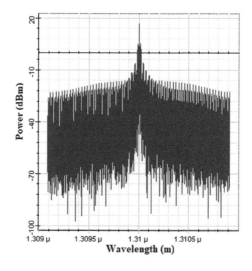

FIGURE 7.44   Signal after amplification for 130 km at 5 Gbps.

Figures 7.52–7.54 illustrate the ED taken at 60 km and Figures 7.55–7.57 show the ED taken for 130 km distance for the three respective data rates. The explanation for the decrease in EH with data rate as well as with distance has been given in Section 3.1. However, the individual EHs in both the sets, at the 1550 nm band are less than those taken at 1310 nm band. The reason can be explained as follows. For the same gain at the 1550 nm band the system provides less power and hence for the detection of signal the threshold increases since higher threshold attributes to lower EHs; therefore, the individual EHs are less at this band.

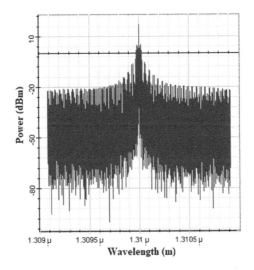

FIGURE 7.45   Signal after amplification for 130 km at 7.5 Gbps.

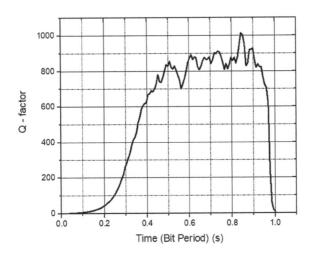

FIGURE 7.46   Q-factor for 60 km at 2.5 Gbps.

Similar variations for signal before amplification are also made. Two plots (Figures 7.58 and 7.59) are provided. The input signal is taken at 0 dBm but due to usage of power during modulation the measured power is a little less than 0 dBm.

Figures 7.60 and 7.61 demonstrate the spectra of the output signal after amplification for the above input signals at both distances of 60 km and 130 km, respectively. It can be observed that the output signal peak powers are nearly around 20 dBm with very little variations for the 60 km set and around 10 dBm for the 130 km distance which is nearly same as those taken at 1310 nm. The reason for that lies in the system

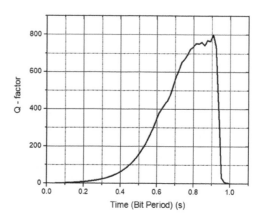

FIGURE 7.47   Q-factor for 60 km at 5 Gbps.

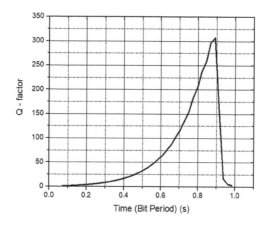

FIGURE 7.48   Q-factor for 60 km at 7.5 Gbps.

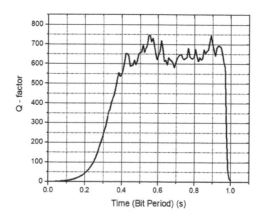

FIGURE 7.49   Q-factor for 130 km at 2.5 Gbps.

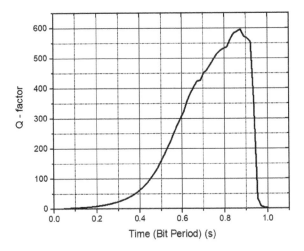

FIGURE 7.50 Q-factor for 130 km at 5 Gbps.

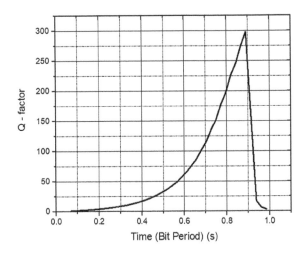

FIGURE 7.51 Q-factor for 130 km at 7.5 Gbps.

parameters. For maximum gain to be achieved the wavelength difference should be 70 nm for 1310 nm band and 100 nm for 1550 nm band. The two systems are simulated with these values; thus, the gain also measures to be same. It goes on to show that with increasing distance the gain decreases and moreover its variation on data rates is very little.

7.4 CONCLUSION

Simulated findings reveal that communication system designed at 1310 nm has potential advantages than the conventional 1550 nm when both the system has equal

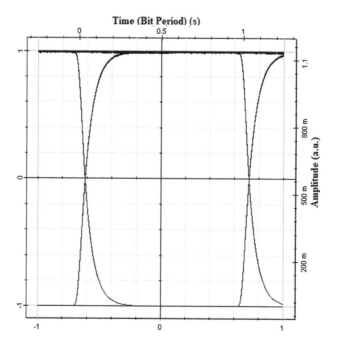

FIGURE 7.52   BER for 60 km at 2.5 Gbps.

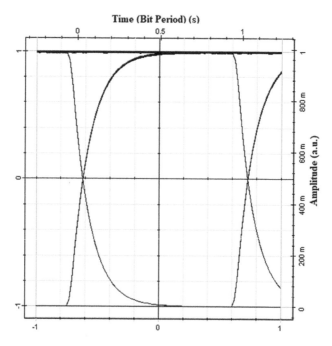

FIGURE 7.53   BER for 60 km at 5 Gbps.

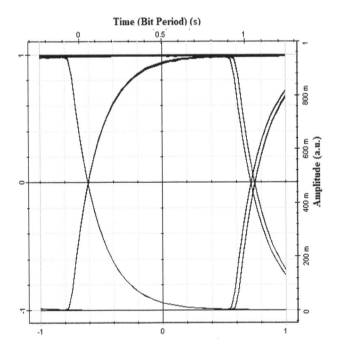

FIGURE 7.54 BER for 60 km at 7.5 Gbps.

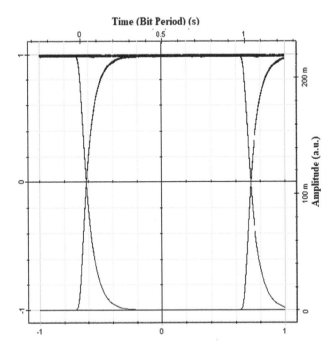

FIGURE 7.55 BER for 130 km at 2.5 Gbps.

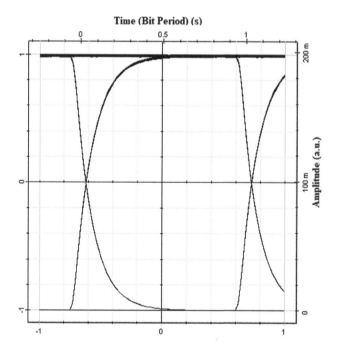

FIGURE 7.56   BER for 130 km at 5 Gbps.

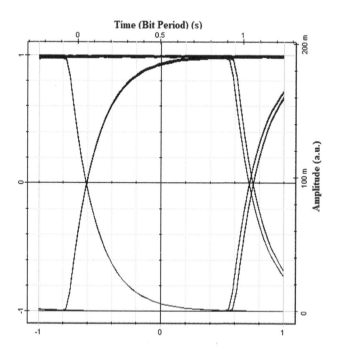

FIGURE 7.57   BER for 130 km at 7.5 Gbps.

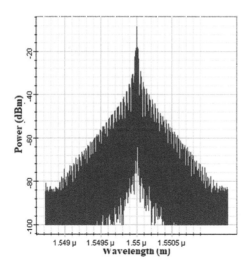

FIGURE 7.58 Signal before amplification for 60 km at 5 Gbps.

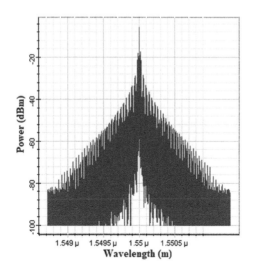

FIGURE 7.59 Signal before amplification for 130 km at 5 Gbps.

input power and embedded with RAMAN amplifier. When both the systems are compared at 5 Gbps data rate for 80 km fibre length, Q-factor for the system operated at 1310 nm is almost three times higher the counterpart. However, eye height is approximately 10% greater for the former, which speaks for less sensitivity. For both the system, gain is reduced with increasing bit rate. Gin is comparatively better for 1310 nm (<1%). Three different data rates are considered from 2.5 to 7.5 Gbps in order to get performance comparison in a wider domain. Output signal level remains

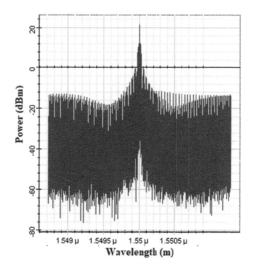

FIGURE 7.60   Signal after amplification for 60 km at 5 Gbps.

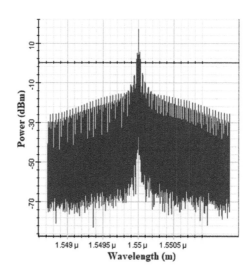

FIGURE 7.61   Signal after amplification for 130 km at 5 Gbps.

almost significant up to 150 km for both the systems, when bit rate is higher. This is one of the novel aspects of the design, and therefore the proposed system can be implemented in real communication system which will provide lower data loss.

REFERENCES

Agarwal, V., Agrawal, M. (2018) *Characterization and optimization of semiconductor optical amplifier for ultra high speed applications: A review, IEEE Conference on Signal Processing And Communication Engineering Systems*

Agrawal, G. P. (2010) *Fiber-optic Communication Systems*, 4th Ed., Wiley, Hoboken, NJ

Agrawal, G. (2013) Nonlinear Fiber Optics, 5th Ed., *Optics and Photonics*, 295–352

Agrawal, G.P. (2016) Optical communication: Its history and recent progress. In Al-Amri M., El-Gomati M., Zubairy M. (eds) *Optics in Our Time*, Springer, Cham, 177199

Al-Taiy, H., Wenzel, N., Preußler, S., Klinger, J., Schneider, T. (2014) Ultra-narrow linewidth, stable and tunable laser source for optical communication systems and spectroscopy, *Optics. Letters*, 39, 5826–5829

Bebawi, J. A., Kandas, I., El-Osairy, M. A., Aly, M. H. (2018) A comprehensive study on EDFA characteristics: Temperature impact, *Applied Science*,8, 1640

Boiyo, D. K., Kuja, S., Kipnoo, E. K. R., Waswa, D., Muguro, K. M., Amolo, G. (2012) Modelling of an optical fibre Raman amplifier, *Proceedings of the Mechanical Engineering Conference on Sustainable Research and Innovation*, 4, 79–81

Bromage, J. (2004) Raman amplification for fiber communications systems, *Journal of Lightwave Technology*, 22(1), 79–93

Chi, R. (2017) *Research on the application of semiconductor optical amplifier in 100G local area network/data center, 16th IEEE International Conference on Optical Communications and Networks*, China

Czyżak, P., Mazurek, P., Turkiewicz, J. P. (2014). 1310 nm Raman amplifier utilizing high-power, quantum-dot pumping lasers, *Optics & Laser Technology*, 64, 195–203

Emori, Y., Kado, S., Namiki, s. (2002) Broadband flat-gain and low-noise Raman amplifiers pumped by wavelength-multiplexed high-power laser diodes, *Optical Fiber Technology*, 8(2), 107–122

Fu, J., Xi, Y. (2016) Narrow spectral width FP lasers for high-speed short-reach applications, Progress in Electromagnetic Research Symposium, China

Garmire, E. (2002) Sources, modulators, and detectors forfiber-optic communication systems. In Bass, M. (ed.), *Fiber Optics Handbook*, McGraw-Hill, New York, 4.1–4. 80

Ghatak, A. K.&Thyagrajan, K. (1989) *Optical Electronics*, 1st Ed., Cambridge University Press, Cambridge

Ghatak, A. K.&Thyagrajan, K. (1998) *Introduction to Fiber Optics*, 1st Ed., Cambridge University Press, Cambridge

Hambali, A., Pamukti, B. (2017) *Performance analysis of hybrid optical amplifier in long-haul ultra-dense wavelength division multiplexing system, International Conference on Control, Electronics, Renewable Energy and Communications*, Yogyakarta, 80–83

Hecht, J. (1999) *City of Light: The Story of Fiber Optics*. Oxford University Press, New York

Hui, R. (2019) *Introduction to Fiber-Optic Communications*, 1st Ed., Academic Press

Huurdeman, A. A. (2003) *The Worldwide History of Telecommunications*. Wiley, Hoboken, NJ

Islam, M. N. (2002) Raman amplifiers for telecommunications, *IEEE Journal of Selected Topics in Quantum electronics*, 8(3), 548–559

Ivaniga, T., Ivaniga, P. (2017) Comparison of the optical amplifiers EDFA and SOA based on the BER and Q-factor in C-band, *Advances in Optical Technologies*, 2017, 9053582

Karimi, M. (2018) Theoretical study of the thermal distribution in Yb-doped double-clad fiber laser by considering different heat sources, *Progress In Electromagnetics Research C*, 88, 59–76

Keiser, G. (2000) *Optical Fiber Communication*, 3rd Ed., McGraw Hill

Miller, S. E., Kaminow, I. P. (1988) *Overview and Summary of Progress', Optical Fiber Telecommunications II*, 1–27, Academic Press

Nayayama, K., Kakui, M., Matsui, M., Saitoh, T., Chigusa, Y. (2002) Ultra-low-loss (0.1484 dB/km) pure silica core fibre and extension of transmission distance, *Electronics Letters*, 38(20), 1168–1169

Nouchi, P., Dany, B., Campion, J. F., de Montmorillon, L. A., Sillard, P., Bertaina, A. (2003) Optical communication and fiber design, *Annales Des Télécommunications*58, 1586–1602

Oh, K., Choi, S., Jung, Y., Lee, J. W. (2005) Novel hollow optical fibers and their applications in photonic devices for optical communications, *Journal of Lightwave Technology*, 23(2), 524–532

Putra, A. W. S., Yamada, M., Ambran, s., Maruyama, T. (2018) Theoretical comparison of noise characteristics in semiconductor and fiber optical amplifiers, *IEEE Photonics Technology Letters*, 30(8), 756–759

Ramaswami, R., Sivarajan, K., Sasaki, G. (2009) *Optical Networks: A Practical Perspective*, 3rd Ed., Morgan Kaufmann

Senior, J. M. (1992) *Optical Fiber Communications*, 2nd Ed., Prentice Hall, Englewood Cliffs, NJ

Seraji, F. E., Kiaee, M. S. (2018) Comparison of EDFA and Raman amplifiers effects on RZ and NRZ encoding techniques in DWDM optical network with bit rate of 80 Gb/s, Physics & Astronomy*International Journal*, 2(1), 116–121

Smith, W. J. (2000) *Modern Optical Engineering*, 3rd Ed., McGraw-Hill

Wasfi, M. (2009) Optical fiber amplifiers-review, *International Journal of Communication Networks and Information Security*, 1(1), 42–47

Zhou, P., Zhan, W., Mukaikubo, M., Nakano, Y., Tanemura, T. (2017) Reflective semiconductor optical amplifier with segmented electrodes for high-speed self-seeded colorless transmitter, *Optics Express*, 25, 28547–28555

Zoiros, K. E. (2018) Special issue on applications of semiconductor optical amplifiers, *Applied Science*,8, 1185

8 Ultra-Narrowband Optical Comb Filter Using Sampled Fibre Bragg Gratings

Somnath Sengupta

8.1 INTRODUCTION

The fibre Bragg gratings (FBGs) are essential devices in the area of optical fibre communication. Nowadays, the FBGs are extensively used in filtering, Add/Drop Multiplexing in WDM systems, multi-wavelength fibre laser, and dispersion compensation in optical network. With the development of dense wavelength division multiplexing (DWDM) system, the number of optical channels in optical fibre has been highly increased. The optical comb filter is the key component in generation of such multiple optical channels or carriers. Optical comb filters are also used in optical networks and systems to process the optical signals and isolate the neighbouring channel signals for reducing the cross-talk. Recently, the optical comb filter is being used in the area of microwave photonics for generation of millimetre wave signal and signal processing for 5G systems.

8.1.1 MOTIVATION

It can be understood that an optical comb filter or multichannel filter is a very crucial device in the field of photonics and optical communication. To generate comb spectrum, several techniques have been developed. However, out of those techniques, it has been found that the sampled-chirped FBG-based method is very advantages as it requires only a single fibre, whereas the other techniques require involvement of many components. There are several factors such as reflectivity, channel-bandwidth, isolation, etc., related to the performance of a comb filter. A multichannel filter with uniform reflectivity, ease of fabrication and ultra-narrow in-channel bandwidth is highly desired for applications mentioned earlier.

In this chapter, we mainly discuss the general techniques for achieving the comb spectrum using sampled-chirped fibre Bragg grating (SCFBG). At the beginning of this chapter, we discuss the different types of FBGs, e.g. apodized, chirped, phase shifted and superstructure gratings. We demonstrate the sampled fibre Bagg gratings as comb filters. The method of achieving the multichannel reflection spectrum using multiple phase shift (MPS) techniques and spectral Talbot effect have been explained and demonstrated.

8.2 OPTICAL COMB FILTER

In long-haul optical communication, to increase the carrier bandwidth/data rate, the multiple optical carrier transmission over a single fibre is highly desired. The wavelength division multiplexing (WDM) is the technology which involves the transmission of multiple channels over a single optical link. In this technology, however, the number of channels or channel spacing are very much limited. To transmit more channels, the DWDM system has been developed. Currently, DWDM systems support up to 192 wavelengths and each wavelength transports up to 100 Gbit/s. To implement WDM/DWDM, generally, the optical filter with comb response or lasers with narrowband lines are used [1–3]. The response of a usual optical comb filter is shown in Figure 8.1.

The comb filter, in its reflection spectrum, generates repetitive spectral lines with certain frequency spacing. These multichannel devices are based on the optical fibre-based grating structures. Several structures [4–7] have been designed for generation of multiple channels with improved reflectivity and narrower channel spacings. Apart

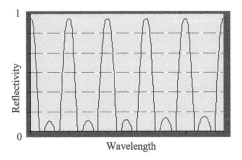

FIGURE 8.1   Response of optical comb filter.

from the use of comb filter as DWDM component, the comb filters have been used in the area of optical signal processing [8], microwave photonics [9,10], etc.

8.3   BASICS OF FIBRE BRAGG GRATING

An FBG is defined as a small section of fibre inscribed with a periodic modulation of refractive index and allows a certain wavelength to get reflected and others are transmitted. The central reflected wavelength of FBG is given by

$$\lambda_B = 2n_{eff}\Lambda \tag{8.1}$$

where Λ is the grating period and n_{eff} is the effective refractive index of fundamental core mode. The coupled-mode theory is used to design and analyse the FBGs.

Simple FBGs are called uniform grating as because the refractive index modulation in these gratings is uniform. However, there are other types of gratings where the refractive index modulation is non-uniform, e.g. apodized gratings, chirped gratings, phase-shifted gratings, and superstructure gratings. These are called non-uniform gratings. In general, the refractive index modulation profile is given by

$$n(z) = n_1 + \Delta n(z) \tag{8.2}$$

where n_1 is the initial core index and $\Delta n(z)$ is the induced index change across the length of the core in the direction z.

The refractive index modulation profile/index change in general is given by [11]

$$\Delta n(z) = \overline{\delta n_{eff}}(z)\left\{1 + \nu\cos\left[\frac{2\pi}{\Lambda}z + \phi(z)\right]\right\} \tag{8.3}$$

where $\overline{\delta n_{eff}}(z)$ is the dc-index change, ν is the fringe visibility of the index change, Λ is the grating period and $\phi(z)$ is the phase function. The FBGs generally have the index change in the range of 10^{-5} to 10^{-2}. The term $\phi(z)$ is zero for uniform gratings, but it is non-zero for non-uniform gratings.

We now discuss the different types of non-uniform gratings with their refractive index profiles. The index profiles of various gratings are shown schematically in Figure 8.2.

8.3.1 APODIZED GRATINGS

The reflection spectrum of FBG contains a significant amount of sidelobes. The sidelobes can be reduced by apodizing the refractive index profile of the grating. The refractive index profile for apodized grating is expressed by

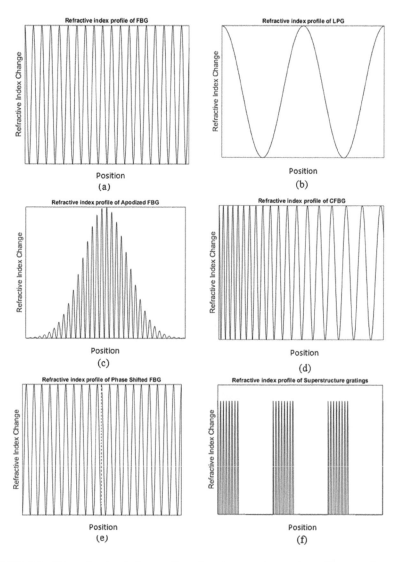

FIGURE 8.2 The refractive index profiles of various gratings; a-b: uniform gratings; c-f: non-uniform gratings.

$$\Delta n(z) = \overline{\delta n_{\text{eff}}} \, S(z) \left\{ 1 + v \cos \left[\frac{2\pi}{\Lambda} z \right] \right\} \tag{8.4}$$

The function S(z) is the apodization envelope of the grating. The commonly used apodization functions are Raised Cosine, Gaussian, Tanh, Blackman and Sinc. [11]. An example of Gaussian apodized grating is shown in Figure 8.2c.

8.3.2 Chirped Gratings

In uniform FBG, the grating period remains constant throughout the length of the fibre, but if the grating period is varied along the length, then a chirped grating or chirped fibre Bragg grating (CFBG) is produced. The varying grating period offers a range of Bragg wavelengths to be reflected from the grating. This causes a wide-band spectrum in the reflectivity of the CFBG. Other advantages of CFBG include the dispersion compensation and linear group delay characteristics that are desired in WDM systems. The refractive index profile of CFBG is expressed by

$$\Delta n(z) = \overline{\delta n_{\text{eff}}} S(z) \left\{ 1 + v \cos \left[\frac{2\pi}{\Lambda(z)} z + \phi(z) \right] \right\} \tag{8.5}$$

where $\Lambda(z)$ is the local grating period and if Λ_0 is the fundamental grating period and C is the chirp coefficient, then $\Lambda(z)$ can be written as

$$\Lambda(z) = \Lambda_0 + Cz \tag{8.6}$$

where C is obtained by differentiating $\Lambda(z)$, with respect to z,

$$C = \frac{d\Lambda(z)}{dz} \tag{8.7}$$

The phase change at any position in the grating can be obtained by subtracting the phase at initial position from the phase at a certain position z, as

$$\phi(z) = \frac{2\pi}{\Lambda(z)} z - \frac{2\pi}{\Lambda_0} z = \frac{2\pi}{\Lambda_0 + Cz} z - \frac{2\pi}{\Lambda_0} z$$

$$= 2\pi z \left(\frac{-Cz}{\Lambda_0^2 + \Lambda_0 Cz} \right) \tag{8.8}$$

Considering the fact that $\Lambda_0^2 \gg \Lambda_0 Cz$ we get, the phase change of the linearly CFBG can be obtained as

$$\phi(z) = -\frac{2\pi C}{\Lambda_0^2} z^2 \tag{8.9}$$

The phase change can be rewritten as

$$\phi(z) = -\frac{2\pi C z^2}{\Lambda_0^2} = -\frac{2\pi \dfrac{d\Lambda(z)}{dz} z^2}{\Lambda_0^2} = -2\pi \frac{\dfrac{d\lambda_D(z)}{dz}}{2n_{eff}} \frac{z^2}{\Lambda_0^2}$$

$$= -2\pi \frac{\dfrac{d\lambda_D(z)}{dz}}{2n_{eff}} \frac{z^2}{\left(\dfrac{\lambda_D}{2n_{eff}}\right)^2} = -\frac{4\pi n_{eff} z^2}{\lambda_D^2} \frac{d\lambda_D}{dz} \tag{8.10}$$

Here, we assume that the local Bragg reflected wavelengths are given by, $\lambda_D(z) = 2n_{eff}\Lambda(z)$ and differentiating it with respect to z, we get

$$\frac{d\lambda_D(z)}{dz} = 2n_{eff}\frac{d\Lambda(z)}{dz} \tag{8.11}$$

The term $\dfrac{d\lambda_D(z)}{dz}$ is called the "Chirp" and is rate of change of the design wavelength with position ingrating [11].

In Equation (8.10), also we have considered the central Bragg reflected wavelength is given by

$$\lambda_D = 2n_{eff}\Lambda_0 \tag{8.12}$$

The differentiation of the phase term $\phi(z)$ with respect to z for a linear chirp can be calculated as

$$\phi(z) = -\frac{4\pi n_{eff} z^2}{\lambda_D^2} \frac{d\lambda_D}{dz}$$

$$\text{or,} \quad \frac{d\phi(z)}{dz} = -\frac{8\pi n_{eff} z}{\lambda_D^2} \frac{d\lambda_D}{dz} \tag{8.13}$$

The other way of defining the phase factor is [12]

$$\frac{d\phi(z)}{dz} = \frac{2Fz}{FWHM^2} \tag{8.14}$$

where FWHM is the full-width-half-maximum of the apodization profile of the grating and F is a parameter defining similar "Chirp" as discussed [12].

Hence, the phase shift can also be written from previous Equation (8.14):

$$\phi(z) = \frac{Fz^2}{FWHM^2} \tag{8.15}$$

Comparing eqns. (10, 15), it can be written as

$$-\frac{4\pi n_{eff} z^2}{\lambda_D^2} \frac{d\lambda_D}{dz} = \frac{Fz^2}{FWHM^2}$$

$$or, F = -4\pi n_{eff} \frac{FWHM^2}{\lambda_D^2} \frac{d\lambda_D}{dz} \qquad (8.16)$$

The refractive index profile of CFBG is shown in Figure 8.2d and explained further.

8.3.3 PHASE SHIFTED GRATINGS

A phase-shifted grating is a grating with a phase shift in the refractive index modulation in the grating region. This phase shift divides the grating into two gratings producing split in the reflection spectrum and narrowband resonance in the transmission spectrum. The location of the resonance in the wavelength band depends on the amplitude and position of the phase shift in the grating [11]. However, there may be many discrete localized phase shifts in the grating which allows extremely narrow transmission resonance in a grating or tailoring the passive filter shape [11]. The refractive index profile of phase-shifted grating is shown in Figure 8.2e. The phase-shifted FBG has been presented later.

8.3.4 SUPERSTRUCTURE GRATINGS

A superstructure grating is an extension of the principle used in phase-shifted grating. A superstructure grating is formed by introducing gaps periodically in the uniform FBG and thus creating a series of grating and non-grating sections. The non-grating sections introduce phase shifts in the light propagating along the fibre. The superstructure grating as earlier stated is also called a sampled grating because it is formed by sampling a uniform FBG with some functions, e.g. rectangular, Gaussian and Sinc. A number of applications of superstructure gratings have been found in the area of fibre optic communication and as well as fibre optic sensors. The principle of superstructure fibre Bragg grating (SFBG) has been discussed in the next section. Here, a schematic of refractive index profile of SFBG is depicted in Figure 8.2f.

8.4 UNIFORM SAMPLED FIBRE BRAGG GRATING

When a uniform FBG is sampled with a long-periodic-rectangular sampling function, a uniform superstructure or sampled fibre Bragg grating results (U-SFBG). The main Bragg coupling in this case will be coupled to a series of side peaks located at the both sides of Bragg resonance.

In Figure 8.3, we represent the refractive index profile of U-SFBG. Figure 8.3a depicts a rectangular sampling profile and Figure 8.3b shows the refractive index

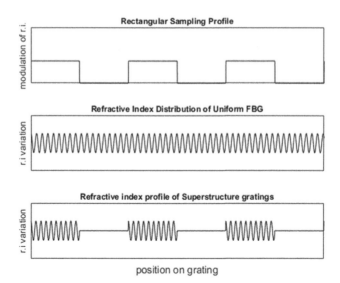

FIGURE 8.3 Illustration of refractive index profile of a SFBG: (a) Rectangular Sampling profile applied on an uniform-FBG; (b) Refractive index profile of an uniform-FBG; (c) Refractive index variation of SFBG.

profile of an uniform FBG. The refractive index profile of SFBG is obtained by multiplying the two r.i. profiles given in Figure 8.3a and 8.3b. Here we will analyse the effect of sampling of a uniform-FBG with the rectangular profile as the rectangular profile is the most common profile used in SFBG.

The rectangular profile given in Figure 8.3.12a can be approximated by a series of pulses. A pulse of width L_g is spatially described by

$$u(z) = \begin{cases} 1 & -\dfrac{L_g}{2} \leq z \leq +\dfrac{L_g}{2} \\ 0 & \text{elsewhere in } z \end{cases} \qquad (8.17)$$

The infinite train of pulses of period L_s is given by

$$p(z) = \sum_{n=-\infty}^{\infty} u(z - nL_s) \qquad (8.18)$$

The spatial variation $p(z)$ can also be written in terms of Fourier series expansion, given by

$$p(z) = \sum_{n=-\infty}^{\infty} P_n e^{jn\omega_0 z} \qquad (8.19)$$

where $\omega_0 = \dfrac{2\pi}{L_s}$ is the angular spatial frequency and P_n is the Fourier series coefficient defined by

$$P_n = \frac{1}{L_s} \int_{-\frac{L_g}{2}}^{\frac{L_g}{2}} u(z) e^{-jn\frac{2\pi}{L_s}z} \, dz$$

$$= \frac{L_g}{L_s} \, sinc\left(\frac{n\pi L_g}{L_s}\right) \tag{8.20}$$

If the index profile of an FBG with grating period Λ_0 is described as

$$x(z) = e^{j\frac{2\pi}{\Lambda_0}z} \tag{8.21}$$

The spatial functions of index variation of a uniform-FBG and the rectangular sampling function with their Fourier spectrum are illustrated in Figure 8.4. Figure 8.4a depicts the index variation of a uniform-FBG, $x(z)$ and Figure 8.4b is the Fourier spectra of $x(z)$, i.e. $X(\omega)$. $X(\omega)$ is nothing but a delta function located at $\omega = \frac{2\pi}{\Lambda_0}$. Figure 8.4c illustrates the rectangular sampling profile $p(z)$ and Figure 8.4d is the schematic of Fourier spectra of $p(z)$, i.e., $P(\omega)$. $X(\omega)$ and $P(\omega)$ are calculated using the Fourier transform and given by

$$X(\omega) = \int_{-\infty}^{\infty} \left\{ e^{j\frac{2\pi}{\Lambda_0}z} \right\} e^{-j\omega z} \, dz = 2\pi\delta\left(\omega - \frac{2\pi}{\Lambda_0}\right) \tag{8.22}$$

$$P(\omega) = \int_{-\infty}^{\infty} \left\{ \sum_{n=-\infty}^{\infty} P_n e^{jn\omega_0 z} \right\} e^{-j\omega z} \, dz = \sum_{n=-\infty}^{\infty} P_n \int_{-\infty}^{\infty} e^{-j(\omega - n\omega_0)z} \, dz$$

[From linearity property of Fourier Transform]

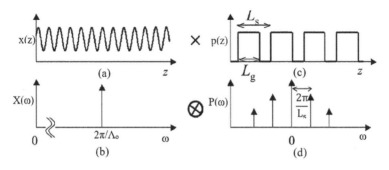

FIGURE 8.4 Principle of sampling technique for constructing a conventional SFBG; (a) R.I. profile of FBG, (b) Fourier spectra of (a); (c) Rectangular sampling profile, (d) Fourier spectra of (c).

$$\text{or}, P(\omega) = \sum_{n=-\infty}^{\infty} \frac{L_g}{L_s} sinc\left(\frac{n\pi L_g}{L_s}\right) 2\pi\delta\left(\omega - n\omega_o\right) \tag{8.23}$$

The index profile of U-SFBG in general is described as

$$w(z) = x(z) \times p(z) \tag{8.24}$$

$$w(z) = e^{j\frac{2\pi}{\Lambda_0}z} \times \sum_{n=-\infty}^{\infty} P_n e^{jn\omega_0 z} \tag{8.25}$$

The Fourier transform of Equation (8.25) becomes

$$W(\omega) = F\left\{e^{j\frac{2\pi}{\Lambda_0}z} \times \sum_{n=-\infty}^{\infty} P_n e^{jn\omega_0 z}\right\}$$

$$= \sum_{n=-\infty}^{\infty} P_n F\left\{e^{j\frac{2\pi}{\Lambda_0}z} \times e^{jn\omega_0 z}\right\} = \sum_{n=-\infty}^{\infty} P_n 2\pi\delta\left(\omega - \frac{2\pi}{\Lambda_0} - n\omega_o\right)$$

$$W(\omega) = \sum_{n=-\infty}^{\infty} \frac{2\pi L_g}{L_s} sinc\left(\frac{n\pi L_g}{L_s}\right)\delta\left(\omega - \frac{2\pi}{\Lambda_0} - n\frac{2\pi}{L_s}\right) \tag{8.26}$$

The Fourier spectra of uniform-SFBG or conventional-SFBG obtained from Equation (8.26) are illustrated in Figure 8.5. It contains the impulses and has envelope of sinc function.

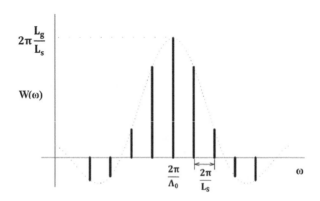

FIGURE 8.5   Spatial Fourier spectra of a conventional SFBG.

The frequency spacing in the spectra can be calculated as

$$\omega = \frac{2\pi}{\Lambda_0} = \frac{2\pi}{\lambda_0 / (2n_{\text{eff}})} = \frac{f \times 2\pi \times 2n_{\text{eff}}}{c}$$

$$\therefore \Delta\omega = \frac{\Delta f \times 2\pi \times 2n_{\text{eff}}}{c} = \frac{2\pi}{L_s}$$

$$\therefore \Delta f = \frac{c}{2n_{\text{eff}}L_s} \tag{8.27}$$

In terms of wavelength,

$$\Delta f = \frac{f_0 \lambda_0}{2n_{\text{eff}}L_s}$$

$$\left|\frac{\Delta f}{f_0}\right| = \left|\frac{\Delta\lambda}{\lambda_0}\right| = \frac{\lambda_0}{2n_{\text{eff}}L_s}$$

$$\therefore \Delta\lambda = \frac{\lambda_0^2}{2n_{\text{eff}}L_s} \tag{8.28}$$

Thus, the frequency or wavelength spacing is inversely proportional to the sampling length.

8.5   SAMPLED-CHIRPED FIBRE BRAGG GRATING

In the previous section, we have seen that using SFBG, a reflection spectrum consisting of a series of nearly identical discrete wavelength channels can be obtained. Such response of SFBG is used to act as periodic comb filter. In these filters, the sampling function and seed grating specification govern the channel spacing and other in-band amplitude or phase characteristics. Uniform-SFBGs have been used in fiusi lasers for achieving multi-wavelength operation [13,14].

We have discussed that, in CFBG, the grating period increases linearly along the fibre length as given in Equation (8.6). If this grating called seed grating is sampled with a rectangular function with period P, then SCFBG is formed. This is demonstrated in Figure 8.6.

Rewriting Equation. (8.6),

$$\Lambda(z) = \Lambda_0 + Cz$$

The refractive index profile of SCFBG can be written as

$$\delta n(z) = p(z)\delta n_0 \cos\left[\frac{2\pi}{\Lambda(z)}z\right]$$

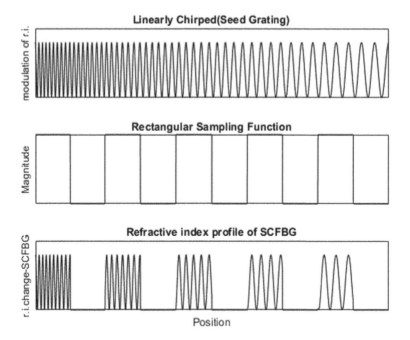

FIGURE 8.6   Illustration of refractive index profile of a SCFBG: (a) Linearly chirped fiber grating acting as seed grating; (b) Rectangular sampling function; (c) r.i. profile of SCFBG obtained by multiplying rectangular sampling function with the seed grating.

$$\delta n(z) = p(z)\delta n_0 \cos\left[\frac{2\pi}{\Lambda_0 + Cz}z\right] \tag{8.29}$$

where δn_0 is the maximum index modulation and $p(z)$ is the rectangular sampling function. Now, the term $\dfrac{1}{\Lambda_0 + Cz}$ can be expanded as

$$\frac{1}{\Lambda_0 + Cz} = \frac{1}{\Lambda_0\left(1 + \dfrac{C}{\Lambda_0}z\right)} = \frac{1}{\Lambda_0}\left[1 - \frac{C}{\Lambda_0}z + \left(\frac{C}{\Lambda_0}z\right)^2 - \left(\frac{C}{\Lambda_0}z\right)^3 + \ldots\ldots\right] \tag{8.30}$$

Hence, Equation (8.29) can be written as

$$\delta n(z) = p(z)\delta n_0 \cos\left[\frac{2\pi}{\Lambda_0}\left(1 - \frac{C}{\Lambda_0}z + \left(\frac{C}{\Lambda_0}z\right)^2 - \left(\frac{C}{\Lambda_0}z\right)^3 + \ldots\ldots\right)z\right]$$

$$\therefore \delta n(z) = p(z)\delta n_0 \cos\left[\frac{2\pi}{\Lambda_0}z + 2\pi\left(-\frac{C}{\Lambda_0^2}z^2\right) + 2\pi\left(\frac{C}{\Lambda_0}z\right)^3 - 2\pi\left(\frac{C}{\Lambda_0}z\right)^4 + \ldots\ldots\right]$$

$$\tag{8.31}$$

In this equation, the terms $2\pi\left(-\dfrac{C}{\Lambda_0^2}\right)$, $2\pi\left(\dfrac{C}{\Lambda_0}\right)^2$, ... are the linear, quadratic and higher order chirp terms. The sampling function can be described in terms of Fourier series expansion as

$$p(z) = \sum_{n=-\infty}^{\infty} P_n e^{jn\frac{2\pi}{P}z} \tag{8.32}$$

where P_n is the Fourier coefficients can be obtained as explained earlier previous section.

The reflection spectrum of SCFBG contains multiple reflection peaks separated by $\Delta\lambda = \dfrac{\lambda_0^2}{2n_{eff}P}$ where λ_0 is the central Bragg wavelength.

8.6   TECHNIQUES FOR OPTICAL COMB SPECTRUM GENERATION

We have discussed the structure and theory of SCFBG. We now discuss the techniques for achieving the comb spectrum using SCFBG. Two basic techniques are used, e.g. (i) MPS and (ii) spectral self-imaging.

8.6.1   MULTIPLE PHASE SHIFT (MPS) TECHNIQUE

The MPS technique can be understood from Equation (8.27), which is the expression for frequency spacing between the individual channels for an SFBG, rewriting it:

$$\Delta f = \frac{c}{2n_{eff}L_s}$$

This equation suggests that, if the sampling period is doubled ($2L_s$), the frequency spacing is halved (i.e. $\Delta f/2$). However, this way of decreasing the channel spacing has the drawback that the total grating length gets doubled. However, instead of increasing the length, an equivalent phase shift can be introduced. This can reduce the length.

When a phase shift between kth and (k+1)th grating section is introduced, given by [15]

$$\Phi_k = \frac{2\pi}{m}k \tag{8.33}$$

the channel spacing becomes

$$\Delta f_{mps} = \frac{\Delta f}{m} = \frac{c}{2mn_{eff}L_s} \tag{8.34}$$

Equation (3.82) implies that the spectrum can be densified by m times. Figure 8.7 shows schematic of the MPS applied on SFBG.

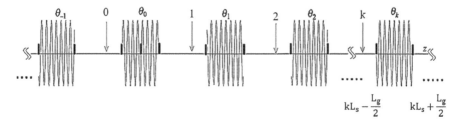

FIGURE 8.7   SFBG with multiple phase shifts.

The term θ_k is the net phase shift at the grating sections.
It can be expressed as follows:

$$\theta_k = \sum_{i=0}^{k} \phi_i \tag{8.35}$$

where ϕ is the phase difference between two sections and given by Equation (8.33).
Thus mathematically, the net phase shift at a particular grating section is given by

$$\theta_k = \frac{\pi}{m} k(k-1) \tag{8.36}$$

The MPS at the grating sections can be inserted using three techniques [16]: (i) using a phase-shifted phase mask, (ii) relative moving of uniform phase mask and (iii) irradiation of uniform UV light as a post-process. The first method requires a complex phase mask and the second one needs critical positioning. The post-process method, however, is the easier one and the amount of phase shift can be monitored from the change of the reflection spectrum.

8.6.2 THE SPECTRAL TALBOT EFFECT

The other way of achieving the densification of spectrum is the self-imaging or Talbot effect. According to the Talbot effect (spatial), if a periodic object is illuminated with a monochromatic light, the exact images of the object are produced at specific distances from the object (integer self-imaging) [17]. At other specific positions, the original periodic pattern reappears but with a spatial period that is an integer fraction of the original period (fractional self-images). The spatial Talbot effect has also been investigated in temporal as well as spectral domain. In the temporal Talbot effect, when the pulses in a periodic pulse train propagate through a first-order dispersive medium, they are reproduced at the same repetition rate or at a multiplied repetition rate, if the following condition is satisfied:

$$|D_v| = \frac{T^2}{m} \tag{8.37}$$

where $|D_\nu|$ is the first-order dispersion coefficient, T is the repetition period of the original pulse train and m is the positive integer. For $m = 1$ the original pulse train is reproduced. This is the so-called integer temporal Talbot effect. For $m > 1$, the output has a repetition rate m times that of the input repetition rate. This is the so-called fractional temporal Talbot effect. The temporal Talbot phenomena is the result of the interference among the identical pulses after dispersion.

The Talbot effect in spectral domain, can similarly be observed if some conditions between chirped grating period and sampling period are satisfied. This will lead to a suitable interference among the individual wavelength channels from the grating sections and result a spectrum consisting of discrete and periodic reflection bands. The spectral Talbot condition in case of linearly SCFBG grating can be achieved from the dispersion relation. The in-band dispersion of a single wavelength channel is given by [18]

$$D_{\nu,grating} = -\left(\frac{\lambda_B}{c}\right)^2 . \frac{1}{C_\Lambda} \tag{8.38}$$

where c is the velocity of light in vacuum and C_Λ is the chirp parameter. However, the spectral dispersion term $D_{\nu,\,grating}$ can also be formulated from its temporal equivalence $(1/D_\nu)$ given in Equation (8.37). Thus, it can be written as

$$D_{\nu,grating} = \frac{m}{\Delta\nu^2} \tag{8.39}$$

where $\Delta\nu$ is the frequency repetition rate (1/T of temporal problem). Comparing Equation (8.38) &(8.39), it can be written as

$$C_\Lambda = -\left(\frac{\lambda_B}{c}\right)^2 . \frac{\Delta\nu^2}{m} \tag{8.40}$$

and from the relation of frequency spacing given in Equation (8.27), we get

$$\Delta\nu = \frac{c}{2n_{eff}P} \tag{8.41}$$

where P is the sampling period of the SCFBG; putting this eqn. over Equation (8.40), we get

$$C_\Lambda = -\left(\frac{\lambda_B}{c}\right)^2 . \left(\frac{c}{2n_{eff}P}\right)^2 . \frac{1}{m}$$

$$= -\left(\frac{\lambda_B}{2n_{eff}P}\right)^2 . \frac{1}{m}$$

$$= -\left(\frac{2n_{\mathrm{eff}}\Lambda_0}{2n_{\mathrm{eff}}P}\right)^2 \cdot \frac{1}{m}$$

$$\therefore |C_\Lambda| = \frac{1}{m}\cdot\left(\frac{\Lambda_0}{P}\right)^2 \tag{8.42}$$

8.6.3 GENERAL CONDITION FOR SPECTRAL SELF-IMAGING

It has been shown that [18] if the phase shift at the grating samples satisfies the following condition:

$$\phi(z) = 2\pi\,\frac{s}{m}\cdot\frac{1}{P^2} \tag{8.43}$$

then both integer and fractional spectral self-imaging can be obtained in SCFBG. The ratio (s/m) in the eqn. is a non-integer and irreducible rational number. The general chirped parameter can be derived from eqns. (3.91) and (3.17). Equating these two equations, we get

$$-2\pi\,\frac{C_\Lambda}{\Lambda_0^2}z = 2\pi\,\frac{s}{m}\cdot\frac{1}{P^2}$$

$$\therefore C_\Lambda = \frac{s}{m}\cdot\frac{\Lambda_0^2}{P^2} \tag{8.44}$$

This is the general condition for spectral Talbot effect.
In summary,

1. In Equation (8.43), if $m = 1$, it represents the integer spectral self-imaging effect. If s is an even integer, the reflection spectrum of the SCFBG will be an exact replica of that of the uniform-SFBG (U-SFBG). It is called direct-integer spectral self-imaging. If s is an odd integer, it will be a frequency-shifted replica of the U-SFBG. It is referred as inverse-integer spectral self-imaging.
2. In Equation (8.43), if $m > 1$, and (s/m) is non-integer and irreducible rational number, the SCFBG will replicate the reflection spectrum of that of the U-SFBG, but the wavelength spacing will be m times narrower than that of the U-SFBG. This is true as long as the reduced wavelength spacing remains larger than the bandwidth of the individual channels in the comb spectrum.

It has been found that, with spectral Talbot effect, in order to obtain clean multi-channel outputs, the duty cycle of each sampling period must be kept small (< 0.2–0.3). In this context, some gratings have been developed, which have full duty cycles [19].

8.7   REFLECTION SPECTRUM OF CFBG, SFBG, SCFBG AND GENERATION OF OPTICAL COMB SPECTRUM

The reflection spectrum of a linearly chirped fibre Bragg grating (LCFBG or normal CFBG) can be calculated using transfer matrix method (TMM)[11]. The different grating parameters are mentioned in Table 8.1.

Figure 8.8a represents the reflection spectrum of the CFBG. Figure 8.8b and 8.8c shows the group delay and the dispersion characteristics of the CFBG. The group delay is calculated from the following relations:

$$\tau_\rho = \frac{d\theta_\rho}{d\omega} = -\frac{\lambda^2}{2\pi C} \cdot \frac{d\theta_\rho}{d\lambda} \tag{8.45}$$

where θ_ρ is the phase angle of the reflection coefficient. The dispersion factor is the rate of change of the group delay with wavelength, given by

$$D = \frac{d\tau_\rho}{d\lambda} \tag{8.46}$$

The Gaussian apodized-CFBG is also shown with the design parameters given in Table 8.2. The parameters assumed are same as those in the case of normal CFBG except additionally FWHM is considered. Figure 8.8d represents the reflection spectrum of the Gaussian apodized CFBG. It can be observed that, due to the apodization, the side-lobes are significantly reduced and the bandwidth is also reduced. Figure 8.8e and 8.8f shows the group delay and the dispersion characteristics of the Gaussian apodized-CFBG. It can be noted that the group delay is linearly decreasing and the dispersion is zero in the central region of the reflection spectrum. This phenomenon is generally used for dispersion compensation in fibre optic communication system.

8.7.1   GENERATION OF OPTICAL COMB SPECTRUM

The U-SFBGs are used to generate comb spectrum, but as discussed, the U-SFBG with MPS technique [15] provides higher channel density in the comb spectrum.

TABLE 8.1
Design Parameters of CFBG

Design Parameters	Values Assumed
λ_D	1550 nm
$\dfrac{d\lambda_D}{dz}$	1 nm/ 1 cm
L	1 cm
$\overline{\delta n_{eff}}$	5×10^{-4}
n_{eff}	1.447

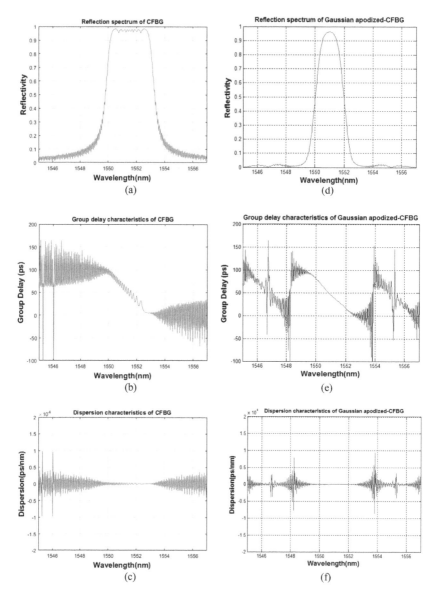

FIGURE 8.8   a: Reflection spectrum of CFBG; b: Group delay of CFBG; c: Dispersion of CFBG; d, e and f are those for Gaussian apodized-CFBG.

Here we first demonstrate optical comb filters based on U-SFBG with and without phase shifts. Further, we discuss the use of SCFBG as comb filter.

8.7.2   U-SFBG BASED OPTICAL COMB FILTER WITH MPS

The parameters for simulation of an 8 cm long U-SFBG are assumed as: n_{eff} = 1.485; the grating length, L_1 = 0.2 mm, non-grating length, L_2 = 1.8 mm; δn = 2 × 10⁻⁴.

TABLE 8.2

Design Parameters: Simulation of Gaussian-Apodized CFBG

Design Parameters	Values Assumed
λ_D	1550 nm
$\dfrac{d\lambda_D}{dz}$	1 nm/ 1 cm
L	1 cm
$\overline{\delta n_{eff}}$	5×10^{-4}
n_{eff}	1.447
FWHM	L/2

The reflection spectrum is depicted in Figure 8.9. It consists of 10 channels in the range of 1548 to 1552 nm. Next, we have applied phase shifts between kth and (k+1) th section as:

$$\Phi_k = \frac{2\pi}{m} k \tag{8.47}$$

With this introduction of MPS, the number of channels is increased by the channel densification factor 'm' as given in Equation (8.34). The reflection spectrum of the same U-SFBG with $m = 2$ is shown in Figure 8.10. It shows that the number channels is doubled, i.e. 20 within the same wavelength range.

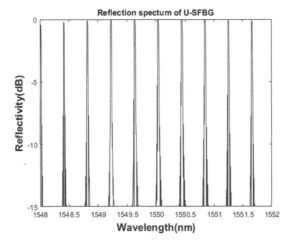

FIGURE 8.9   The reflection spectrum of a U-SFBG without MPS.

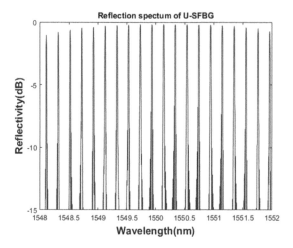

FIGURE 8.10 The reflection spectrum of a U-SFBG with MPS.

8.7.3 SCFBG Based Optical Comb Filter Using Spectral Talbot Effect

The optical comb filter has also been designed using an SCFBG. In [17,18], it has been shown that, at specific conditions between the grating chirp and sampling period, the channel spacing can be reduced compared to the value obtained using conventional SFBG. The channel density has been multiplied without needing to increase the sampling period. In SCFBG, the spectral Talbot effect or the spectral self-imaging discussed earlier section, has been used for channel multiplication. The wavelength spacing has been reduced by changing the chirp-parameter while keeping the sampling period same. According to Equation (8.42), rewriting the general condition for spectral Talbot effect as:

$$C_\Lambda = \frac{s}{m} \cdot \frac{\Lambda_0^2}{P^2} \tag{8.48}$$

C_Λ is the chirp parameter, Λ_0 is the fundamental grating period and P is the sampling period of the SCFBG. The ratio (s/m) is a non-integer and irreducible rational number. By varying the chirp parameter along with the phase shifts at the grating sections, the various channel spacings can be achieved. We use the following relation for introducing the phase shifts in SCFBG between kth and (k+1)th section as:

$$\Phi_k = \pi \frac{s}{m} k \tag{8.49}$$

For the SCFBG, we apply different spectral Talbot conditions as described in earlier section. First, we simulate the SCFBG with same parameters defined in the case mentioned for Figure 8.9 (USFBG-without MPS) except the integers for C_Λ are as s = 1 and m = 1 ($C_\Lambda = 6.809 \times 10^{-8}$). The reflection spectrum obtained is depicted in Figure 8.11. This shows the number of channel as 10. If we increase the value of

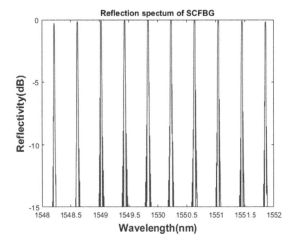

FIGURE 8.11   The reflection spectrum of a SCFBG (s = 1, m = 1).

m to 2 (i.e. $C_\Lambda = 3.405 \times 10^{-8}$), the number of channels will be doubled. It is shown in Figure 8.12.

The group delay characteristics of the reflection bands of SCFBG exhibit similar in-band characteristics as in an equivalent U-SFBG [18]. We have calculated the group delay of SCFBG ($s = 1$, $m = 2$) of the previous case (Figure 8.12) and plotted in Figure 8.13.

8.8   ULTRA-NARROW BAND OPTICAL COMB FILTERS

So far, we have seen the all major techniques for achieving optical comb spectrum. Those techniques have been widely utilized to obtain narrowband filtering. Different

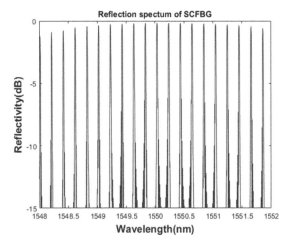

FIGURE 8.12   The reflection spectrum of a SCFBG (s = 1, m = 2).

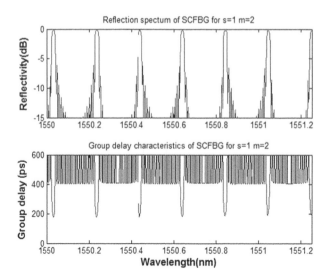

FIGURE 8.13   The reflection spectrum (a) and its group delay (b) of a SCFBG (s = 1, m = 2).

techniques have been reported for design of ultra-narrowband comb filter [14–24]. However, it is beyond our scope to describe each of this technique in detail. Some of those current works are highlighted here. Li and Chen [20] have shown an ultra-narrow band filter using a triply sampled fiber Bragg grating. In that work, good reflectivity has been obtained. Sengupta et al. [21] have proposed a design of optical comb filter using Gaussian-sampled-periodically-chirped fiber Bragg grating (GS-PCFBG) for ultra-narrow band applications. In the work, the periodic chirp effect and the Gaussian sampling have been used for generation of ultra-narrowband comb spectrum. Firstly, the grating period is varied periodically (periodic-chirp effect) and then the grating is sampled with Gaussian pulses along the length of the fibre. To obtain high channel counts, the spectral Talbot effect or spectral self-imaging has been utilized. Apart from the SFBG, the comb filters are also designed using some other techniques. Guo et al. [22] have demonstrated an ultra-narrowband optical comb filter based on the sampled Brillouin grating. This involves the process of stimulated Brillouin scattering. Recently, Mengmeng et al. [23] have shown an optical comb filter with tunable channel spacing based on Mach-Zehnder interferometer (MZI) with tapered fibre in one arm. The transmission spectrum wavelength can be continuously tuned by rotating the polarization controller placed in the other arm of the MZI. The channel spacings from 0.2 to 3.0 nm have been experimentally achieved. Heesuk et al. [24] reported a multi-channel optical frequency synthesizer to generate stable continuous-wave laser from the optical comb of an Er-doped fibre oscillator.

Here we illustrate one efficient technique for obtaining the ultra-narrowband peaks in the comb spectrum using GS-PCFBG filter [21]. The working of ultra-narrowband comb filter has been pictorially demonstrated in Figure 8.14. The effect of Gaussian apodization on a conventional CFBG has been shown in Figure 8.14a and 8.14b. Figure 8.14a is the r.i. profile of a conventional-CFBG and its reflection spectrum. Figure 8.14b is the r.i. profile of the same CFBG but Gaussian apodized and its

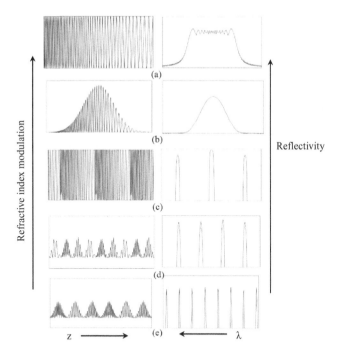

FIGURE 8.14 Illustration of GS-PCFBG: (a) r.i. profile of CFBG and its reflection spectrum; (b) Gaussian apodized-CFBG with its reflection spectrum; (c) Multiple phase shifts based Periodic-chirped r.i. profile with and its reflection spectrum; (d) r.i. profile of Gaussian-sampled-periodically-chirped fiber Bragg grating with its reflection spectrum; (e) the r.i. variation of same GS-PCFBG with increased sampling length and its reflection spectrum showing ultra-narrowband spectrum.

reflection spectrum. The reflection spectrum depicts that the Gaussian apodization reduces the bandwidth significantly. This helps to reduce the bandwidth and also lower the side-lobes. When a grating is periodically Gaussian apodized, the bandwidth of individual channels also gets decreased. Further, low index modulation may be used to provide the ultra-narrowband peaks in the reflection spectrum. In addition, it is seen that, if channel spacing is reduced, the channel-bandwidth is decreased. It is also seen that the in-channel bandwidth can be highly reduced if sampling length is increased. In Figure 8.14c, the r.i. variation of a periodically chirped grating and its reflection spectrum are shown. The reflection spectrum is actually a comb spectrum where MPS technique has been used. Further, in Figure 8.14d, it is shown that the grating is sampled periodically with a Gaussian function and its reflection spectrum. It is observed that the spectrum has the lower channel spacing and narrower channel-bandwidth. The channel bandwidth can be further lowered if the sampling length is increased. It is depicted in Figure 8.14e. The length of the individual grating section is increased and ultra-narrowband peaks have been obtained. The reflection spectrum obtained using GS-PCFBG is shown in Figure 8.15. It provides channel spacing of 13.9 GHz. The reflection spectrum of a single peak of Figure 8.15 is given in Figure 8.16, showing the 3-dB bandwidth of 450 MHz.

A quantitative comparison of the performance of the several reported comb filters has been carried out. It is presented in Table 8.3. The parameters such as channel spacing, channel isolation from adjacent channels and 3-dB bandwidth have been considered for their characteristic evaluation.

8.9   CONCLUSION

The optical comb filters have been used for long time as multichannel devices and have shown impact in the areas of WDM/DWDM system, multi-wavelength lasers and optical signal processing. To have better performance in those specific

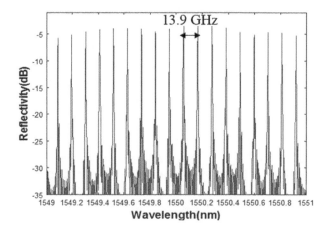

FIGURE 8.15   Reflection Spectrum of GS-PCFBGhavinginter-channel spacing of 13.9 GHz.

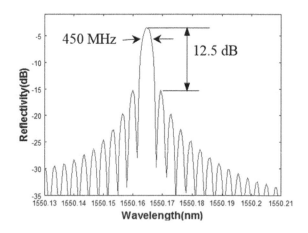

FIGURE 8.16   Single peak reflection spectrum of Figure 15 having the 3-dB bandwidth of 450 MHz.

TABLE 8.3
The Performance-Characteristics of Multichannel Filters

Ref. No.	Multichannel Filter Configurations	Channel Spacing	Channel Isolation	3-dB Bandwidth
[4]	Sinc-sampled fiber Bragg gratings	200 GHz	10 dB	200 GHz
[5]	phase-shift phase-only sampled FBG	–	–	5 GHz
[6]	Dammann fibre Bragg gratings	50 GHz	15 dB	43.75 GHz
[7]	Phased-only sampled FBG	–	–	43.75 GHz
[17]	Spectral fractional Talbot effect based sampled chirped FBGs	100 GHz	10 dB	5 GHz
[19]	Full-duty-cycle sampled fibre Bragg grating	50 GHz	30 dB	5.75 GHz
[20]	Triply sampled Fibre Bragg grating	25 GHz	80 dB	1.5 GHz
[21]	Gaussian-sampled periodically chirped fibre Bragg grating	13.9 GHz–50 GHz	12.5 dB	450 MHz–2.7 GHz
[22]	Sampled Brilouin dynamic gratings	200 MHz–50 GHz	10 dB	12.5 MHz–1.0 GHz
[24]	Er-doped fibre oscillator	100 MHz	50 dB	1.5 Hz

applications, these comb filters are required to use in ultra-narrow band operations. Different techniques for achieving comb spectrum have been described. Detailed theoretical analysis and results have been shown. A recent development in this area is also highlighted. A specific design of optical comb filter using Gaussian-sampled periodically CFBG has been demonstrated schematically. The various properties of fibre grating like chirp-effect, Gaussian apodization, sampling, MPSs, etc., can be well understood from the design. Further, the method of achieving ultra-narrowband comb spectrum has also been explained.

REFERENCES

1. H. P. Li, M. Li, Y. L. Sheng, J. E. Rothenberg, "Advances in the design and fabrication of high-channel-count fibre Bragg gratings," *Journal of Lightwave Technology*, vol. 25 no. 9, pp. 2739–2750, 2007.
2. X. M. Liu, "Tunable ultranarrow dual-channel filter based on sampled FBGs," *Journal of Lightwave Technology*, vol. 26, no. 13, pp. 1885–1890, 2008.
3. A. V. Buryak, K. Y. Kolossovski, D. Y. Stepanov, "Optimization of refractive index sampling for multichannel fibre Bragg gratings", *IEEE Journal of Quantum Electronics*, vol. 39, no. 1, pp. 91–98, 2003.
4. Ibsen, M., Durkin, M. K., Cole, M. J., Laming, R.I., "Sinc-sampled fisam Bragg gratings for identical multiple wavelength operation", *IEEE Photonics Technology Letters*, vol. 10, pp. 842–844 1998.
5. M. Li, H. Li, and Y. Painchaud, "Tunable high channel-count notch filter based on a phase-shift phase-only sampled FBG and its application to multi-wavelength fihig laser," in Proc. OFC 2009, San Diego, CA, 2009.
6. J. E. Rothenberg, H. Li, Y. Li, J. Popelek, Y. Sheng, Y. Wang, R. B. Wilcox, and J. Zweiback, "Damman fiber Bragg gratings and phase-only sampling for high-channel counts," *IEEE Photonics Technology Letters*, vol. 14, pp. 1309–1311, Sept. 2002.
7. Li, H., Sheng, Y., Li, Y., Rothenberg, J.E., "Phased-only sampled fibre Bragg gratings for high channel counts chromatic dispersion compensation," *Journal of Lightwave Technology*, vol. 21, 2074-2083, 2003.
8. T. Udem, R. Holzwarth, T. W. Hansch, "Optical frequency metrology," *Nature*, vol. 416, no. 14, pp. 233-237, 2002.
9. J. P. Yao, "Microwave photonics," *Journal of Lightwave Technology*, vol. 27, no. 3, pp. 314-335, 2009.
10. J. Capmany, B. Ortega, and D. Pastor, "A tutorial on microwave photonic filters," *Journal of Lightwave Technology*, vol. 24, no. 1, pp. 201-229, 2006.
11. T. Erdogan, "Fibre grating spectra", *Journal of Lightwave Technology* vol. 15, no. 8, pp. 1277-1294, 1997.
12. H. Kogelnik, "Filter response of nonuniform almost-periodic structures," *Bell Sys. Tech. J.*, vol. 55, pp. 109-126, 1976.
13. J. Chow, G. Town, B. Eggleton, M. Ibsen, K. Sugden, I. Bennion, "Multiwavelength generation in an erbium-doped fiber laser using in-fiber comb filters," *IEEE Photonics Technology Letters*, vol. 8, pp. 60-62, 1996.
14. M. Li, H. Li, and Y. Painchaud, "Multi-channel notch filter based on a phase-shift phase-only sampled fiber Bragg grating," *Opt. Express*, vol. 16, pp. 19388–19394, 2008.
15. Y Nasu, and S Yamashita, "Densification of sampled fibre Bragg gratings using Multiple-Phase-Shift (MPS) technique," *Journal of Lightwave Technology*, vol. 23, no. 4, pp. 1808-1817, 2005.

16. Y Dai, X Chen, X Xu, C Fan, and S Xie, "High Channel-count comb filter based on chirped sampled fibre Bragg grating and phase shift," *IEEE Photonics Technology Letters*, vol. 17, no. 5, pp. 1040–1042, 2005.

17. C. Wang, J. Azaña, and L. R. Chen, "Efficient technique for increasing the channel density in multiwavelength sampled fiber Bragg grating filters," *IEEE Photonics Technology Letters*, vol. 16, no. 8, pp. 1867-1869, 2004.

18. J. Azaña, C. Wang, and L. R. Chen, "Spectral self-imaging phenomena in sampled Bragg gratings," *J. Opt. Soc. Amer. B*, vol. 22, no. 9, pp. 1829–1841, 2005.

19. L Deng, and F Luo, "Novel optical comb filter based on full-duty-cycle sampled fibre Bragg grating with spectral Talbot effect," *Optics Communication*, vol. 285, pp. 5132-5137, 2012.

20. H. Li, and X. Chen, "High Channel-Count Ultra-Narrow Comb-Filter Based on a Triply Sampled Fibre Bragg Grating," *IEEE Photonics Technology Letters*, vol. 26, pp. 1112-1115, 2014.

21. S. Sengupta, and S. K. Ghorai, "Ultra-narrow band optical comb filter using Gaussian-sampled fiber Bragg grating with periodic chirp effect," *Optical and Quantum Electron*, vol. 48, p. 482 (13pp.), 2016.

22. J-J. Guo, M. Li, Y. Deng, N. Huang, J. Liu, and N. Zhu, "Multichannel optical filters with an ultranarrow bandwidth based on sampled Brillouin dynamic gratings," *Optics Express*, vol. 22, pp. 4290-4300, 2014.

23. Han, Mengmeng, et al. "Tunable and channel spacing precisely controlled comb filters based on the fused taper technology." *Optics Express* 26.1 (2018): 265-272.

24. Jang, Heesuk, et al. "Comb-rooted multi-channel synthesis of ultra-narrow optical frequencies of few Hz linewidth." *Scientific reports* 9.1 (2019): 1-8.

9 A Real-Time and Wireless Structural Health Monitoring Scheme for Aerospace Structures Using Fibre Bragg Grating Principle

Ahsan Aqueeb, Mijia Yang, Benjamin Braaten, Ellie Burczek and Sayan Roy

CONTENTS

9.1 INTRODUCTION

The population of artificial satellites and near-Earth objects in space has massively increased since the last decade. With an advancement of space industries all over the world, accumulations of expired and malfunctioning satellites along with natural space debris are presenting a rising concern for current and prospective aerospace structures. In the past, critical damages were reported multiple times due to impacts between rogue and victim aerospace structures leading to severe threats of catastrophic accidents arising from structural failures [1,2]. In this book chapter, a wireless remote monitoring scheme was presented to detect, identify, and locate such unwanted and sudden impacts in real-time using fibre Bragg grating sensors. The developed scheme was first analytically modelled, then implemented, and finally validated through in-lab measurements. For the sake of the discussion, the overall work was divided into two major sections. The first section presented identifying the location and estimation of any impact on a two-dimensional surface using fibre Bragg grating sensors on-board, while the second section presented a wireless monitoring scheme for transmitting the sensor data in realtime to a remote location. It was envisioned that such critical mission information will be extremely useful in (1) making real-time decisions on flight management information system (for example, continuing flight or returning to the base) and (2) understanding the effects of impacts on aerospace structures. Now, location estimations and characterizations of impact loads (IL) for various types of structures are already well established in current literatures [3]. On the contrary, simultaneous identifications of not only the IL locations but also analysing the impact load time history (ILTH) in realtime for space applications have not been properly discussed yet. In this research, a non-stochastic inverse analysis was used to address the uncertainty in detecting, identifying and locating ILs on a simply supported two-dimensional plate to obtain and simplify ILTH using a sinusoidal function and Whitney and Pagano's solution [3]. To do so, an effective search method was utilized to reduce the objective function along with a novel and simplified least square method in detecting the maximum limits of the ILTH and IL. Finally, a real-time monitoring system was proposed through optical sensor interrogators and wireless transmitters.

To further investigate the efficiency of the proposed algorithm, variable distances between sensors could be tested. Also, a study could be done on the minimum distance between sensors required to ensure the algorithm would work properly. Simultaneous multi-point and distributed impacts could also be taken into consideration; however, such studies were beyond the scope of the current discussion.

9.2 IL AND ILTH ANALYSIS

9.2.1 Existing Techniques

Passive sensing is a well-established method for health monitoring of structural bodies experiencing mechanical stress over the time in civil and aerospace engineering. Analysis of IL and its identification constitute the primary step to correctly attribute the health of a structure in operation by locating potential damage. On the other hand, ILTH provides the information regarding the sequence of values as a time-varying quantity which is measured at a set of fixed intervals. Now, different techniques of inverse analysis exist for IL and ILTH assessment [4,5]. Inverse analysis can be executed using two methods; either by reducing the error between predicted data and measured data acquired using model-based forward solving of a mathematical model or by using a neural network that requires extensive training of any given dataset. The model-based approach is well accepted among researchers due to its convenience of implementation and simplicity of calculation compared to a neural network-based approach. Another advantage of the model-based method is that ILTH and IL can be obtained either separately by transfer matrix or simultaneously by back analysis [4]. Now, IL and ILTH can be determined from local sensor measurements by using different methods of inverse analysis. However, inverse (deconvolution) analysis is challenging to implement as the objective function at the minimum point does not always converge to zero as the process is noisy and measured responses result in instability. This challenge can be eradicated by the regularization of the measured responses. To do this, additional conditions are typically imposed to optimize the objective function by implicitly filtering the noisy measured data. For non-linear structure, Kalman filter and recursive least square estimator are used. In the case of linear structural systems Tikhonov regularization, truncation, and singular value decomposition can be used. In general, one common drawback of such classical regularization methods is the requirement of determining appropriate regularization parameters. Additionally, the presence of noise further degrades the overall performance of these algorithms. Unlike the deterministic methods, the stochastic analysis considers the uncertainties associated with structural properties, like measurement error and elastic modulus, this method also accounts for the noise in the measured data. Inverse analysis and stochastic analysis can work together, where the stochastic analysis will discover an affected area and inverse analysis will then work to detect the area of the impact location and the IL magnitude. The drawback for such statistical distribution as prior information on the data needs to be known. An alternative to stochastic analysis method is interval analysis where to rectify this disadvantage, interval analysis can be introduced instead of stochastic analysis. Interval analysis takes into consideration the minimum and maximum boundaries of the system response. However, it does not provide the insight into the distribution of ILTH and IL. Also, interval analysis alone or combined with the regularization method it is a costly process for real-time detection.

In this research, the IL and ILTH were concurrently obtained, and the uncertainty was addressed by the inverse analysis which is non-stochastic. To minimize the objective function, particle swarm optimization (PSO) technique was adopted as an efficient search method. A novel method was created to detect the maximum limits

of the IL and ILTH which is based on the least square method. The ILTH was simplified by a sinusoidal function.

9.2.2 Three-Layer Identification Process

A three-layer identification process for IL and ILTH analysis is introduced in this
section. It is an effective and simple scheme. These layers can be integrated to effectively detect IL and ILTH in an intelligent structure, which contains a network of at
least three sensors. These layers are designed in such a way that, it will be able to
differentiate IL and ILTH of a structure. Now for this design to work, the network
must have at least three sensors. At first, strain time histories are collected from all
sensors and only three least times of arrivals related to the nearest sensors to the IL
are selected for further analysis. Time of arrival (TOA or ToA) is the travel time of a
(surface) wave from the point of impact to the sensor. IL can then be evaluated by
using the triangulation method [4]. Here, IL is decided by minimizing the error
between measured and the predicted TOAs. The output from the first layer helps in
defining the sampling space. One significant challenge in this approach is finding
TOAs under a noisy response. In addition, measurement of the speed of the surface
wave is a complex process. For an anisotropic plate, that also depends on the wave
frequency and propagation direction.

The next layer of the identification process contains detecting IL and reconstruct-
ing ILTH by reducing the predicted and measured strain time histories. Now the
noisy measured data from the sensor locations were filtered using the moving average method. Those values are later applied to determine IL and ILTH. By subtracting
the filtered strain time history from the noisy one, the error is measured. Finally, the
identification process approaches the third layer based on the maximum observed
error in the measured data and establishes the extreme ILs and ILTHs.

9.2.3 Theories

9.2.3.1 Layer 1: Estimation of Location

At first, TOAs were analysed by differential TOA (Figure 9.1) as described in [6]
because the effect of noise must be reduced:

$$\Delta t_i = t_i - t_1 \tag{9.1}$$

where t_i is the TOA at the ith sensor and t_i is the TOA at the firstsensor location. As
three sensors are employed, i can be of values of two and three (for second and third
sensors, respectively) when the reference was the first sensor and so on. The objective function was then defined as the following:

$$Err = \sqrt{\left(\Delta t_{2m} - \Delta t_{2c}\right)^2 + \left(\Delta t_{3m} - \Delta t_{3c}\right)^2} = \left\{ \left[\Delta t_{2m} - \frac{1}{V} \left(\sqrt{\left(x_2 - \xi\right)^2 + \left(y_2 - \eta\right)^2} \right. \right. \right.$$
$$\left. \left. - \sqrt{\left(x_1 - \xi\right)^2 + \left(y_1 - \eta\right)^2} \right) \right]^2 + \left[\Delta t_{3m} - \frac{1}{V} \left(\sqrt{\left(x_3 - \xi\right)^2 + \left(y_3 - \eta\right)^2} - \sqrt{\left(x_1 - \xi\right)^2 + \left(y_1 - \eta\right)^2} \right) \right]^2 \right\}^{\frac{1}{2}} \tag{9.2}$$

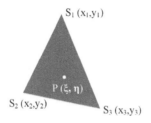

FIGURE 9.1   Time of arrival (TOA) based impact identification..

where

Δt_{im}: measured value of differential TOA,

Δt_{ic}: predicted value of differential TOA,

x_i: x-coordinate of the *i*th sensor,

y_i: y-coordinate of the *i*th sensor,

ξ: x-coordinate of the IL,

η: y-coordinate of the IL,

V: wave speed and

S_1, S_2 and S_3: three sensors located closest to the point of impact (ξ, η).

In Equation (9.2), the plate was considered to be isotropic. For an anisotropic plate, a direction-dependent wave speed must be incorporated. In a plate structure, the bending wave's speed *V* can be calculated as follows:

$$V = \sqrt[4]{\frac{Eh^2\omega^2}{12\left(1-v^2\right)\rho}} \tag{9.3}$$

where

E = elastic modulus,

h = plate's thickness,

ρ = unit of weight,

ω = wave's frequency and

v = Poisson's ratio.

Fast Fourier transform (FFT) on the sensor measurements can help measure wave frequency [4]. After calculating TOAs and wave speed at the sensor locations, the steepest descent method was used to minimize the objective function (Equation (9.2)).

9.2.3.2   Layer 2: Deterministic Identification of Impact Location and Load Characteristics

This layer consists of a plate dynamics model. This model is used for forward solving method. Layer 2 tries to make a prediction of the strain time history. Now, two classical methods can be used here to understand the plate problem: Kirchhoff–Love's solution and Mindlin's solution. Kirchhoff–Love's formulation only considers the in-plane shear. In the case of Mindlin's solution, it also considers the out-of-plane shear. In this research, a different method known as Whitney and Pagano's approach [4] that also considers anisotropy was utilized for a simply supported rectangular plate. Now, to derive an expression for strain along the x-direction (ε_x), Soares [4] as well as Carvalho and Dobyns [4] methodology can be adapted. The relationship between modal amplitudes, the time-dependent function of bending deflection would be used for the expression[4], which is

$$\varepsilon_x = z \frac{\delta \psi_x(x,y,t)}{\delta x} \tag{9.4}$$

where z is the respective coordinate to mid-plane, and Ψ_x is the bending curvature respective to the x-axis.

Next, strain respective to the x-axis can be derived using the following equation [4]:

$$\varepsilon_x = K_c \sum_i \sum_j K_{ij} \sin\left(\frac{i\pi\xi}{a}\right) \sin\left(\frac{j\pi\eta}{b}\right) \sin\left(\frac{i\pi x}{a}\right) \sin\left(\frac{j\pi y}{b}\right) \int_0^t F(\tau) \sin\left[\omega_{ij}(t-\tau)\right] d\tau \tag{9.5}$$

where

i and j = mode numbers,

a = length of plate,

b = width of plate,

η = x-coordinate of IL,

ξ = y-coordinate of IL and

ω_{ij} = modal frequency of plate.

K_c and K_{ij} are next defined as follows:

$$K_c = -\frac{8}{a\rho\pi uv} \tag{9.6}$$

$$K_{ij} = \frac{K_A}{j\omega_{ij}} \sin\left(\frac{i\pi u}{2a}\right) \sin\left(\frac{j\pi v}{2b}\right) \tag{9.7}$$

where u is the length and v is the width of the location (impact area) and K_A is a transverse shear stiffness bending factor. Although small values of u and v were chosen to symbolize pointwise IL, they can be assumed as unknown quantities. Now, $F(\tau)$ that symbolizes the ILTH can then be reduced using a half-cycle sinusoidal function as in (9.8):

$$F(\tau) = \begin{cases} 0, \tau < t_0 \\ F_0 \sin\left[\omega_L(\tau - t_0)\right], t_0 < \tau < t_1 \\ F_0 \sin\left[\omega_U(\tau - t_3)\right], t_1 \leq \tau < t_2 \\ 0, t_2 \leq \tau \end{cases} \tag{9.8}$$

where

F_0 = IL amplitude,

ω_U = unloading frequency,

ω_L = loading frequency,

t_0 = impact start time,

t_1 = time of impact and

t_2 = impact end time.

It should be noted here that if the measurement starts while impact occurs, t_0 is then considered to be zero. In practice, this is different since there is always a time lag as the strain recording is initiated ahead of the impact occurrence for measurement. Hence, whenever the first signal was received, the time was then set to zero. The time lag was equalled to TOA of the closest sensor. Then, t_1, t_2, and t_3 can be defined as

$$t_1 = t_0 + \frac{1}{4\omega_L} \tag{9.9a}$$

$$t_2 = t_0 + \frac{1}{4\omega_L} + \frac{1}{4\omega_U} \tag{9.9b}$$

$$t_3 = t_2 - \frac{1}{2\omega_U} \tag{9.9c}$$

Equation (9.5) can be solved analytically by replacing Equation (9.8) as an integral part of the equation, which leads to

$$\int_0^t F(\tau)\sin\left[\omega_{ij}(t-\tau)\right]d\tau = \cos(\omega t_0)\sin(\omega_{ij}t)Z_1(\tau) - \cos(\omega t_0)\cos(\omega_{ij}t)Z_2(\tau)$$
$$-\sin(\omega t_0)\sin(\omega_{ij}t)Z_3(\tau) + \sin(\omega t_0)\cos(\omega_{ij}t)Z_4(\tau) \tag{9.10}$$

where

$$Z_1(\tau) = -\frac{\cos\left[(\omega-\omega_{ij})\tau\right]}{2(\omega-\omega_{ij})} - \frac{\cos\left[(\omega+\omega_{ij})\tau\right]}{2(\omega+\omega_{ij})} \tag{9.11}$$

$$Z_2(\tau) = \frac{\sin\left[(\omega-\omega_{ij})\tau\right]}{2(\omega-\omega_{ij})} - \frac{\sin\left[(\omega+\omega_{ij})\tau\right]}{2(\omega+\omega_{ij})} \tag{9.12}$$

$$Z_3(\tau) = \frac{\sin\left[(\omega-\omega_{ij})\tau\right]}{2(\omega-\omega_{ij})} + \frac{\sin\left[(\omega+\omega_{ij})\tau\right]}{2(\omega+\omega_{ij})} \tag{9.13}$$

$$Z_4(\tau) = \frac{\cos\left[(\omega-\omega_{ij})\tau\right]}{2(\omega-\omega_{ij})} - \frac{\cos\left[(\omega+\omega_{ij})\tau\right]}{2(\omega+\omega_{ij})} \tag{9.14}$$

the objective function can then be expressed as follows:

$$Err = \sqrt{\sum_{i=1}^{3}\sum_{j=1}^{N}\left[\varepsilon_{ic}\left(t_j\right)-\varepsilon_{im}\left(t_j\right)\right]^2} \qquad (9.15)$$

where $\varepsilon_{ic}(t_j)$ is the predicted strain and $\varepsilon_{im}(t_j)$ is the measured strains at the ith sensor, and the number of time steps is N. The above function is expressed using the forward solving model. The variables of the objective function are a 6-dimensional space in terms of X, expressed as

$$X = X\left(\xi,\eta,F_0,t_0,\omega_L,\omega_U\right) \qquad (9.16)$$

The layer 1 output was used to shorten the range of ξ and. The objective function was computed iteratively to converge the sample vectors and in computing the destination vector. The following equation represents how the particle speed and position updated after each iteration:

$$\vec{V}_{k+1} = \vec{A}\otimes\vec{V}_k + \vec{B}_1\otimes r_1\otimes\left(\vec{P}_1-\vec{X}_k\right)+\vec{B}_2\otimes r_2\otimes\left(\vec{P}_2-\vec{X}_k\right) \qquad (9.17)$$

$$\vec{X}_{k+1} = \vec{C}\otimes\vec{X}_k + \vec{D}\otimes\vec{V}_{k+1} \qquad (9.18)$$

where
   X_k = particle position after kth iteration,
   X_{k+1} = particle position after the $(k+1)^{th}$ iteration,
   V_k = the particle's speed after the kth iteration,
   V_{k+1} = the particle's speed after the $(k+1)^{th}$ iteration,
   P_1 = the particle's past best position,
   P_2 = the finest position experienced by other particles in the herd,
   B_1, B_2 = two factors relates to the personal influence and social influence of
      particles,
   A = momentum factor,
   r_1 and r_2 = stochastic variables
   C, D = the weight factors for updating the particle's position, considered as
      equal to one, and
   $\otimes$ represents the element by element multiplication of the two matrices.

Each term of Equation (9.17) represents a different aspect of the computation like the first term points out the momentum of motion of the particles in the search space. The second term indicates the speed of the particles based on their personal experience, and the third term updates, relying on the values of A, B_1 and B_2. The long-term trend of the particles can show convergent (to the optimal position), harmonic oscillatory, and zigzagging behaviour [4]. To make the algorithm more efficient, these parameters can be tuned.

9.2.4 VERIFICATION

At Layers 1 and 2, finite-element modelling (FEM) was adopted to verify the inverse model and forward solving mathematical model. Thus, a comparison was being done between the actual IL and ILTH with detected IL and reconstructed ILTH.

9.2.4.1 Verifying the Forward Solving Model

To verify the proposed mathematical model, finite-element model (FEM) was used for a virtual experiment using ABAQUS. An aluminium plate of size $88 \times 88 \times 0.25$ cm³ was modelled. A sinusoidal IL was applied. This aluminium plate was given an IL at the centre at time zero. The IL had an amplitude of 1000N, with an unloading frequency of 300 Hz and a loading frequency of 100 Hz. The simulation was done using a quadrilateral shell element with reduced integration for small strains (S4RS). The FEM consisted of 17,161 nodes and 16900 shell elements. Only 1000μS vibration time was used, and vibrating duration was divided into 100-time intervals. It is worth to mention that due to 1000μS vibration time, a significant damping effect may not be witnessed.

Figure 9.2 shows a comparison between the suggested mathematical model and FEM result at the impact location. The results obtained from both methods showcased a suitable match.

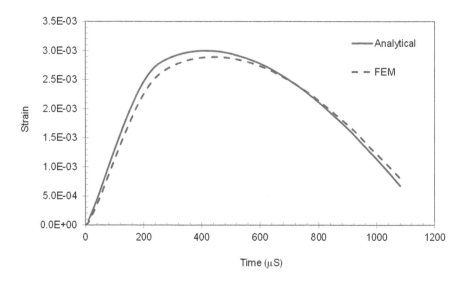

FIGURE 9.2 Comparison of the mathematical model and FEM's strain time histories is shown at impact location at the center of the plate [3].

TABLE 9.1

Plate Structure's Mechanical and Geometrical Properties [3]

Length of plate	88 cm
Width of plate	88 cm
Thickness of plate	0.25 cm
Elastic modulus	69GPa
Poisson's ratio	0.34
Unit weight	2700kg/m³
Stiffness properties	$D_{11} = D_{22} = 101.59$Pa; $D_{12} = 34.54$Pa; $D_{66} = 33.52$Pa; $A_{22} = 195$ MPa; $A_{44} = A_{55} = 64.4$MPa

9.2.4.1.1 Layer 1 Verifying Process

Table 9.1 shows the exact plate with the properties. The exact IL that was used for IL and ILTH identification through Layers 1 and 2 was then used upon the plate. For the strain time histories and impact location, three different points were selected at nine different areas that were collected for IL and ILTH identification (Figure 9.3), and in Table 9.2, coordinates of impact and sensor locations were reported.

At the first layer, TOA needs to be determined at different sensor locations. The threshold method was used at each sensor location. It was expected that the level of threshold strain would be two times the maximum amount of noise. To find the dominant frequency content, FFT analysis was applied to the time histories from the three closest sensor locations. To calculate the wave speed, corresponding parameters such as the frequency of plate properties were provided to Equation (9.3). After

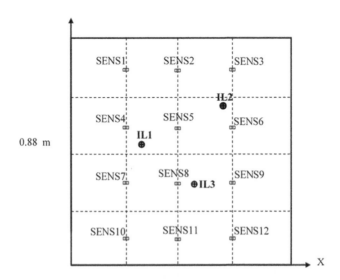

FIGURE 9.3   Impact locations and Sensor configuration set for identification (represents sensor location and impact location. Please refer to Table 9.3 for coordinate information) [3].

TABLE 9.2

Impact Locations and Sensor Locations of X- and Y-Coordinates Shown in Figure 9.2 [3]

Points	X (cm)	Y (cm)
IL 1	25	50
IL 2	64	62
IL 3	53	30
SENS 1	22	77
SENS 2	44	77
SENS 3	66	77
SENS 4	22	55
SENS 5	44	55
SENS 6	66	55
SENS 7	22	33
SENS 8	44	33
SENS 9	66	33
SENS 10	22	11
SENS 11	44	11
SENS 12	66	11

having the parameters, Equation (9.2) was reduced to evaluate the IL. Table 9.2 shows the results obtained of IL estimation for the three ILs. As an example, from the TOA, it can be inferred that SENS 4, SENS 5 and SENS 7 were the nearest sensors to IL 1 (Figure 9.3), and X- and Y-coordinates of IL were estimated by 20% and 9% error (origin was at the bottom left corner of the plate), respectively. In all those cases, it was observed that there was a relatively slight anomaly in identifying the exact impact location. This information helped to adjust the analysis of the second layer.

9.2.4.1.2 Layer 2 Verification

In this step, IL was refined and ILTH was calculated. From the previous layer, analysis of X- and Y-coordinates of impact locations were obtained and they were utilized to define the sampling space. As the number of particles increases, computation costs also increase because of the large number of particles needs a lot of calculation of the objective function, so it helped to maintain a minimum number of particles with lower cost. Surrounding sensors formed the boundary and helped to refine the sampling space. As can be seen, when the TOA in Table 9.3 show that IL 1 was contained within a boundary formed by SENS 4, SENS 5, SENS 7, and SENS 8, it implied that the upper bound for ξ was 0.40. After estimating $\xi = 0.31$ m, the closest vertical gridline had X = 0.22 m. Hence, $\xi = 0.22$ was considered as the lower bound of the sampling space. Hence, $\xi = 0.31$ was the mean of $\xi = 0.22 - 0.40$. Analogous with previous, $\eta = 0.37$ and 0.55 were used as the sampling space.

TABLE 9.3

Layer 1 IL Detection Results Summary [3]

TOA (mS)

	Impact locations (IL)		
Sensor	**IL 1**	**IL 2**	**IL 3**
SENS 1	0.34	0.48	0.59
SENS 2	0.40	0.31	0.51
SENS 3	0.53	0.29	0.49
SENS 4	0.13	0.48	0.40
SENS 5	0.25	0.29	0.31
SENS 6	0.42	0.30	0.34
SENS 7	0.25	0.57	0.34
SENS 8	0.33	0.44	0.13
SENS 9	0.48	0.59	0.25
SENS 10	0.60	0.67	0.41
SENS 11	0.45	0.60	0.41
SENS 12	0.58	0.54	0.29
Dominant wave frequency			
Dominant frequency* (Hz)	2600	3900	4160
Impact location estimation			

Impact locations coordinates (ILC)	IL 1		IL 2		IL 3	
	Predicated	Target	Predicated	Target	Predicated	Target
zeta (cm)	31	25	56	64	52	53
eta (cm)	46	50	65	62	25	30

*For Table 9.3, note that the dominant frequency is reported as the average of the frequencies at each sensor location due to impact at the indicated point [3].

After parameter tuning, *A, B1,* and *B2* matrices were as follows:

$$A = \begin{bmatrix} 0.7 \\ 0.7 \\ 0.2 \\ 0.9 \\ 0.1 \\ 0.2 \end{bmatrix} \text{ and } B_1 = B_2 = \begin{bmatrix} 0.5 \\ 0.5 \\ 0.1 \\ 0.5 \\ 0.1 \\ 0.1 \end{bmatrix} \tag{9.19}$$

Figure 9.4 illustrates the fitness convergence of the three analyses corresponding to the three impact locations. For each particle at each iteration fitness is calculated, fitness was determined. Convergence occurred for the analyses of IL 1 and IL 3 at 70 iterations and for IL 2 at about 80 iterations.

The identified impact locations for IL 1, IL 2, and IL 3 and the corresponding reconstructed load time histories are, respectively, demonstrated in Figure 9.5. First,

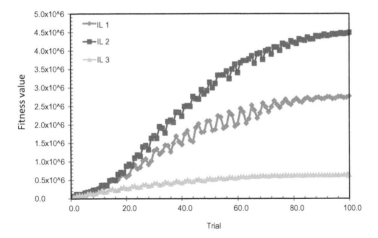

FIGURE 9.4 At three impact locations, fitness value vs. number of iterations for particle swarm optimization (PSO) is shown here (please refer to Figure 9.2 for impact locations) [3].

an approximated location with an error of 1.9% – 24% was obtained through the triangulation method (Table 9.3), and then in the second layer of analysis, it was adjusted with an error ranging between 0.02% and 2.7%. ILTH showed an error for the amplitude of about 5.3%, 3.1% and 1.7%, for unloading and loading frequencies, respectively, and 3.4% for t_0 at the IL2 impact location, so it matched well with the actual time history.

9.2.5 SUMMARY

In this discussion, a numerical scheme based on the inverse analysis technique for a simply supported plate is proposed for IL and ILTH identification. The model proposed here offers an economic and computationally efficient approach. It includes a simplified interval analysis and heuristic algorithm for optimization so that noise effect can be considered. Even though the suggested approach is based on a mathematical forward solving model for a simply supported plate, it can also be used for other types of structures, but a suitable mathematical model might have to be created for such structures.

For verification, a $88 \times 88 \times 0.25$ cm³ aluminium plate was used to implement the numerical scheme after being subjected to an IL, which also established the accuracy and efficiency of the proposed IL and ILTH identification model.

9.3 A REAL-TIME WIRELESS REMOTE MONITORING SCHEME

9.3.1 REAL-TIME MONITORING OF STRAIN

A reliable communication system with remote locations is essential for the assurance of the structural health monitoring (SHM) of aerospace structures. This is especially true in monitoring real-time impact measurements caused by debris colliding with aerospace structures. A simple but a robust wireless monitoring system is thus

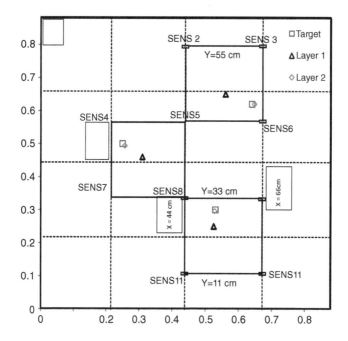

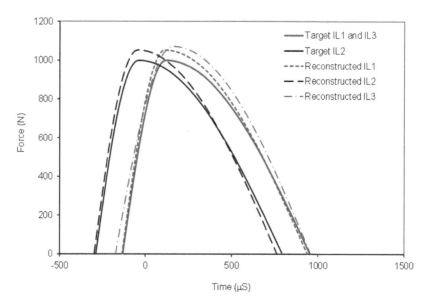

FIGURE 9.5   (a) from layer 1Deterministic IL identification results obtained and particle swarm optimization (PSO) methods (Layer 2), (b) at different ILs comparison of the target and reconstructed impact load time history (ILTH) [3].

required to estimate the residual strength and appraise the signs for crack propagation on the surface. Due to the remote location of the user, wireless communication is naturally chosen for this project. As per the project regulation, only the Federal Communication Commission (FCC) approved radios, signal processing, and electronic modules have been considered [7].

9.3.2 METHODOLOGY

Any non-destructive damage assessment (NDDA) monitoring system combined with wireless communication usually includes four major steps:

(i) Choice of sensor and deployment
(ii) Wireless transceiver selection and implementation
(iii) Signal acquisition and processing on local site
(iv) Authentication of acquired signal and identification analysis on remote sites.

Details of these steps will be discussed as functional blocks later in Section 3.3. Considering the above steps, a real-time monitoring approach is proposed in detailed steps below. A schematic block diagram is provided in Figure 9.6.

Step 1: On aerospace glass in space, real-time measurement of strain needs a duplex communication arrangement between the remote location and local (aerospace). At first, transparent fibre optics sensors will be set on the aerospace glass. In the event of an impact, the embedded fibre optic sensors will sense the strain using a fibre Bragg grating technique using an optical sensor interrogator.

Step 2: Information of magnitude and time of the impact with optical sensor interrogator's output will be encoded and then sent using a Transmitter (wireless transceiver) to the remote location to monitor.

Steps 3 and 4: Receiver (identical wireless transceiver) set will be used in the far end (remote location) to acquire the transmitted signal.

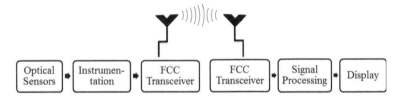

FIGURE 9.6 Proposed Block diagram of structural health monitoring scheme with functional units [3].

Step 5: A signal processing unit will not only decode the received signal but also apply logical processing and arithmetic to extract information.

Step 6: For understanding the received signal with necessary information would be displayed via a Graphical User Interface (GUI).

9.3.3 Functional Blocks

9.3.3.1 Fibre Bragg Grating

Fibre Bragg gratings (FBG) using fibre optical sensors [3] for NDDA is a well-received approach among researchers such as reported in [4]. FBG is a recurring disruption of the refractive index which is created by exposure of the core to an intense optical interference pattern along the fibre length [8]. It is a small length of optical fibre that consists of a pattern of many reflection points, that forges a reflection of wavelengths of incident light [9,10]. These reflection points have equal distance between them. If a wavelength matches with a distance between two reflection points, then it is reflected by the grating. The interrogator determines the individual reflection peak's wavelength. Thus, as strain is applied to an FBG, the distance between the reflection point changes, as a result, a different wavelength is reflected, thus the determination of Bragg wavelength variation is done.

9.3.3.1.1 Sensor Selection and Deployment

A fibre optic sensor system has numerous advantages over conventional electrical or mechanical strain sensor systems especially for space applications and hence fibre optic sensor system was preferred by the authors in this project.These systems exhibit that insusceptibility to any electromagnetic interference is lightweight, and can easily be incorporated onto the glass. As fibre optics are hardly opaque, these sensors perfectly suit the skin of the aerospace including such areas covered with glasses without any compromising of visibility. In this work, for testing and verification of the proposed monitoring technique, a set of four sensors were used.

9.3.3.1.2 Wireless Transceiver Selection and Deployment

There are numerous self-governing FCC-approved wireless transceiver components available to fulfil the needs of this project. These wireless transceiver modules need to be operated in real-time with a desired data transmission rate of 10 samples/sec as per the user-defined specification. This step is described in two parts in detail for better understanding:

Task 9.1: While considering necessary features such as simpler design, low profile, lightweight, integration ease, reliable performance, inexpensive, and compatibility with in-lab experiment - the choice was made to use the IEEE 802.15.4 XBee RF Module (FCC ID: OUR-XBEE, IC: 4214A-XBEE, Series: 0013A200) manufactured by Digi International [3]. An additional unit of SparkFun Electronics [3] Standard RS232 to Serial Base Explorer was used for each XBee RF module, An embedded trace antenna able to perform duplex communication which also has one mW of transmitted output power and an indoor range of 100 feet (30 meters). The chosen XBee module operates in the ISM 2.4 GHz frequency band with a supply voltage of 3.3V DC and an input current of 50 mA. While integrated to the RS232 to Serial Base Explorer module, the Xbee wireless transceiver can be programmed to get the desired operation. An image of the Xbee wireless transceiver (shown in colour blue) integrated with the RS232 to Serial Base Explorer (shown in colour red) is given in Figure 9.8. An identical pair of these modules were used for both transmission and reception purposes in local and remote locations.

Task 9.2: Now, both the hardware and software deployment of the RF transceiver requires a careful selection of optimum settings parameters. These have been described in detail in the following two sections:

• Hardware deployment

The hardware deployment of the RF module can be further divided into three subsections:

■ CXBee RF Module (Series: 0013A200, Manufacturer: Digi International Inc.)

FIGURE 9.7   Wireless transceiver module with necessary attachments.

FIGURE 9.8   NI PXIe-4844 OSI with FBG optical sensors.

- The XBee RF module can connect to any serial device using a level translator while connected with an interface to a host device through a logic-level asynchronous serial port. This 20-pin user-ready module is integrated with an on-chip antenna and configured to be used within a peer-to-peer network. Hence no additional Master/Slave topology is required. Only additional CMOS logic to transfer the wireless signal into serial data is therefore required.
- XBee Explorer Serial (Series: WRL – 09111, Manufacturer: SparkFun Electronics)
  The XBee Explorer Serial is integrated with CMOS logic that allows the user to communicate with the XBee RF module in an asynchronous serial fashion through an RS232 interface. This module is compatible with Data Terminal Ready (DTR) communication and thus can be reprogrammed and configured using Digi International X-CTU software, available on the manufacturer's website.
- Generic Serial to USB Adapter (Series: DA-70119, Manufacturer: Assmann WSW Components)
- generic serial to USB adapter is required for each integrated pair of the above two modules for communication and accessing purposes through a computer system. It is expected that the user has been equipped with Prolific PL2303 USB to Serial/UART [11] device driver in an environment of Windows XP or higher version.
- Software deployment

The X-CTU software from Digi International Inc. allows users to configure the integrated RF module to achieve the user-defined performance. The COM port properties and firmware details of both the transceivers located in the local and remote end are mentioned next. COM port properties (Driver: Prolific USB-to-Serial Communication Port): Bits per second: 9600; Data bits: 8; Parity: None; Stop bits: 1; Flow control: None; Firmware Information: Product Family: XB24; Function Set: XBEE 802.15.4; Firmware Version: 10ec.

9.3.3.1.3   Signal Acquisition and Processing on Local Site

The proposed wireless transceiver system is expected to receive the residual strain data from deployed optical sensors and then transmit the data to the remote location. The task of signal processing and identification can therefore be divided into three subtasks:

- Signal Acquisition:
  The deployed FBG optical sensors manufactured by Optilab, LLC [3] sense the changes enforced by strain on the test platform and shift the associated grating reflections proportional to the applied strain. The NI PXIe-4844 Optical Sensor Interrogator (OSI) for FBG by National Instruments [3] secures change in grating reflections while connected to the FBG optical sensors. It can adapt for processing purposes and display, as shown in Figure 9.9 In this work, 10 samples/sec of data rate and a nominal optical wavelength of 1539.1 nm have been picked for the purpose of signal collection.

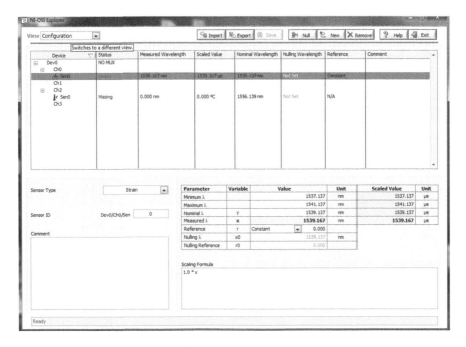

FIGURE 9.9 A screenshot of NI-OSI Explorer.

- Signal Processing:

  For signal processing, these following tools were used:

  (1) National Instruments (NI) - Optical Sensor Interrogator (OSI) Explorer to identify and configure sensors connected to the hardware module and

  (2) NI LabVIEW 2010 for signal processing of the received sensor data to wirelessly transmit through RF module.

  Next, the details of each tool are described.

- Receiving optical data from sensors

  The optical signal from sensors was received by NI PXIe-4844 OSI, as shown in Figure 9.8. The NI-OSI Explorer interface, loaded in the OSI system, transfers the received optical signal into an equivalent digital signal for data processing purposes. Figure 9.9 is a screenshot of NI-OSI Explorer, while the optical sensors were connected and working as a strain sensor with a measured nominal optical wavelength of 1539.167 nm on Channel 0.

- Processing the received data for wireless transmission

  Next, the digital signal from NI-OSI Explorer was fed to NI LabVIEW 2010 for processing purposes. Figure 9.10 shows the block diagram of the designed NI LabVIEW interface to acquire the digital data. Once acquired, the data can be viewed in realtime on a graph window in the NI LabVIEW front panel, as shown in Figure 9.11.

In the following stage, the received data were mathematically processed in realtime to excerpt the sensor data from the raw sensor input. This was done by designing a

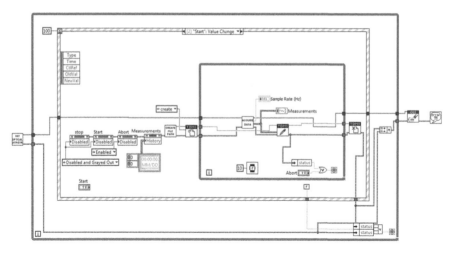

FIGURE 9.10   NI LabVIEW block diagram to acquire the FBG sensor data.

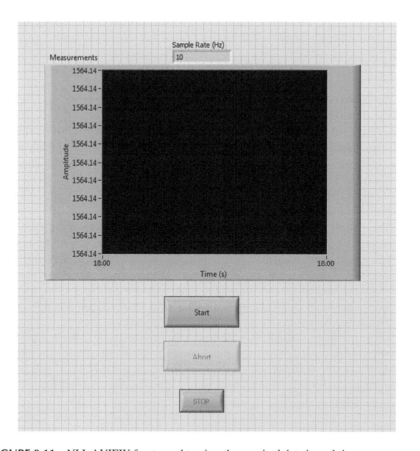

FIGURE 9.11   NI LabVIEW front panel to view the acquired data in real time.

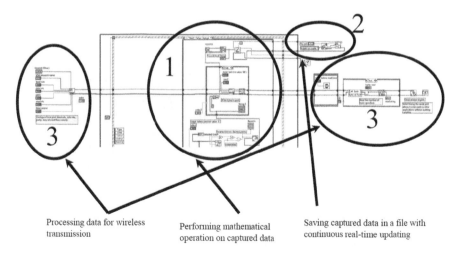

Processing data for wireless transmission

Performing mathematical operation on captured data

Saving captured data in a file with continuous real-time updating

FIGURE 9.12 NI LabVIEW block diagram for signal processing task.

functional block diagram in NI LabVIEW 2010, shown in Figure 9.12. The raw data from the block diagram, shown in Figure 9.10, were first captured in a buffer memory of the system, and then mathematical operations were done on the captured data for the extraction with a data rate of 10 samples/sec. While the process is going, the value wavelength of the FBG sensor is recorded at each 100 msec and then a comparison is done with the nominal wavelength of the sensor to understand the relative change of stress that happened due to the applied strain at any particular moment. The encircled block numbered as '1' in Figure 9.12 was designed to obtain the described operation. The real-time data with timestamp were also updated in a CSV (comma-separated values) formatted file. The encircled block numbered as '2' in Figure 9.12 was designed to perform this operation. After successful processing, the obtained data need to be wirelessly transmitted by the connected RF module through serial communication. For that, NI-Virtual Instrument Software Architecture (NI-VISA) was used here. NI-VISA is an industrial standard programming instrumentation system used to communicate with any generic serial/USB interface. The encircled blocks numbered as '3' in Figure 9.12 perform this operation on the processed data. After successful processing, the serial data were fed to the RF module for transmitting to the remote side. Additionally, the locations of sensors in a two-dimensional format on the test platform were also transmitted to the remote side.

The resulting Ni LabVIEW front panel is shown in Figure 9.13. At first, the user must set the NI-VISA resource values, shown on the left side of the panel, according to the COM port properties of the connected RF module to avoid any compatibility error. While initiated by the user, it asks to create a file in the 'file path' window, on the top left corner in Figure 9.13, to update the log record of input data in runtime. If the user pauses or stops the program for a moment, the system always remembers its last position and continues to update the data-log upon resuming. The 'Measurements' window on the top right side shows all the sensor's data with a current timestamp.

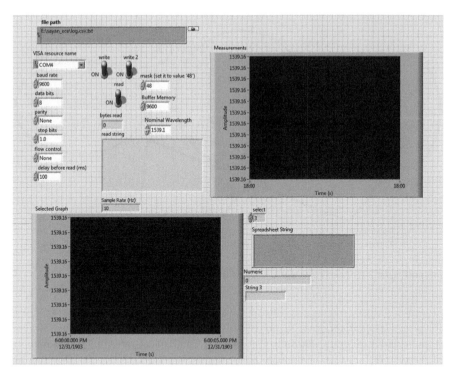

FIGURE 9.13   NI LabVIEW front panel for signal processing task.

The controllers located between the 'VISA resource' and the 'Measurements' window are used to enable/disable 'write' and 'read' operations of the serial communication and 'write' operation of the data record operation. Also, an option of adjusting the required buffer memory size is provided for achieving a higher speed of communication, if required in the future. Buffer memory of 9600 bytes is set for current operation with a data rate of 10 samples/sec. A mask value of 48 must be provided by the user before initiating the program to ensure that the buffer memory can be used during both reading and write operations. The 'Read-String' window, located below the controllers, is activated only when there is an alert message from the remote side and displays the alert. The 'Spreadsheet String' window below the 'Measurements' window displays the real-time numeric value of the obtained sensor data. While working with multiple sensors, the 'Selected Graph' window enables the feature of distinctly displaying each sensor's output when the associated input channel is properly selected through the 'Select' controller located just to the right side of the 'Selected Graph' window.

9.3.3.1.4   Identification Analysis on Remote Site and Received Signal Verification

The transmitted sensor data were received by the RF transceiver module located on the remote site and fed to the MathWorks MATLAB R2012a [3] for information retrieving purposes. It was expected that the remote system was equipped with

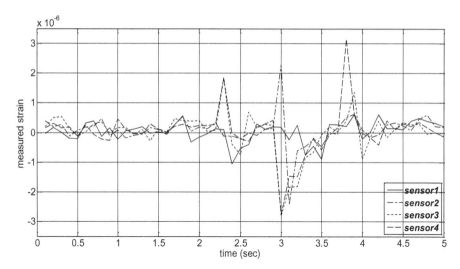

FIGURE 9.14 Normalized FBG strain sensor output in a five second window, displayed at the remote location.

Prolific PL2303 USB to Serial/UART [11] device driver software in an environment of Windows XP or higher version. After retrieval of the received data through serial communication from the RF module, the mathematical computation was performed to display the measured strain for each sensor distinctly. As shown in Figure 9.14, the vertical axis shows the magnitude of measured strain and the horizontal axis shows the time in seconds which was being shown in a window of five (5) seconds to the remote user, and the display was designed to refresh regularly to update with the real-time input. 50 successive data points were displayed to the remote user. During the initial test setup, at four preselected locations, four (4) FBG optical sensors were planted on the test platform.

Experimentally, an industrial standard impact hammer was used to impact multiple events at different locations of the test platform. Under standard circumstances, a change in the value should be experienced by the measurement of residual strain when impact occurs, otherwise, it should stay at zero. However ideal condition is rare and as a result, the noise was witnessed during the rest condition of the test platform. For the first couple of seconds, it can be seen near the normalized zero lines on the vertical axis in Figure 9.14. A meticulous inspection of Figure 9.14 shows the existence of three peaks, between the open interval of two (2) to four (4) seconds. These values are separable from the normalized zero lines by a value of ±1.5E–6 in the vertical axis. The event of impact during the experiment can be represented by each of those cumulative peak responses of sensors. It can also be noticed that each sensor was not generating a similar kind of output with respect to the outputs of other sensors. The dissimilar outputs are expected, as each sensor registered a different amount of residual strain because of varied distances between impact locations and sensor locations. An average impact force calculation from the outputs of all sensors' data was also performed and displayed at the remote location, as shown in Figure 9.15.

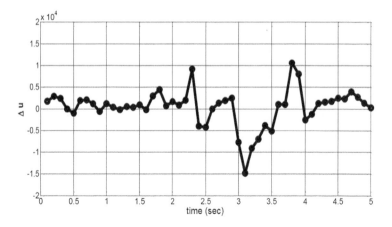

FIGURE 9.15 Average magnitude of impact force on the test platform with time, displayed at the remote location.

9.3.4 Summary

Here, a wireless communication system that enables us to retrieve, process, and transmit data from an array of FBG optical sensors is discussed. Additionally, a remote system able to receive, process, and display the transmitted intelligence in realtime is also presented. Only FCC approved instruments were used as per the requirement of the project. Industrially available FBG interrogator system and wireless radio frequency modules working at 2.45 GHz band were used. It has been expected that the local and remote site will be integrated with NI LabVIEW 2010 or higher version and MathWorks MATLAB 2012A or higher version, respectively. The entire system is shown in Figure 9.16.

9.4 CONCLUSION

In this book chapter, a system has been proposed where a location estimation technique with a wireless remote monitoring system is used for non-destructive damage assessment (NDDA) of impact response on aerospace structures in realtime. To identify the coordinates of an applied impact, the estimation theory accounts for the sensor information and physical properties of the surface. The proposed model presents an inexpensive method that is also computationally efficient and also includes an interval analysis approach. This analysis is required to address for the noise in the system response and a heuristic algorithm for optimization. The numerical scheme indicated efficiency and accuracy of the proposed method. Any deviation found because of the noise was successfully foreseen with the least square regression and linear approximation. In the second phase, a wireless remote monitoring system that enables us to retrieve, process, and transmit information from an array of Fibre Bragg Grating optical sensors is reported along with a remote system capable of receiving, processing, and display the transmitted data in realtime. Here, only FCC approved modules were taken into consideration. The overall system has been assessed on a

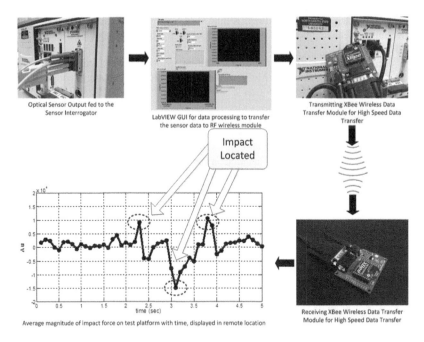

Optical Sensor Output fed to the
Sensor Interrogator

LabVIEW GUI for data processing to transfer
the sensor data to RF wireless module

Transmitting XBee Wireless Data
Transfer Module for High Speed Data
Transfer

Average magnitude of impact force on test platform with time, displayed in remote location

Receiving XBee Wireless Data Transfer
Module for High Speed Data Transfer

FIGURE 9.16   The complete system with job descriptions of individual modules.

laboratory scale with detail experimental measurement to remotely monitor the strain information in realtime due to an event of multiple sudden impacts on a surface.

ACKNOWLEDGEMENT

This research is supported by the National Aeronautics and Space Administration (NASA) of the United States (Grant Number 80NSSC18M0022).

REFERENCES

1. Jenny Howard. *"Sensor to Monitor Orbital Debris Outside Space Station"* https://www.nasa.gov/mission_pages/station/research/news/sensor_to_monitor_orbital_debris_outside_ISS(accessed May. 27, 2020).
2. James L.Hyde, and Ronald P. Bernhard. *"STS-111 (OV-105 Flight 18) Meteoroid/Orbital Debris Postflight Assessment."* (2006).
3. Sayan Roy, Benjamin D. Braaten and Mijia Yang, "An integrated remote monitoring system for impact responses of aerospace structures," 2015*IEEE International Conference on Wireless for Space and Extreme Environments (WiSEE), Orlando, FL*, 2015, pp. 1–6, doi: 10.1109/WiSEE.2015.7392984.
4. Mijia Yang. An Integrated Real-Time Health Monitoring and Impact/Collision Detection System for Bridges in Cold Remote Regions, MPC-15-282. North Dakota State University - Upper Great Plains Transportation Institute, Fargo: Mountain-Plains Consortium, 2015.
5. Mijia Yang, Saeed Ahmari, and Majura Selekwa. (2016). Impact event identification in thin plates through real strain measurements: Impact event identification in thin plates. *Structural Control and Health Monitoring*. 10.1002/stc.1933.

6. Talieh Hajzargerbashi, Tribikram Kundu, and Scott Bland 2011. "An improved algorithm for detecting point of impact in anisotropic inhomogeneous plates." *Ultrasonics* 51 317–324.

7. Federal Communications Commission [Online.. Available: www.fcc.gov

8. Kenneth O. Hill and Gerald Meltz, "Fiber Bragg grating technology fundamentals and overview," *in Journal of Lightwave Technology*, vol. 15, no. 8, pp. 1263–1276, Aug. 1997, doi: 10.1109/50.618320.

9. Accessed: Aug. 26, 2020. [Online]. Available: https://www.hbm.com/en/1629/fibre-bragg-grating-technology-explained/#c254072.

10. Accessed: Aug. 26, 2020. [Online]. Available: https://hittech.com/en/portfolio-posts/noria-the-fiber-bragg-grating-manufacturing-solution/ .

11. Prolific [Online]. Available: www.prolific.com

10 Gap Solitons in Photorefractive Optical Lattices

Aavishkar Katti and Priya Singh

CONTENTS

10.1 INTRODUCTION

Photorefractive (PR) materials exhibit a rich diversity of nonlinear optical phenomena, especially in the field of optical spatial solitons[1–5]. Optical spatial solitons have wide applications in diverse areas such as optical switching, routing, navigation, storage, waveguiding, etc. In PR materials, the refractive index changes with increase in intensity. Nonuniform illumination induces photogenerated charge carriers. These carriers can drift and diffuse and consequently, set up an induced space charge field within the crystal. The light is self-trapped as an index waveguide is created which guides the propagating light beam in a stable fashion. PR materials are a very convenient medium for investigating optical spatial solitons because they exhibit a saturable nonlinearity and the solitons can be realised at relatively low laser powers [6]. Different types of PRsolitons can form based on the drift mechanism of the charge carriers and the type of nonlinearity (linear, quadratic or both) present in the PR crystal. Optical solitons are supported by noncentrosymmetric PR crystals such as strontium barium niobate (SBN) and lithium niobate(LN), centrosymmetric PR crystals such as potassium lithium tantalateniobate (KLTN) and potassium tantalateniobate (KTN), certain PR crystals exhibiting the linear and quadratic nonlinearity simultaneouslysuch as lead magnesium niobate-lead titanate (PMN-0.33PT). There has been extensive research on the various characteristics of PRsolitonssuch as multiple soliton interactions, dynamical evolution, fission, fusion, coupling, waveguiding and oscillations[1,2].

Photonic crystals or photonic lattices are optical structures made of periodically modulated or alternating refractive index layers. The existence of optical spatial solitons in such photonic lattices has been investigated extensively [7–11]. Similar to the energy bandgap observed in solids, we can observe a photonic bandgap in periodic refractive index structures. Light beams should not have their frequency within the photonic bandgap for stable propagation. But gap solitons, as the name suggests are light beams whose frequency lies within the photonic bandgap of the Floquet Bloch lattice spectrum. They exist as defect nonlinear modes and the nonlinearity in the optical crystal is the actual cause for the existence of gap solitons. The photonic lattice may be induced optically by interference or incorporated during the fabrication of the crystal itself. Inducing the lattice pattern optically preferred as it is reversible. It can be done either by interference of two coherent optical beams, or by an amplitude modulation mask.

In this chapter, we shall discuss the existence and stability of gap solitons in different types of PR optical lattices. We shall consider optical lattices in noncentrosymmetric PR crystals, centrosymmetricPRs and pyroelectric PRs. A theoretical foundation using the Helmholtz equation has been laid which serves as a general framework for PR crystals having different nonlinearities and configurations. A particularly interesting case which we consider is an optical lattice embedded in a conventional PR crystal exhibiting the pyroelectric effect. The transient pyroelectric field substitutes for the external electric field with clear advantages changing the nonlinearity drastically.

10.2 THEORETICAL FOUNDATION

The theoretical model's foundation is the paraxial Helmholtz equation for the electric field envelope $\vec{E} = \hat{x}A(x,z)\exp(ikz)$ [12],

$$\left[i\frac{\partial}{\partial z} + \frac{1}{2k}\frac{\partial^2}{\partial x^2} + \frac{k}{n_e}(\Delta n_{PR} + \Delta n_G) \right]A(x,z) = 0 \quad \text{where} k = k0ne = (2\pi/\lambda 0)ne \quad (10.1)$$

Δn_{PR} is the change in the refractive index due to the PR nonlinearity. It can be described by linear electro-optic effect, the quadratic electro-optic effect or simultaneous linear and quadratic electro-optic effect. In general[6,13,14],

$$\Delta n_{PR} = -\frac{1}{2}\left[a n_e^3 r_{eff} E_{sc} + b n_e^3 g_{eff} \epsilon_0^2 \left(\epsilon_r - 1 \right)^2 E_{sc}^2 \right] \qquad (10.2)$$

Here, $a = 1$, $b = 0$ for a PR crystal with the linear electro-optic effect; $a = 0$, $b = 1$ for a PR crystal which shows the quadratic electro-optic effect; $a = 1$, $b = 1$ for a PR crystal demonstrating both electro-optic effects simultaneously.

$$\Delta n_G \alpha \ pR(x) \qquad (10.3)$$

is the refractive index pattern due to the periodic optical lattice where $R(s) = \cos(2\pi s/T)$ signifies the optical lattice having a modulation period T and p is the scaled lattice depth.

We shall consider four cases of different kinds of nonlinear responses of refractive index. We consider a PR crystal with the linear electro-optic effect, quadratic electro-optic effect, and simultaneous linear and quadratic and linear electro-optic effects along with an applied external electric field.

We shall firstly investigate the band structure of the corresponding linear system. As per the Floquet Bloch theory, linearized version of Equation (10.1), along with (10.2) and (10.3) has periodic solutions with an identical period T. Analogous to the E-k curve, i.e. the dispersion relation in atomic lattices, the dispersion relation for the photonic lattice encompasses infinite branches within the first Brillouin zone. The Bloch wave solutions for (10.1) will be obtained by solving the differential eigenvalue problem numerically in each branch.

We will then solve for gap soliton solutions to the dimensionless equivalent of (10.1) using a suitable ansatz depending upon the axial propagation constant and transverse field profile. Since stable solitons are of interest in real-world applications, we intend to investigate the stability of these gap solitons which will be done by the perturbation theory. A relevant crystal having the desired nonlinearity will be taken (PMN-0.33PT, KLTN, LN, etc.) to illustrate our results.

10.2.1 Non-centrosymmetric Photorefractive Lattices

We shall consider a noncentrosymmetric PR crystal like SBN which exhibits the linear electro-optic effect. Substituting (10.2) and (10.3), (10.1)becomes[6]

$$iU_\xi + \frac{1}{2} U_{ss} + pR(s)U - \frac{\beta U}{1 + |U|^2} = 0 \qquad (10.4)$$

where $\beta = \left(k_0 x_0 \right)^2 n_e^4 r_{33} E_0 / 2$ signifies the strength of nonlinearity which depends on the external biasing voltage. $R(s) = \cos(2\pi s/T)$ defines the lattice pattern with T as a time period and p is the lattice depth. The following dimensionless coordinates shall be used throughout this chapter, $\xi = z/\left(k x_0^2 \right)$, $s = x/x_0$, $A = \left(2\eta_0 I_d / n_e \right)^{1/2} U$. On substituting the gap soliton ansatz $q(s,\xi) = w(s,\mu)\exp(i\mu\xi)$ and Bloch wave function $w(x) = w(x + a) = e^{ikx} w(x)$ in (4), we obtain,

$$\frac{1}{2}\frac{\partial^2 w}{\partial s^2} + pR(s)w - \frac{\beta w}{\left(1+|w|^2\right)} = 0 \qquad (10.5)$$

In the limit of small powers of the gap solitons, we can linearize (10.5) and then solve numerically to find the photonic bandgap structure of the ensuing linear photonic lattice. (10.5) is then solved with a frequency lying within the photonic bandgap to obtain the gap soliton spatial profile.

10.2.1.1 Bandgap Structure

If we assume the total power of the soliton to be low, i.e. if we are in the low power limit, the soliton solutions become linear modes which can then be computed numerically. Hence we linearize Equation (10.5) and then obtain its solution. The optical lattice has a modulation period T, total power P and a lattice depth p. If a wave propagates in periodic media, usually there are multiple forbidden bandgaps in the transmission spectrum. We shall consider $p = 7$, $T = 2$ and obtain the bandgap structure of the optical lattice. This can be found by solving the differential equation numerically as an eigenvalue problem. With reference to the Floquet Bloch theory, the linearized Equation(10.4) can be seen to have periodic solutions with period T. Bloch wave solutions for (10.3) exist in each branch while no periodic solutions exist between adjoining branches due to the bandgap. An infinite no. of finite bandgaps along with a semi-infinite bandgap constitutes the Bloch band spectrum.

For the linear photonic lattice, the band structure is illustrated in Figure 10.1 Figure 10.2 portrays the same thing but now the band structure is plotted versus lattice depth p. We can see that bandgap increases with increasing lattice depth. The

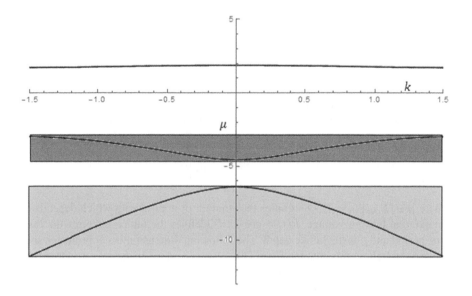

FIGURE 10.1   Band structure w.r.t. Bloch wave vector for the periodic lattice.

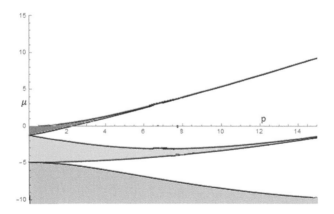

FIGURE 10.2   Band structure w.r.t lattice depth for the periodic lattice.

semi-infinite gap starts from $\mu \geq 1.91$, the first bandgap is in between $-2.945 \leq \mu \leq 1.761$, and the second bandgap is $-6.410 \leq \mu \leq -4.520$.

10.2.1.2   Gap Solitons

Now, on satisfying the modified NLS equation (5) with the gap solitonansatzalong with the Bloch wave function, we numerically obtain the spatial profile of gap soliton supported solely by linear electro-optic effect, or the screening gap soliton.

The spatial profiles of the gap solitons in the first finite gap are plotted at high, moderate and low powers ($\mu = 1.6, -0.5, -2.8$) in Figure 10.3(a)–(c). The soliton field and intensity profiles are seen to be symmetric, single hump and wholly positive.

The spatial profiles of gap solitons in the second finite gap are also plotted at high, moderate and low power ($\mu = -4.4, -5.4, -6.3$) in Figure 10.4(a)–(c). They show that solitons are asymmetric and double hump. The powers of the solitons increase with an increase in the propagation constant

10.2.1.3   Stability

The stability of solitons is an important issue since only stable solitons can be observed experimentally. Stable gap solitons have potential practical applications and hence it is imperative to examine the stability of such gap solitons. Usually, we can study the stability by the Vakhitov–Kolokolov (VK) stability criterion. The VK stability criterion states that the gap solitons so formed in the photonic bandgaps are stable if $dP/d\mu > 0$, where $P = \int_{-\infty}^{\infty} ww * dx$ is the conserved total energy flow or power. But the VK stability criterion suffers from the drawback that it is a necessary but not sufficient condition. Hence, we shall evaluate the *linear* stability of the gap solitons using the perturbation theory. The steady-state solution is considered to be perturbed through the functions $u(s)$ and $v(s)$,

$$U(s,\xi) = \left(w(s,\xi) + \left[u(s) - v(s) \right] \exp\left[i\delta\xi \right] + \left[u^*(s) + v^*(s) \right] \exp\left[i\delta^*\xi \right] \exp\left[i\mu\xi \right] \right) \quad (10.6)$$

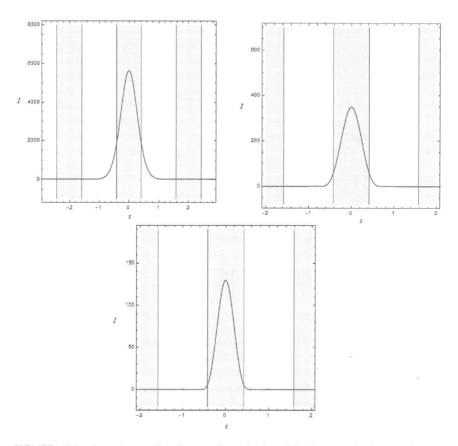

FIGURE 10.3 Intensity profile of gap solitons in first finite bandgap in the (a) high, (b) moderate and (c) low-power regime.

The asterisk indicates a complex conjugate, while v and u are the two components of the perturbation. $\delta = \delta_r + i\delta_i$ is the growth rate of the perturbation which is complex. Substituting (10.6) in (10.5), we get an eigenvalue problem from the resulting linearized version of (10.5),

$$\begin{bmatrix} 0 & L_1 \\ L_2 & 0 \end{bmatrix}\begin{bmatrix} u \\ v \end{bmatrix} = \delta \begin{bmatrix} u \\ v \end{bmatrix} \tag{10.7}$$

with

$$L_1 = -\frac{1}{2}\frac{d^2}{ds^2} - pR + \frac{\beta}{\left(1+w^2\right)} + \mu \tag{10.8}$$

$$L_2 = -\frac{1}{2}\frac{d^2}{ds^2} - pR + \frac{\beta\left(1-w^2\right)}{\left(1+w^2\right)^2} + \mu \tag{10.9}$$

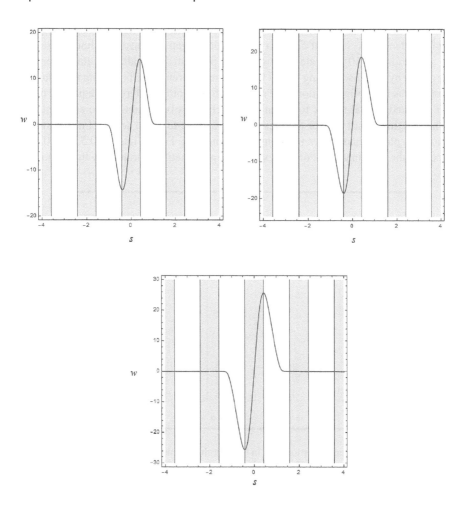

FIGURE 10.4   Field profile of gap solitons in second finite bandgap in the (a) high, (b) moderate and (c) low-power regime.

We note here that (10.8) and (10.9) represent two interdependent eigenvalue problems and can evidently be reduced to a single equation,

$$L_1 L_2 u = \delta^2 u \qquad (10.10)$$

From (10.8) and (10.9), (10.10) becomes,

$$\left(-\frac{1}{2}\frac{d^2}{ds^2} - pR + \frac{\beta}{\left(1+w^2\right)} + \mu\right)\left(-\frac{1}{2}\frac{d^2}{ds^2} - pR + \frac{\beta\left(1-w^2\right)}{\left(1+w^2\right)^2} + \mu\right) = \delta^2 u \quad (10.11)$$

From (10.6), we infer that any perturbation will exponentially increase if δ has an imaginary component, i.e. $\delta_i \neq 0$. If the imaginary component is zero, $\delta_i = 0$ and there exists only a real component and we can interpret the gap solitons to be completely stable. Hence, the value of δ^2 has to be real for the gap solitons to be linearly stable.

Equation (10.10) indicates an eigenvalue problem. We shall solve this eigenvalue problem by first discretizing (10.11) by a finite difference method using the boundary conditions for a bright soliton. The resultant fourth-order differential equation results in a pentadiagonal matrix formulation, which we solve to obtain the eigenvalue δ^2. The imaginary part of δ for the gap solitons in both first and second finite bandgaps has been plotted and shown in Figures 10.5 and 10.6. From Figures 10.5 and 10.6, we see that the imaginary part of δ does not follow any uniform trajectory throughout the

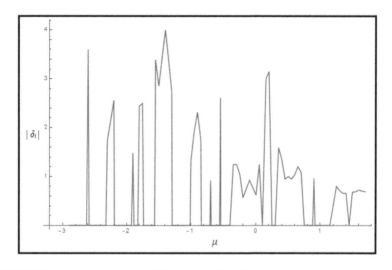

FIGURE 10.5   Imaginary portion of δ in the first finite bandgap.

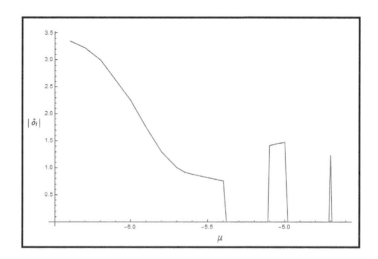

FIGURE 10.6   Imaginary portion of δ in the second finite bandgap.

first or second finite bandgap. There are certain frequencies where the imaginary part of δ is zero, while there are certain frequencies where it spikes to a finite value. This shows that there is a mixture of stability and instability throughout the first and second finite bandgap.

10.2.2 PYROELECTRIC PHOTOREFRACTIVE LATTICES

Pyroelectricity is a phenomenon exhibited in certain materials where an electrical potential is generated due to a finite temperature change resulting in heating or cooling. These effects can be observed in crystals which possess a spontaneous polarization. We can define the pyroelectric coefficient as the change in the electric charge per unit of surface area due to a unit temperature change. Mathematically, the pyroelectric coefficient can be expressed as $p = - dP_S / dT$ where P_S is the spontaneous polarization.

Certain PR crystals have been found to possess a large and finite pyroelectric coefficient. Recent studies have reported formation of optical spatial solitons supported by the pyroelectric effect in PR crystals[15–18]. Spontaneous polarization of a crystal controlled by the temperature variation produces a transitory pyroelectric field. This transient pyroelectric field results in the formation of space charge field which in turn results in a change in refractive index due to the (Pockel's) electro-optic effect. The index waveguide then self-traps a solitary wave which is known as a pyroelectricsoliton or a pyroliton. The transient pyroelectric field can be written as[18]

$$E_{py} = -\frac{1}{\varepsilon_0 \varepsilon_r} \frac{\partial P}{\partial T} \Delta T \qquad (10.12)$$

where ΔT is the change in temperature, ε_0 is the permittivity of free space and ε_r is the relative permittivity. This large pyroelectric field results in the formation of a space charge field and hence a change in refractive index and hence can support the soliton. For a stable soliton to be formed at steady state, the pyroelectric field should sustain for time longer as compare to the time of formation of soliton[18],

$$\tau = \frac{\varepsilon_0 \varepsilon_r}{\sigma_d} \qquad (10.13)$$

where σ_d is the dark conductivity of a PR crystal.

The question arises as to how the transient pyroelectric field is different from the so-called screening external electric field and what are the ensuing changes in the self-trapping mechanism. There are multiple clear advantages of pyroelectric field over the external bias field: firstly, we do not have to orient the c-axis along x-direction like in the earlier case as the pyroelectric field is automatically induced parallel to the c-axis; secondly, we do not require any external connection or electrodes and any external source in this case as our objective is to just induce a temperature change which then results in the pyroelectric field.

Consider an optical beam which is allowed to propagate along the z-axis in a PR crystal, and is allowed to diffract only in x-direction. Here, the optical beam is linearly polarized along the x-axis and the crystal is placed between an insulating plastic cover and metallic plate whose temperature is controllable.

The incident optical beam is articulated in terms of slowly varying envelope A, i.e. E_{opt} $= \hat{x} A(x,z) \exp(ikz)$, where $k = k_0 n_e k_0 = 2\dfrac{\pi}{\lambda_0}$ and λ_0 is the wavelength of free space, n_e is the unperturbed refractive index. In this case, Equation (10.1) now reads as[18]

$$iA_z + \frac{1}{2k} A_{xx} - \frac{k_0}{2}\left(n_e^3 r_{eff} E_{pysc}\right)A = 0 \tag{10.14}$$

where r_{eff} is the effective electro-optic coefficient and E_{pysc} is the space charge field resulted due to the pyroelectric field.

We shall follow the approach of Ref. [18] to derive the expression for the space charge field which has been resulted due to the pyroelectric effect. Recall the following expressions[18]:

$$j = \sigma E \quad \left(\text{Ohms law in differential form}\right) \tag{10.15}$$

$$\frac{\partial \rho}{\partial t} + \nabla.j = 0 \quad \left(\text{Continuity equation}\right) \tag{10.16}$$

$$\nabla.D = \rho \quad \left(\text{Gaussian theorrem}\right) \tag{10.17}$$

where j is total current, $\sigma = kI + \sigma_d$ is the total conductivity, ρ is the space charge density and k is the specific photoconductivity. E is the total electric field and D is the displacement electric vector. The light intensity is given by $I = \dfrac{n_e}{2\eta_0}|A|^2$ and can be expressed as a function of x as, $I(x) = I_0 \exp\left(-2x/x_1^2\right)$ where x_1 is the beam radius and I_0 is the intensity at the beam centre. Total conductivity can be written as, $\sigma = \kappa I_0\left(\exp\left(-2x/x_1^2\right) + I_d/I_0\right)$

Considering that the region of illumination is smaller in comparison to the thickness H of the crystal, $x_1 \ll H$. Now the total current j will be,

$$j = j_d = \sigma_d V/H = \sigma_d E_{py}$$

where j_d is the divergence-less current and V is the voltage. Solving Equations (10.15)–(10.17), we have[18],

$$\nabla.\left[\varepsilon_0\varepsilon_r \frac{\partial E}{\partial t} + \sigma E\right] = 0 \tag{10.18}$$

On neglecting the diffusion and photovoltaic effect and taking the boundary condition under consideration, we have

$$\varepsilon_0\varepsilon_r \frac{\partial E}{\partial t} + \sigma E = j_d \tag{10.19}$$

Solving the differential Equation (10.19), we get

$$E\left(t,x_1\right)=\frac{V}{H}\left\{\frac{\eta}{\exp\left(-\dfrac{2x^2}{x_1^2}\right)+\eta}+\frac{\exp\left(-\dfrac{2x^2}{x_1^2}\right)}{\exp\left(-\dfrac{2x^2}{x_1^2}\right)+\eta}\exp\left[-\frac{\sigma_0 t}{\varepsilon_0\varepsilon_r}\left(\exp\left(-\frac{2x^2}{x_1^2}\right)+\eta\right)\right]+\eta\right\} \tag{10.20}$$

The total electric field can be expressed as a summation of the transient pyroelectric fieldand the induced space charge field,

$$E = E_{py} + E_{pysc} \tag{10.21}$$

The homogeneous pyroelectric field E_{py} is caused by the homogenous heating of a crystal, which results in an inhomogeneous alteration in a refractive index. The soliton forms due to the change in refractive index made by the inhomogeneous space charge field E_{pysc} [18]

$$E_{pysc}=E-E_{py}=E_{py}\frac{I}{I+I_d}\left\{\exp\left[-\frac{\sigma_0 t}{\varepsilon_0\varepsilon_r}\left(\exp\left(-\frac{2x^2}{x_1^2}\right)+\eta\right)\right]-1\right\} \tag{10.22}$$

For steady-state condition, i.e. $t \gg 1$, the space charge field becomes,

$$E_{pysc}=-E_{py}\frac{I}{I+I_d} \tag{10.23}$$

where I is the intensity of light and I_d is the dark irradiance.

The expression (10.23) is quite similar to the space charge field obtained by the photovoltaic effect in unbiased photovoltaic PR crystal. However, in the present case, changing the temperature results in varying values and sign of the space charge field. Substituting (10.23) in (10.1) and then normalizing to dimensionless coordinates,

$$\xi=\frac{z}{kx_0^2}, s=\frac{x}{x_0}, A=\left(\frac{2\eta_0 I_d}{n_e}\right)^{1/2}U, \eta_0=\left(\frac{\mu_0}{\varepsilon_0}\right)^{1/2}, \alpha=\left(k_0 x_0\right)^2 n_e^4 r_{eff} E_{py}/2 \text{ where } x_0 \text{ is}$$

an arbitrary spatial width, we deduce that the envelope U obeys the following evolution equation in absence of any photonic lattice [18]:

$$iU_\xi+\frac{1}{2}U_{ss}+\alpha\frac{|U|^2}{1+|U|^2}U=0 \tag{10.24}$$

The bright soliton solution can be attained from (10.24) by use of the following ansatz along with the relevant boundary conditions:

$$U=r^{\frac{1}{2}}y(s)\exp\left(iv\xi\right) \tag{10.25}$$

where v represents: nonlinear shift and $y(s)$ is the normalized real bound function with boundary condition $0 \leq y(s) \leq 1$ and $r = \dfrac{I_0}{I_d} = \dfrac{1}{\eta}$ which is defined as the ratio of maximum beam power density to the dark irradiance. On substituting these values and solving the Equation (10.24), we obtain the intensity profile and FWHM of bright soliton with the boundary conditions $y(0) = 1$, $\dot{y}(0) = 0$ and $y(s \rightarrow \pm\infty) = 0$. Figures 10.7 and 10.8 show the pyroelectricsoliton profile and the existence curve, respectively.

In presence of the photonic lattice and the photo-induced nonlinearity, the dynamical evolution equation will be modified as

$$iU_\xi + \frac{1}{2}U_{ss} + pR(s)U + \frac{\alpha|U|^2}{1+|U|^2}U = 0 \tag{10.26}$$

where p is the lattice depth and $R(s) = cos\,(2\pi s/T)$defines the lattice pattern with T as a time period.

On substituting the gap soliton ansatz $q(s,\xi) = w(s,\mu)exp(i\mu\xi)$ and Bloch wave function $w(x) = w(x + a) = e^{ikx}w(x)$in (26), we obtain

$$\frac{1}{2}\frac{\partial^2 w}{\partial s^2} + pR(s)w + \frac{\alpha w^3}{\left(1+|w|^2\right)} - \mu w = 0 \tag{10.27}$$

(10.27) will be the equation governing the existence of gap solitons in pyroelectric PR media.

10.2.2.1 Bandgap Structure

On linearizing the modified NLS equation (27), we get the basic properties and the photonic band structure of optical lattice. Since the linearized version of (10.27) is

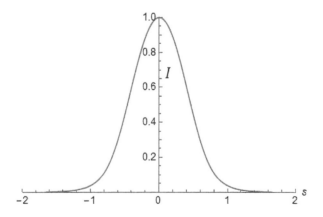

FIGURE 10.7 Normalized intensity profile of bright spatial soliton.

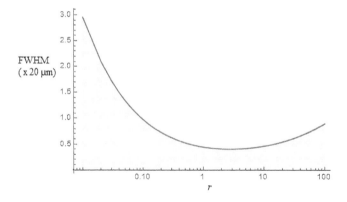

FIGURE 10.8 Graph of the soliton width of bright soliton as a function of r.

same as that of (10.9), the bandgap structure will be the same as shown in Figures 10.3 and 10.4. Figure 10.3 shows the band structure clearly with respect to the Bloch wave vector k. The bandgap structure with respect to the lattice depth is portrayed in Figure 10.4 and shows that bandgap increases with increasing lattice depth.

10.2.2.2 Gap Solitons

The intensity profiles of the gap solitons in the first finite bandgap are plotted at high, moderate and low power ($\mu = 1.75, 0, 2.95$) in Figure 10.8. They show that the soliton field and intensity profile is symmetric and single hump.

The intensity profiles of the gap solitons in the second finite bandgap are plotted at high, moderate and low power ($\mu = -6.4, -5.3, -4.5$) in Figure 10.9. They show that soliton field profile is asymmetric and intensity profile is multi-hump (Figure 10.10).

10.2.2.3 Stability

Perturbation theory is going to be used to calculate the *linear* stability of the gap solitons. Small perturbations $u(s)$ and $v(s)$ in the steady-state solution are considered,

$$U(s,\xi) = \left(w(s,\xi) + \left[u(s) - v(s)\right]\exp\left[i\delta\xi\right] + \left[u^*(s) + v^*(s)\right]\exp\left[i\delta^*\xi\right]\exp\left[i\mu\xi\right]\right) \quad (10.28)$$

v and u are the two components of the perturbation while the asterisk indicates a complex conjugate. $\delta = \delta_r + i\delta_i$ is the growth rate of the perturbation which is complex. Substituting (10.28) in (10.27), we get an eigenvalue problem from the resulting linearized version,

$$\begin{bmatrix} 0 & L_1 \\ L_2 & 0 \end{bmatrix}\begin{bmatrix} u \\ v \end{bmatrix} = \delta\begin{bmatrix} u \\ v \end{bmatrix} \quad (10.29)$$

with

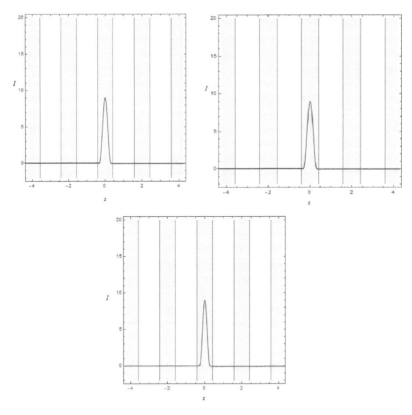

FIGURE 10.9 Intensity profiles of single-humped gap solitons in the first finite bandgap at (a) low (b), moderate and (c) high power.

$$L_1 = -\frac{1}{2}\frac{d^2}{ds^2} - pR - \frac{\alpha w^3}{\left(1+w^2\right)} + \mu \tag{10.30}$$

$$L_2 = -\frac{1}{2}\frac{d^2}{ds^2} - pR - \frac{\alpha\left(w^3 - w^5\right)}{\left(1+w^2\right)^2} + \mu \tag{10.31}$$

The individual eigenvalue problems in (10.29) can be reduced to a single equation,

$$L_1 L_2 u = \delta^2 u \tag{10.32}$$

From (10.30) and (10.31), (10.32) becomes,

$$\left(-\frac{1}{2}\frac{d^2}{ds^2} - pR - \frac{\alpha w^3}{\left(1+w^2\right)} + \mu\right)\left(-\frac{1}{2}\frac{d^2}{ds^2} - pR - \frac{\alpha\left(w^3 - w^5\right)}{\left(1+w^2\right)^2} + \mu\right)u = \delta^2 u \tag{10.33}$$

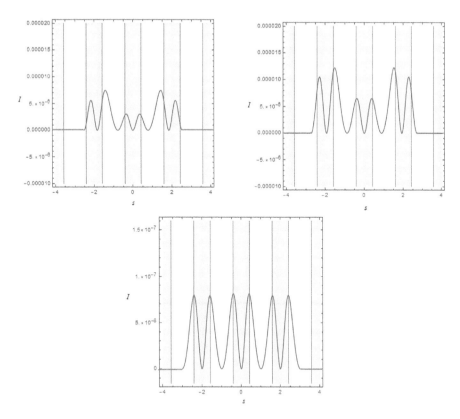

FIGURE 10.10 Intensity profiles of multi-humped gap solitons in the second finite bandgap at (a) low, (b) moderate and (c) high power.

Again, any perturbation will exponentially increase if δ has an imaginary component, i.e.$\delta_i \neq 0$. As discussed before, if the imaginary component is zero, $\delta_i = 0$, then the gap solitons are stable. Hence, the value of δ^2 has to be real for the gap solitons to be linearly stable.

The imaginary portion of δ is plotted in Figures 10.11 and 10.12 for the first and second finite bandgaps. We see that the imaginary portion of δ is homogeneously zero throughout the two finite bandgaps. This shows that the single hump pyroelectric gap solitons in the first finite bandgap are stable. Also, the multi-hump solitons in the second finite bandgap are found to be stable when subjected to perturbations and this is quite an interesting result as we would usually expect multi-hump solitons to be unstable, even though exceptions exist as in the present case[19,20].

10.2.3 CENTROSYMMETRIC PHOTOREFRACTIVE LATTICES

A centrosymmetric crystal is one which possesses inversion symmetry or an inversion centre as a symmetry element. The first discovery of stable optical spatial solitons in PR materials was in noncentrosymmetric crystals like SBN. The mediating nonlinearity was because of the linear electro-optic effect or the Pockels' effect. The

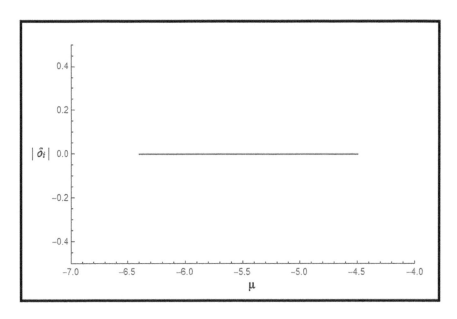

FIGURE 10.11   Imaginary portion of δ in the first finite bandgap.

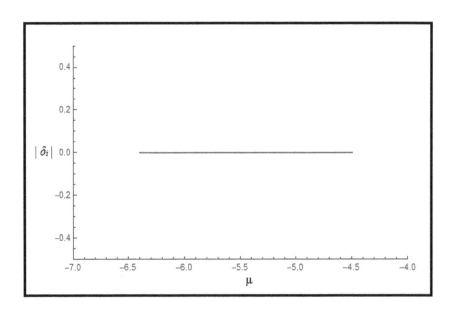

FIGURE 10.12   Imaginary portion of δ in the second finite bandgap.

Pockels' effect mandates the change in refractive index is proportional to the electric field. Optical spatial solitons in centrosymmetric PR crystals were later predicted by Segev et al.[13]. Centrosymmetric PR crystals exhibit quadratic electro-optic effect or the dc Kerr effect in which the refractive index changes as the square of field, i.e.$\Delta n \propto E^2$. The gap solitons in such centrosymmetricPR materials have been studied in [12] and we shall follow the same approach to introduce the topic to the reader.

Consider a centrosymmetric PR crystal with an imprinted optical lattice where an optical beam is allowed to propagate along the z-axis in a PR crystal, which is permitted to diffract only in x-direction. Here, the optical beam is linearly polarized along the x-axis and the bias field is applied in the same direction resulting in a space charge field (E_{sc}).

The incident optical beam is stated in terms of the slowly varying envelope, $E = \hat{x}A(x,z)\exp(ikz)$, where $k = k_0 n_e$, $k_0 = 2\pi/\lambda_0$ and λ_0 is the free space wavelength. The paraxial diffraction equation (10.1) now reads as,

$$iA_z + \frac{1}{2k}A_{xx} - \frac{k_0}{2}(\Delta n_{PR} + \Delta n_G)A = 0 \qquad (10.34)$$

where [4], [13], $\Delta n_{PR} = -n_e^3 g_{eff}\varepsilon_0^2(\varepsilon_r - 1)^2 E_{SC}^2/2$ and Δn_G is the periodic optical lattice as defined before. On satisfying (10.34) using dimensionless co-ordinates defined before, the dynamical evolution equation becomes,

$$iU_\xi + \frac{1}{2}U_{ss} + pR(s)U - \beta\frac{U}{(1+|U|^2)^2} = 0 \qquad (10.35)$$

where $\beta = (k_0 x_0)^2 n_e^4 g_{eff}\epsilon_0^2(\epsilon_r - 1)^2 E_{sc}^2/2$ is the strength of nonlinearity which depends on the external bias voltage, p is the lattice depth and $R(s) = cos(2\pi s/T)$ defines the lattice pattern with T as the time period.

On substituting the gap soliton ansatz $q(s,\xi) = w(s,\mu)\exp(i\mu\xi)$ and Bloch wave function $u(x) = u(x+a) = e^{ikx}u(x)$ in Eq,(10.23), we obtain[12],

$$\frac{1}{2}\frac{\partial^2 w}{\partial s^2} + pR(s)w - \frac{\beta w}{(1+|w|^2)^2} - \mu w = 0 \qquad (10.36)$$

Equation(10.36) serves as the central equation to solve for gap solitons in centrosymmetric PR media.

10.2.3.1   Bandgap Structure

On linearizing the modified NLS equation (36), we get the basic properties and photonic bandgap structure of optical periodic lattice $pR(s)$. As the linearized version of (10.36) is same as that of (10.9), the bandgap structure will be the same as shown in Figures 10.3 and 10.4. Figure 10.3 shows that we will get the first and second bandgap and will focus on the soliton solutions in these gaps only. The band -gap structure

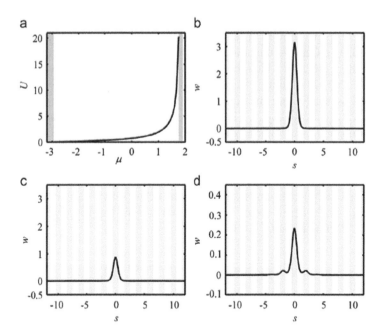

FIGURE 10.13 (a) Normalized power curve, Profile of gap solitons in first finite bandgap for (b) $\mu = -4.4$,(c) $\mu = -5.4$ and (d) $\mu = -6.3$. (Reprinted from Optics Communications, 285, Kaiyun Zhan and Chunfeng Hou, Gap solitons supported by optical lattices in biased centrosymmetric photorefractive crystals, 3649-3653, 2012, with permission from Elsevier).

of periodic lattices versus lattice depth is portrayed in Figure 10.4 and shows that bandgap increases with increasing lattice depth.

10.2.3.2 Gap Solitons

Substituting the gap solitonansatz and Bloch wave function in (10.36), we get the spatial profile of gap solitons in the respective bandgaps supported by quadratic electro-optic effect or the dc Kerr effect.

The field profiles of the solitons in the first finite gap are plotted at high, moderate and low power ($\mu = 1.6, -0.5, -2.8$) in Figure 10.13 (b)–(d). We see that the solitons are symmetric and single hump.

The field profiles of solitons in the second finite gap are plotted at high, moderate and low power ($\mu = -4.4, -5.4, -6.3$) in Figure 10.14 (b)–(d). They show that solitons field profile is asymmetric and the intensity profile is double hump.

10.2.3.3 Stability

As before, we use perturbation theory. Small perturbations in the steady-state solution are considered,

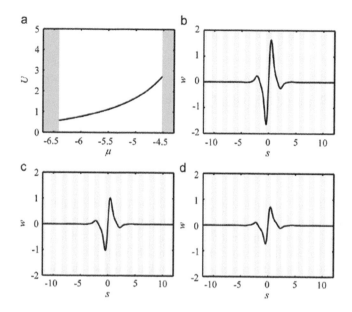

FIGURE 10.14   (a) Normalized power curve, Profile of gap solitons in second finite bandgap for (b) $\mu = -4.4$,(c) $\mu = -5.4$ and (d) $\mu = -6.3$. (Reprinted from Optics Communications, 285, Kaiyun Zhan and Chunfeng Hou, Gap solitons supported by optical lattices in biased centrosymmetric photorefractive crystals, 3649-3653, 2012, with permission from Elsevier).

$$U(s,\xi)=\left(w(s,\xi)+\left[u(s)-v(s)\right]\exp\left[i\delta\xi\right]+\left[u^{*}(s)+v^{*}(s)\right]\exp\left[i\delta^{*}\xi\right]\exp\left[i\mu\xi\right]\right) \quad (10.37)$$

The asterisk shows a complex conjugation, v and u are the components of the perturbation. $\delta = \delta_{r} + i\delta_{i}$ illustrates how the complex perturbation grows. Substituting (10.37) in (10.35), we get an eigenvalue problem from the resulting linearized version of (10.35),

$$\begin{bmatrix} 0 & L_{1} \\ L_{2} & 0 \end{bmatrix}\begin{bmatrix} u \\ v \end{bmatrix}=\delta\begin{bmatrix} u \\ v \end{bmatrix} \quad (10.38)$$

with[12]

$$L_{1}=-\frac{1}{2}\frac{d^{2}}{ds^{2}}-pR+\frac{\beta}{\left(1+w^{2}\right)^{2}}+\mu \quad (10.39)$$

$$L_{2}=-\frac{1}{2}\frac{d^{2}}{ds^{2}}-pR+\frac{\beta\left(1-3w^{2}\right)}{\left(1+w^{2}\right)^{3}}+\mu \quad (10.40)$$

The two individual eigenvalue problems in (10.39) and (10.40) can be reformulated to signify a single eigenvalue problem as follows:

$$L_1 L_2 u = \delta^2 u \qquad (10.41)$$

From (10.39) and (10.40), (10.38) becomes,

$$\left(-\frac{1}{2}\frac{d^2}{ds^2} - pR + \frac{\beta}{\left(1+w^2\right)^2} + \mu\right)\left(-\frac{1}{2}\frac{d^2}{ds^2} - pR + \frac{\beta\left(1-3w^2\right)}{\left(1+w^2\right)^3} + \mu\right)u = \delta^2 u \qquad (10.42)$$

From (10.37), we infer that any perturbation will exponentially increase if δ has an imaginary component, i.e. $\delta_i \neq 0$. If the imaginary component is zero, $\delta_i = 0$ and there exists only a real component then the gap solitons are stable. Hence, the value of δ^2 has to be real for the gap solitons to be linearly stable.

Equation (10.42) signifies an eigenvalue problem and we can obtain the value of δ by solving it. Figure 10.15 (a)–(b) show the plot of the imaginary part of δ in the first and second finite bandgaps. From Figure 10.15, we can clearly see that the imaginary part of δ is zero through the first finite bandgap, while it does not follow any uniform trajectory in the second finite bandgap. There are certain frequencies where the imaginary part of δ is zero while there are certain frequencies where it spikes to a finite value within the second finite bandgap. This shows that there is a mixture of stability and instability in the second finite bandgap.

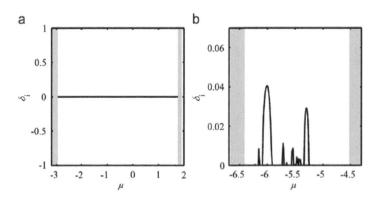

FIGURE 10.15 Imaginary portion of δ in the (a) first finite bandgap and (b) second finite bandgap. (Reprinted from Optics Communications, 285, Kaiyun Zhan and ChunfengHou, Gap solitons supported by optical lattices in biased centrosymmetric photorefractive crystals, 3649-3653, 2012, with permission from Elsevier).

10.2.4 COMPARATIVE STUDY

Gap soliton properties	Centrosymmetric PR lattice	Non-centrosymmetricPR lattice	Pyroelectric PR lattice
Field profile- First finite gap	Symmetric, single hump and entirely positive.	Symmetric, single hump and wholly positive	Symmetric and single hump
Field profile- Second finite gap	Asymmetric and double hump	Asymmetric and double hump	Asymmetric and multi hump
Stability- First finite gap	Stable at all frequencies $(\delta_i = 0)$	Mixture of stability and instability in the whole frequency range	Stable at all frequencies $(\delta_i = 0)$
Stability- Second finite gap	Mixture of stability and instability in the whole frequency range	Mixture of stability and instability in the whole frequency range	Stable at all frequencies $(\delta_i = 0)$

10.3 CONCLUSIONS

We have undertaken a comprehensive discussion about the existence of gap solitons in different types of PR optical lattices, i.e.noncentrosymmetricPRs, centrosymmetricPRs and pyroelectricPRs. A theoretical foundation using the Helmholtz equation has been laid which can function as a general framework for PR crystals having different nonlinearities and/or configurations. We have obtained the bandgap structure and analyzed the first and second finite bandgap. We have observed the gap soliton profiles in the first and second finite bandgap for each of the three configurations of the PR crystal mentioned above. The gap solitons are either single-humped and symmetric or double-humped or even multi-humped and asymmetric. Generally, the first finite bandgap supports entirely positive single-humped solitons and the second finite bandgap supports asymmetric and double-humped or multi-humped solitons. The stability of gap solitons is examined in all three configurations by linear stability analysis. The growth rate of perturbation has been studied by formulating a general eigenvalue problem which can be modified to suit each configuration of PR crystal. Solving the eigenvalue problem numerically can be achieved by first discretizing the Sturm Liouville system by finite differences and then finding out the eigenvalues of the resultant pentadiagonal matrix. While the pyroelectric gap solitons are found to be stable across the two finite bandgaps, gap solitons in other configurations are found to be both stable and unstable for different values of frequency within the two finite bandgaps.

REFERENCES

1. G. I. Stegeman, "Optical spatial solitons and their interactions: Universality and diversity," *Science*, vol. 286, no. 5444, pp. 1518–1523, 1999, doi: 10.1126/science.286.5444.1518.
2. S. Trillo and W. E. Torruellas, Eds., *Spatial Solitons, Springer Series in Optical Sciences*, vol. 31. Springer, Berlin, 2001.

3. A. Katti, "Bright screening solitons in a photorefractive waveguide," *Opt Quant Electron*, vol. 50, no.6, p. 263, Jun. 2018, doi: 10.1007/s11082-018-1524-y.

4. A. Katti, R. A. Yadav, and D. P. Singh, "Theoretical investigation of incoherently coupled solitons in centrosymmetricphotorefractive crystals," *Optik - International Journal for Light and Electron Optics*, vol. 136, pp. 89–106, 2017, doi: 10.1016/j.ijleo.2017.01.099.

5. A. Katti, R. A. Yadav, and A. Prasad, "Bright optical spatial solitons in photorefractive waveguides having both the linear and quadratic electro-optic effect," *Wave Motion*, vol. 77, pp. 64–76, Mar. 2018, doi: 10.1016/J.WAVEMOTI.2017.10.002.

6. D. N. Christodoulides and M. I. Carvalho, "Bright, dark, and gray spatial soliton states in photorefractive media," *Journal of the Optical Society of America B*, vol. 12, no. 9, p. 1628, Sep. 1995, doi: 10.1364/JOSAB.12.001628.

7. A. S. Desyatnikov, E. A. Ostrovskaya, Y. S. Kivshar, and C. Denz, "Composite bandgap solitons in nonlinear optically induced lattices," *Physical review letters*, vol. 91, no. 15, p. 153902, 2003.

8. A. B. Aceves, "Optical gap solitons: Past, present, and future; theory and experiments," *Chaos: An Interdisciplinary Journal of Nonlinear Science*, vol. 10, no. 3, pp. 584–589, 2000.

9. Y. J. He, W. H. Chen, H. Z. Wang, and B. A. Malomed, "Surface superlattice gap solitons," *Optics Letters*, vol. 32, no. 11, pp. 1390–1392, 2007.

10. D. L. Mills and S. E. Trullinger, "Gap solitons in nonlinear periodic structures," *Physical Review B*, vol. 36, no. 2, p. 947, 1987.

11. H. Sakaguchi and B. A. Malomed, "Gap solitons in quasiperiodic optical lattices," *Physical Review E*, vol. 74, no.2, p. 026601, 2006.

12. K. Zhan and C. Hou, "Gap solitons supported by optical lattices in biased centrosymmetric photorefractive crystals," *Optics Communications*, vol. 285, no.17, pp. 3649–3653, 2012.

13. M. Segev and A. J. Agranat, "Spatial solitons in centrosymmetric photorefractive media.," *Optics Letters*, vol. 22, no.17, pp. 1299–1301, 1997, doi: 10.1364/OL.22.001299.

14. L. Hao, Q. Wang, and C. Hou, "Spatial solitons in biased photorefractive materials with both the linear and quadratic electro-optic effects," *Journal of Modern Optics*, vol. 61, no.15, pp. 1236–1245, Sep. 2014, doi: 10.1080/09500340.2014.928379.

15. J. Safioui, F. Devaux, and M. Chauvet, "Pyroliton: Pyroelectric spatial soliton.," *Optics express*, vol. 17, no. 24, pp. 22209–22216, 2009, doi: 10.1364/OE.17.022209.

16. Q. Jiang, Y. Su, and X. Ji, "Pyroelectric photovoltaic spatial solitons in unbiased photorefractive crystals," *Physics Letters, Section A: General, Atomic and Solid State Physics*, vol. 376, no. 45, pp. 3085–3087, 2012, doi: 10.1016/j.physleta.2012.08.043.

17. A. Katti and R. A. Yadav, "Spatial solitons in biased photovoltaic photorefractive materials with the pyroelectric effect," *Physics Letters, Section A: General, Atomic and Solid State Physics*, vol. 381, no. 3, pp. 166–170, 2017, doi: 10.1016/j.physleta.2016.10.054.

18. Y. Su, Q. Jiang, and X. Ji, "Photorefractive spatial solitons supported by pyroelectric effects in strontium barium niobate crystals," *Optik*, vol. 126, no.18, pp. 1621–1624, 2015, doi: 10.1016/j.ijleo.2015.04.053.

19. E. A. Ostrovskaya and Y. S. Kivshar, "Multi-hump optical solitons in a saturable medium," *Journal of Optics B: Quantum and Semiclassical Optics*, vol. 1, no. 1, p. 77, 1999.

20. E. A. Ostrovskaya, Y. S. Kivshar, D. V. Skryabin, and W. J. Firth, "Stability of multi-hump optical solitons," *Physical Review Letters*, vol. 83, no. 2, p. 296, 1999.

11 Real-Time Numerical Analysis of Photonic Bandgap Structures Using Finite Difference in Time-Domain Method

Rajorshi Bandyopadhyay and Rajib Chakraborty

CONTENTS

11.1   INTRODUCTION

According to Professor Richard Feynman, the development of Maxwell partial differential equations of electrodynamics, the fundamentals of electromagnetism, is one of the best findings of the 19th-century science. During World War II, electromagnetism was studied because of its connection to military defence. However, at present the field of electromagnetism is also important in the field of high-speed computing and biomedicine other than its applications in electromagnetic wave guiding, radiation and scattering phenomena. In 1966, finite difference algorithm of Maxwell's equation was first proposed by Kane S. Yee in his paper "Numerical solutions of initial boundary value problems involving Maxwell's equations in isotropic media" where he first introduced the concept of Yee cell and the scheme of finite difference in time domain (FDTD). Due to the deficient computational capacity of earlier computers, his paper was not given due consideration. However, in the mid-seventies, when the computational capabilities of computers had been developed enough, importance of implementation of Yee's algorithm was realized for broad range of important problems. In 1975, Taflove and Browdin wrote a paper demonstrating the importance of the FDTD method in electrodynamics and gave the first absorbing boundary conditions [1].That was the beginning of computational electrodynamics as we know it today.

During the period we have just discussed, frequency-domain solution techniques were used to find the solution of Maxwell's equations using the powerful language FORTRAN. This technique is useful for analysing the scattering properties resulting from wave matter interaction but analysis becomes difficult for objects with complex shape and volume. Even, the sophisticated frequency-domain technique which is used nowadays is sometimes not capable of providing the correct analysis of complex structures. The concept of direct time-domain numerical method for solving Maxwell's equations was realized in the decade of 1970s and '80s. FDTD is the first time-domain numerical method for numerical analysis. Modern communication demand systems with increased transmission capacity at a faster rate and communication system in the electronic domain have reached a bottleneck because of its bandwidth limitations.This increasing demand of transmission capacity forced the scientist and researcher to think about a new kind of devices where data transmission will be in optical domain and that has given birth to a new field of integrated photonics. In photonics, photons are used instead of electrons and in this domain, the bandwidth is much more than the bandwidth obtained in electrical transmission system. Chip-to-chip or even on-chip integration compatibility is another problem[2] associated with high-speed electronics. On the contrary, photonic devices can be designed with wavelength-scale components, which make it well suited for compact integrated photonic system design with an added advantage of low power requirement for its operations. At this point, photonic bandgap structure (PBGS) is playing an important role in manipulating light energy in both temporal and spatial domain [3] and this manipulation of light energy is one of the key functionalities for developing optical integrated circuits. The discovery of PBG materials opens a new and exciting era in the fields of physical and material science, electrical and optical engineering[3–5].

11.2   RELATED WORKS

Dielectric or dielectric/metallic materials having a suitable refractive index (r.i.) contrast are periodically arranged to form a PBGS which is an optical equivalent of conventional crystal. An electron propagating in a crystal is characterized by its energy bandgap. Similarly, a PBGS is recognized by its photonic bandgap, i.e. for certain wavelength range, no light, irrespective of polarization can propagate through it. It is because of the dispersion relation, i.e. the relation between the wavelengths (λ) and wave vector (κ) of the electromagnetic (EM) wave propagating through the PBGS. This characteristic was first noticed by Lord Rayleigh in an one-dimensional (1D) optical periodic structure [6,7]. He also derived a general solution of 1D periodic media. The generalized concept of omni-directional photonic gap for two-dimensional (2D) and three-dimensional(3D) PBGS was made clear by Yablonovitch and John after a century [4,5] and those structures were named as photonic crystal (PhC) by Yablonovitch. However, nowadays, conventionally, all the 1D, 2D, 3D periodic structures are popularly known as PhCs. The basic 1D, 2D and 3D PhC structures are shown in Figure 11.1. 1D PBGSs also known as distributed Bragg reflectors (DBR) are widely used as optical filters, surface emitting lasers, anti-reflection coating,etc.[8–14]. 2D PhC exhibits permittivity periodicity along two directions. However, the permittivity of the medium is uniform in the third direction. Porous silicon where the pores are periodically arranged or periodically arranged dielectric rods in air are the examples of 2D PhC. In 3D PhCs, the permittivity modulations are found in all the three directions. Stone opal is the natural example of 3D PhC having unique optical properties of showing different colours from different angles, which is because of the typical micro-structure of the opal. In a PBGS with perturbed refractive index pattern (defective PBGS), photons can be localized within the perturbed area [15]. This photon localization results in a transmission peak or a reflection dip within the wider stopband. This type of doped PBGS finds potential applications in lasing [16], sensing [9], space application [17], biophotonics [18],resonator [19], etc. PBGSs with defect only in one dimension are known as Bragg grating and find huge applications in optical communication system as transmission filter [20–23]. Dielectrics, metals, semiconductors, organic materials, etc., are used depending on the desired wavelength of filter, etc. [24–27]. A single resonance mode (or group of resonance modes) may be found at one wavelength (or some wavelengths) within the PBGS depending on the width of the defect layer [28]. But the resonance modes are found to be fixed in those cases. Tunable filters are also studied [29–34].

However, for wavelength division multiplexing (WDM) system, used substantially in modern fibre optic communication, requires selection of several wavelengths simultaneously. For C-band communication system (1530-1565 nm), CWDM

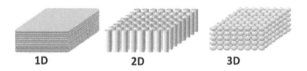

1D   2D   3D

FIGURE 11.1   Schematic of (a) 1D PhC (b) 2D PhC (C) 3D PhC.

(coarse WDM) operates with 8 channels and DWDM (dense WDM) system uses 40 channels at 100 GHz spacing or 80 channels with 50 GHz spacing. For WDM applications, multiband filters are also designed [35–38] but there, either the lattice parameters or the defect layer width are changed by trial and error method to obtain the filtering effect at different wavelengths. Thermo-optically tunable filters using 1D PBGSs for DWDM applications have also been reported [39]. However, filters tuned thermo-optically are not very reliable or stable and are slow too. [40]. Optically tunable narrowband filters by changing the angle of incidence of the incident light were also proposed in [41,42]. However, the required mechanical arrangement is troublesome for these filters. Here, in this chapter it is shown that PBGS can be used to design optical filters with predetermined materials widths and a study is also carried out to know that how the change of layer widths affects the shift of peak of the transmission band. These type of information are highly beneficial for the designer.

There are a number of different methods to study the characteristics of PBGS. Some of these methods are finite element method (FEM) [43,44], plane wave expansion method [45–48], transfer matrix method (TMM) [49–51], finite difference method [52,53], tight-binding formulation [54,55], multiple-scattering theory (Korringa–Kohn–Rostoker method) [56–59], generalized Rayleigh identity method [60], averaged field approach [61–63], etc. With these methods, PBGSs have been analysed to study different characteristics in a number ways. Each method has its own pros and cons and methods are generally chosen depending on the particular problem. In this chapter, a brief discussion on FDTD method is presented first and thereafter simple 1D periodic structures as optical transmission filter is analysed by the application of FDTD method. The main advantage of FDTD method is that it gives a direct time domain solution for this the structure is discretized into a number of small meshes. So the source of error is well known. However, as the object and its surroundings are required to be implemented in the FDTD algorithm, the execution time is more.

11.3 FINITE DIFFERENCE AND MAXWELL'S EQUATION

From elementary calculus, the standard mathematical expression for derivative can be written as

$$u'(x) = \lim_{\Delta x \to 0} \frac{u(x + \Delta x) - u(x)}{\Delta x} \tag{11.1}$$

where Δx is finite and small. This is the Forward Euler approximation. Similarly, Backward Euler approximation and the central difference approximation can be written as

$$u'(x_i) = \lim_{\Delta x \to 0} \frac{u(x_i) - u(x_i - \Delta x)}{\Delta x} \cong \frac{u_i - u_{i-1}}{\Delta x} \tag{11.2}$$

$$u'(x_i) = \lim_{\Delta x \to 0} \frac{u(x_i + \Delta x) - u(x_i - \Delta x)}{2\Delta x} \cong \frac{u_{i+1} - u_{i-1}}{2\Delta x} \tag{11.3}$$

For continuous cases, all these definitions are equivalent but for discrete cases, the approximations of these equations are different. By applying the Taylor series analysis, it can be shown that central difference approximation is the most accurate one (second-order accurate).

We know that 1D scalar wave equation in an infinite homogeneous space can be represented as

$$\frac{\partial^2 u}{\partial t^2} = c^2 \frac{\partial^2 u}{\partial t^2} \tag{11.4}$$

where $u = u(x,t)$ and c is the velocity of light. By applying the central difference approximation for both time and space, the explicit time marching solution for u can be written as

$$
u_i^{n+1} = \left(c\Delta t\right)^2 \frac{u_{i+1}^n - 2u_i^n + u_{i-1}^n}{\left(\Delta x\right)^2} + 2u_i^n - u_i^{n-1}
$$
$$
+ O\left(\left(\Delta x\right)^2\right) + O\left(\left(\Delta t\right)^2\right) \tag{11.5}
$$

If a region is free from source, Maxwell's equation can be written as
Gauss' law for electric field

$$\nabla.\mathbf{D} = 0 \qquad \oiint_A \mathbf{D}.d\mathbf{A} = 0 \tag{11.6}$$

Gauss' law for magnetic field

$$\nabla.\mathbf{B} = 0 \qquad \oiint_A \mathbf{B}.d\mathbf{A} = 0 \tag{11.7}$$

Faraday's law

$$\nabla \times \mathbf{E} = -\frac{\partial \mathbf{B}}{\partial t} \qquad \int_l \mathbf{E}.d\mathbf{l} = -\frac{\partial}{\partial t} \oiint_A \mathbf{B}.d\mathbf{A} \tag{11.8}$$

Ampere's law

$$\nabla \times \mathbf{H} = \frac{\partial \mathbf{D}}{\partial t} \qquad \oiint_l \mathbf{H}.d\mathbf{l} = \frac{\partial}{\partial t} \oiint_A \mathbf{D}.d\mathbf{A} \tag{11.9}$$

Considering most of the materials as non-magnetic and by applying the constitutive relationship, the curl equations can be rewritten as

$$\nabla \times \mathbf{E} = -\mu \frac{\partial \mathbf{H}}{\partial t} \tag{11.10}$$

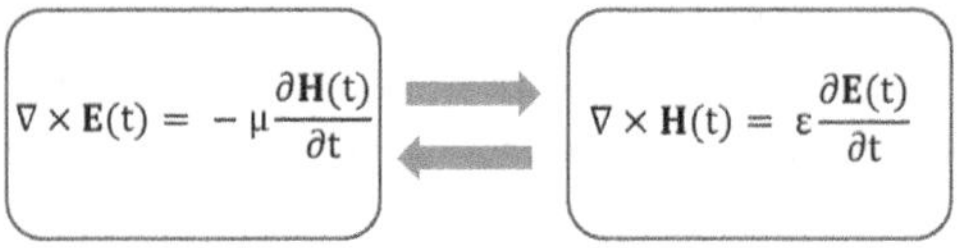

$$\nabla \times E(t) = -\mu \frac{\partial H(t)}{\partial t} \quad \Longleftrightarrow \quad \nabla \times H(t) = \varepsilon \frac{\partial E(t)}{\partial t}$$

FIGURE 11.2   Maxwell curl equations in a linear, isotropic, non-dispersive material.

$$\nabla \times \mathbf{H} = \quad \varepsilon \frac{\partial \mathbf{E}}{\partial t} \tag{11.11}$$

In FDTD, all the Maxwell's equations are considered as a sequence of events. When an EM wave propagates in a linear, isotropic, non-dispersive material, it is assumed that electric fields are continuously updated from the previous values of magnetic field and magnetic fields are continuously updated from the previous values of electric field as shown in Figure 11.2. Now time derivative of Equations 11.10 and 11.11 can be written as

$$\nabla \times \mathbf{E}(\mathbf{t}) = -\mu \frac{\mathbf{H}\left(t + \Delta t/2\right) - \mathbf{H}\left(t - \Delta t/2\right)}{\Delta t} \tag{11.12}$$

$$\nabla \times \mathbf{H}\left(\mathbf{t} + \Delta \mathbf{t}/\mathbf{2}\right) = \varepsilon \frac{\mathbf{E}\left(t + \Delta t\right) - \mathbf{E}\left(t\right)}{\Delta t} \tag{11.13}$$

Here, $\mathbf{E}$ and $\mathbf{H}$ are staggered in such a way that $\mathbf{E}$ exists at $0, \Delta t, 2\Delta t$.... and $\mathbf{H}$ exists at $\Delta t/2, t + \Delta t/2, 2t + \Delta t/2$.... . This is done so that each term of Equations 11.10 and 11.11 exists at the same point of time. Arranging Equations 11.12 and 11.13, we get

$$\mathbf{H}\left(t + \Delta t/2\right) = \mathbf{H}\left(t - \Delta t/2\right) - \frac{\Delta t}{\mu} \nabla \times \mathbf{E}\left(t\right) \tag{11.14}$$

$$\mathbf{E}\left(t + \Delta t\right) = \mathbf{E}\left(t\right) + \frac{\Delta t}{\varepsilon} \nabla \times \mathbf{H}\left(t + \Delta t/2\right) \tag{11.15}$$

These are the update equations for $\mathbf{E}$ and $\mathbf{H}$. Here, all the symbols are having their usual meaning.

Maxwell's divergence equations (Equations 11.6 and 11.7) are automatically satisfied in Yee's algorithm [64] in which $\mathbf{E}$ and $\mathbf{H}$ are positioned in time following leapfrog arrangement. In this, the $\mathbf{E}$ values are computed using the previously stored $\mathbf{H}$ data and stored in memory for a particular time. After that, the $\mathbf{H}$ values are computed using the $\mathbf{E}$ values just computed. To make the right hand and left hand side of Maxwell curl equation compatible, $\mathbf{E}$ values are computed in integral time steps and $\mathbf{H}$ values are computed in half time steps. This process continues till it reaches the final time step. But while dealing with Maxwell's curl equations (Equations 11.10 and 11.11), it is found that in any medium, electric and magnetic fields are related to the impedance of the medium. For example, while dealing with Maxwell's equations in free space, it is seen that electric and magnetic fields are related to the free space impedance which is around 377 Ω. Therefore, to maintain the order of magnitude of electric and magnetic field, the magnitude of magnetic fields are generally normalized ($\tilde{H}$). Considering the vector component of the above curl operators, following six coupled scalar equations are obtained:

$$-\frac{\mu_x}{c_0}\frac{\partial \tilde{H}_x}{\partial t} = \frac{\partial E_z}{\partial y} - \frac{\partial E_y}{\partial z} \tag{11.16}$$

$$-\frac{\mu_y}{c_0}\frac{\partial \tilde{H}_y}{\partial t} = \frac{\partial E_x}{\partial z} - \frac{\partial E_z}{\partial x} \tag{11.17}$$

$$-\frac{\mu_z}{c_0}\frac{\partial \tilde{H}_z}{\partial t} = \frac{\partial E_y}{\partial x} - \frac{\partial E_x}{\partial y} \tag{11.18}$$

$$\frac{\varepsilon_x}{c_0}\frac{\partial E_x}{\partial t} = \frac{\partial \tilde{H}_z}{\partial y} - \frac{\partial \tilde{H}_y}{\partial z} \tag{11.19}$$

$$\frac{\varepsilon_y}{c_0}\frac{\partial E_y}{\partial t} = \frac{\partial \tilde{H}_x}{\partial z} - \frac{\partial \tilde{H}_z}{\partial x} \tag{11.20}$$

$$\frac{\varepsilon_z}{c_0}\frac{\partial E_z}{\partial t} = \frac{\partial \tilde{H}_y}{\partial x} - \frac{\partial \tilde{H}_x}{\partial y} \tag{11.21}$$

Figure 11.3 shows a unit Yee cell. This cell can be used to show the field vectors around H_x. With reference to this figure, equivalent finite difference equation for Equation. 11.16 can be written as

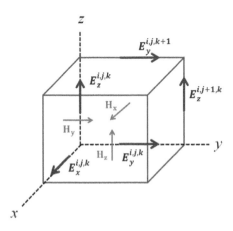

FIGURE 11.3   Fields around Hx in a unit Yee cell.

$$\frac{\left(E_z\right)^t_{i,j+1,k} - \left(E_z\right)^t_{i,j,k}}{\Delta y} - \frac{\left(E_y\right)^t_{i,j,k+1} - \left(E_y\right)^t_{i,j,k}}{\Delta z} =$$

$$- \frac{\left(\mu_x\right)_{i,j,k}}{c_0} \cdot \frac{\left(\tilde{H}_x\right)^{t+\frac{\Delta t}{2}}_{i,j,k} - \left(\tilde{H}_x\right)^{t-\frac{\Delta t}{2}}_{i,j,k}}{\Delta t} \qquad (11.22)$$

This equation can be rearranged and can be rewritten as

$$\left(\tilde{H}_x\right)^{t+\frac{\Delta t}{2}}_{i,j,k} = \left(\tilde{H}_x\right)^{t-\frac{\Delta t}{2}}_{i,j,k} + \frac{c_0 \Delta t}{\left(\mu_x\right)_{i,j,k}} \left(\frac{\left(E_y\right)^t_{i,j,k+1} - \left(E_y\right)^t_{i,j,k}}{\Delta z} \right.$$

$$\left. - \frac{\left(E_z\right)^t_{i,j+1,k} - \left(E_z\right)^t_{i,j,k}}{\Delta y} \right) \qquad (11.23)$$

This is the 3D update equation for H_x in FDTD time marching loop following the Yee scheme. Similar figures can also be drawn for H_y, H_z, E_x, E_y and E_z and their update equations can also be obtained in a similar fashion.

However, there are many cases where the problem spaces are 2D and even 1D. In the case of 2D problem space, problem geometry or field distribution doesnot have any variation in one direction. For those cases, update equations are simplified and can be obtained from the Maxwell Curl's equations (Equations 11.16–11.21). If the problem space is independent of z-direction, then the same set of equations can be rewritten as

$$- \frac{\mu_x}{c_0} \frac{\partial \tilde{H}_x}{\partial t} = \frac{\partial E_z}{\partial y} \qquad (11.24)$$

$$- \frac{\mu_y}{c_0} \frac{\partial \tilde{H}_y}{\partial t} = - \frac{\partial E_z}{\partial x} \qquad (11.25)$$

$$- \frac{\mu_z}{c_0} \frac{\partial \tilde{H}_z}{\partial t} = \frac{\partial E_y}{\partial x} - \frac{\partial E_x}{\partial y} \qquad (11.26)$$

$$\frac{\varepsilon_x}{c_0} \frac{\partial E_x}{\partial t} = \frac{\partial \tilde{H}_z}{\partial y} \qquad (11.27)$$

$$\frac{\varepsilon_y}{c_0} \frac{\partial E_y}{\partial t} = - \frac{\partial \tilde{H}_z}{\partial x} \qquad (11.28)$$

$$\frac{\varepsilon_z}{c_0} \frac{\partial E_z}{\partial t} = \frac{\partial \tilde{H}_y}{\partial x} - \frac{\partial \tilde{H}_x}{\partial y} \qquad (11.29)$$

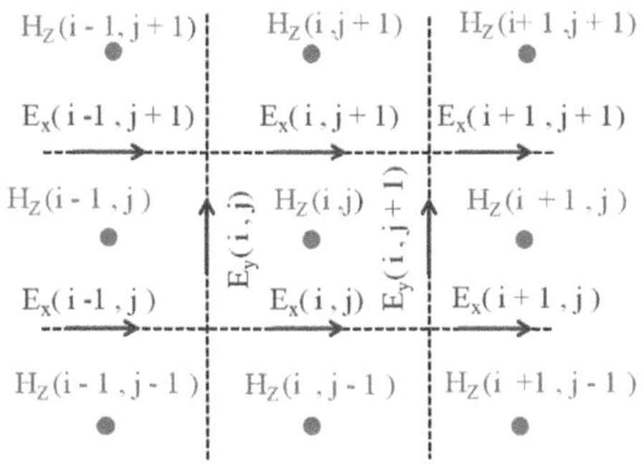

FIGURE 11.4 Position of field components in 2D FDTD (TE_z mode).

Out of these above-mentioned equations, Equations 11.24, 11.25 and 11.29 involve H_x, H_y and E_z whereas Equations 11.26, 11.27 and 11.28 involve H_z, E_x, E_y. Therefore, there are two separate sets of equations. In the first set (Equations 11.24, 11.25 and 11.29), all the magnetic field components are transverse to the reference direction z and this set of equations represent the transverse magnetic mode with respect to z(**TM$_z$**). In the second set (Equations 11.26, 11.27 and 11.28), all the electric field components are transverse to the reference direction z; hence, this set of equations represent the transverse electric mode with respect to z(**TE$_z$**). Figure 11.4 shows the position of the field components for TE_z mode in 2D FDTD. By applying central difference approximation to the set of equations mentioned above, update equations for TE_z mode can be written as

$$\left(E_x\right)_{i,j}^{t+\Delta t} = \left(E_x\right)_{i,j}^{t} + \frac{c_0 \Delta t}{\left(\varepsilon_x\right)_{i,j}} \left(\frac{\left(\tilde{H}_z\right)_{i,j}^{t+\frac{\Delta t}{2}} - \left(\tilde{H}_z\right)_{i,j-1}^{t+\frac{\Delta t}{2}}}{\Delta y} \right) \tag{11.30}$$

$$\left(E_y\right)_{i,j}^{t+\Delta t} = \left(E_y\right)_{i,j}^{t} + \frac{c_0 \Delta t}{\left(\varepsilon_y\right)_{i,j}} \left(-\frac{\left(\tilde{H}_z\right)_{i,j}^{t+\frac{\Delta t}{2}} - \left(\tilde{H}_z\right)_{i-1,j}^{t+\frac{\Delta t}{2}}}{\Delta x} \right) \tag{11.31}$$

$$\left(\tilde{H}_z\right)_{i,j}^{t+\frac{\Delta t}{2}} = \left(\tilde{H}_z\right)_{i,j}^{t-\frac{\Delta t}{2}} + \frac{c_0 \Delta t}{\left(\mu_y\right)_{i,j}} \left(\frac{\left(E_x\right)_{i,j+1}^{t} - \left(E_x\right)_{i,j}^{t}}{\Delta y} \right.$$

$$\left. -\frac{\left(E_y\right)_{i+1,j}^{t} - \left(E_y\right)_{i,j}^{t}}{\Delta x} \right) \tag{11.32}$$

In a similar fashion, field components for TM_z mode can also be shown and update equations for H_x, H_y, E_z for that mode can be written.

When there are no variations of the problem geometry and field distribution in two directions, the problem space is considered as 1D. Considering no variation in y- and z-directions, Maxwell's curl equations can be written as

$$-\frac{\mu_x}{c_0}\frac{\partial \tilde{H}_x}{\partial t} = 0 \tag{11.33}$$

$$-\frac{\mu_y}{c_0}\frac{\partial \tilde{H}_y}{\partial t} = -\frac{\partial E_z}{\partial x} \tag{11.34}$$

$$-\frac{\mu_z}{c_0}\frac{\partial \tilde{H}_z}{\partial t} = \frac{\partial E_y}{\partial x} \tag{11.35}$$

$$\frac{\varepsilon_x}{c_0}\frac{\partial E_x}{\partial t} = 0 \tag{11.36}$$

$$\frac{\varepsilon_y}{c_0}\frac{\partial E_y}{\partial t} = -\frac{\partial \tilde{H}_z}{\partial x} \tag{11.37}$$

$$\frac{\varepsilon_z}{c_0}\frac{\partial E_z}{\partial t} = \frac{\partial \tilde{H}_y}{\partial x} \tag{11.38}$$

Equations 11.33 and 11.36 do not represent any progressive fields because the equations do not contain any space derivatives, whereas the other four equations represent progressive fields. Here also, two independent set of equations are found. Equations 11.34 and 11.38 contains only E_z and H_y but Equations 11.35 and 11.37 contain only E_y and H_z. Both represent 1D plane wave, where electric and magnetic fields are transverse to x-direction (**TEM$_x$**). Figure 11.5 shows the position of the field components for 1D problem space (**TEM$_x$**). Applying central difference approximation (based on position of the fields in 1D space), 1D update equations can be written as

$$\left(E_y\right)_i^{t+\Delta t} = \left(E_y\right)_i^t + \frac{c_0 \Delta t}{\left(\varepsilon\right)_i}\left(-\frac{\left(\tilde{H}_z\right)_i^{t+\frac{\Delta t}{2}} - \left(\tilde{H}_z\right)_{i-1}^{t+\frac{\Delta t}{2}}}{\Delta x}\right) \tag{11.39}$$

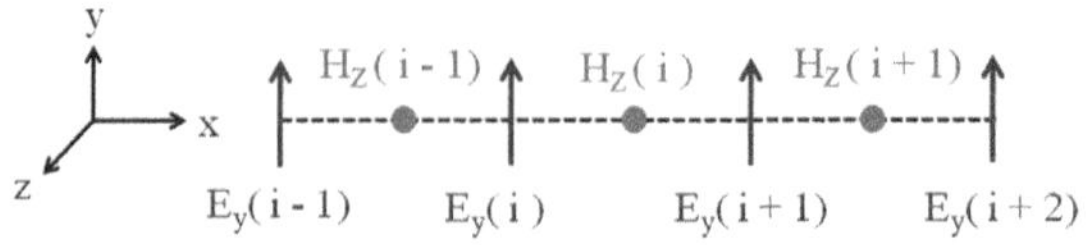

FIGURE 11.5   Position of field components in 1D FDTD (*TEM$_x$* mode).

$$\left(\tilde{H}_z\right)_i^{t+\frac{\Delta t}{2}} = \left(\tilde{H}_z\right)_i^{t-\frac{\Delta t}{2}} + \frac{c_0 \Delta t}{(\mu)_i}\left(-\frac{\left(E_y\right)_{i+1}^t - \left(E_y\right)_i^t}{\Delta x}\right) \qquad (11.40)$$

The equations mentioned for 3D,2D and 1D problem space are the main update equations, which are required to be coded for FDTD time marching loop to find the numerical solutions of the 3D, 2D and 1D problem geometry, respectively.

11.4 SOURCE WAVEFORM FOR FDTD SIMULATION

For FDTD simulation, sources are one of the necessary and important parameters which demand special attention. It is the source that excites electric and magnetic fields as a function of time. In general, waveform is selected depending on the problems under consideration. The wavelength spectrum of the source waveform must include all the wavelengths of interest for the simulation. In this chapter, either modulated Gaussian or Gaussian pulses are used for simulations to investigate the scattering characteristics of periodic structures. The field in the time domain is captured and frequency-domain response can be obtained by the Fourier transform of the captured field.

11.4.1 GAUSSIAN WAVE

FDTD simulations should provide the numerical results for all the wavelengths present in the spectrum. For simulation, the cell size of the problem space and the highest wavelength of the source waveform plays an important role. One may set the highest wavelength larger than 20 cell size to obtain a reasonably accurate result (to be discussed later in numerical dispersion section). A Gaussian wave as a function of time can be expressed as

$$x(t) = e^{-\frac{t^2}{\tau^2}} \qquad (11.41)$$

Here τ determines the width of the Gaussian pulse and related with maximum frequency component as [65]

$$\tau \cong \frac{0.5}{f_{max}} \qquad (11.42)$$

Once τ corresponding to the maximum frequency is known, it is possible to construct the Gaussian wave for the FDTD simulation. One thing to be noted here is that for FDTD simulation, field values are initialized as zero. Therefore the source should also be zero at t = 0. For doing this, time shifted Gaussian wave is to be constructed so that the value of the waveform is zero at zero instant of time. A time-shifted Gaussian wave is represented as

$$x(t) = e^{-\frac{(t-t_0)^2}{\tau^2}} \tag{11.43}$$

Here, t_0 is the time shift. Considering a negligible value of $x(t)(e^{-20})$ at zero instant of time, it can be shown that [65,66]

$$t_0 \cong 4.5 \; \tau \tag{11.44}$$

Considering the above equation for constructing a Gaussian waveform, the required waveform with a centre wavelength 1550 nm (frequency 193 *THz*) is constructed and shown in Figure 11.6. To make the value of the waveform zero at zero instant of time, the required amount time shift is 0.011658 ps also shown in the figure.

11.4.2 MODULATED GAUSSIAN WAVE

A time-shifted cosine modulated Gaussian wave can be represented as

$$x(t) = \cos(\omega_c t) \; e^{-\frac{(t-t_0)^2}{\tau^2}} \tag{11.45}$$

In FDTD simulation, this type of modulated Gaussian is required to find the spectral response for frequency band centred at ω_c ($\omega_c = 2\pi f_c$). To construct a modulated Gaussian wave, a Gaussian wave with required bandwidth is constructed first and then the modulating function with frequency f_c is multiplied with Gaussian waveform as shown in Equation. 11.45. τ can be found from the relation [11.65]

$$\tau = \frac{0.966}{\Delta f} \tag{11.46}$$

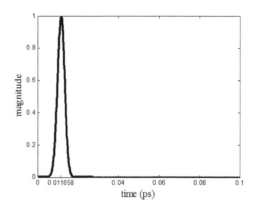

FIGURE 11.6 Gaussian waveform for FDTD simulation at 1550 nm.

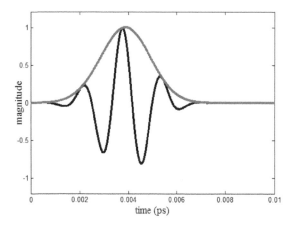

FIGURE 11.7 Modulated Gaussian waveform with Gaussian envelop for FDTD simulation.

Figure 11.7 shows a modulated Gaussian wave along with the Gaussian envelope. Here, the central frequency of the modulating signal is 600 THz. Throughout this chapter, these type of Gaussian and modulated Gaussian waveforms are used as the source for FDTD simulations.

11.4.3 TOTAL FIELD/SCATTER FIELD CORRECTION OF THE FDTD SOURCE

FDTD method computes the numerical solution of the wave matter interaction located within the computational domain. For that, plane wave propagating in one direction is to be generated. This was quite challenging because FDTD domain is finite whereas plane wave is infinite in the direction perpendicular to the propagation. Total Field/Scattered Field (TF/SF) method (also known as the Huygens surface method), first reported in [67], and later in [68,69] is used to solve this problem. The concept of TF/SF is based on the linearity of Maxwell's equations where total electric field (E_{total}) and total magnetic field (H_{total}) is considered as

$$E_{total} = E_{inc} + E_{sca} \qquad\qquad H_{total} = H_{inc} + H_{sca} \qquad\qquad (11.47)$$

These values result from the wave matter interaction in space. Figure 11.8 shows the 1D FDTD lattice with TF/SF regions. Referring Figure 11.8, Equations 11.39 and 11.40 can then be written in terms of TF/SF zoning as

$$\left(E_{y(tot)}\right)_{i_L}^{t+\Delta t} = \left(E_{y(tot)}\right)_{i_L}^{t} + \frac{c_0 \Delta t}{(\varepsilon)_i} \left(\frac{\left(\tilde{H}_{z(tot)}\right)_{i_L+\frac{1}{2}}^{t+\frac{\Delta t}{2}} - \left(\tilde{H}_{z(sca)}\right)_{i_L-\frac{1}{2}}^{t+\frac{\Delta t}{2}}}{\Delta x} \right) \qquad (11.48)$$

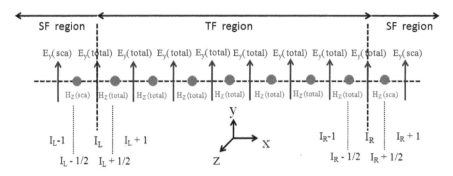

FIGURE 11.8   1D FDTD lattice with TF/SF regions.

$$\left(\tilde{H}_{z(sca)}\right)^{t+\frac{\Delta t}{2}}_{i_L-\frac{1}{2}} = \left(\tilde{H}_{z(sca)}\right)^{t-\frac{\Delta t}{2}}_{i_L-\frac{1}{2}} + \frac{c_0 \Delta t}{(\mu)_i}\left[\frac{\left(E_{y(tot)}\right)^{t}_{i_L} - \left(E_{y(sca)}\right)^{t}_{i_L-1}}{\Delta x}\right] \qquad (11.49)$$

These equations are inconsistent and incorrect because in the same equations, some fields are stored as total filed and some fields are stored as scatter fields in the memory of the computer. However as

$$-\tilde{H}_{z(sca)}\Big)^{t+\frac{\Delta t}{2}}_{i_L-\frac{1}{2}} - \tilde{H}_{z(inc)}\Big)^{t+\frac{\Delta t}{2}}_{i_L-\frac{1}{2}} = -\tilde{H}_{z(tot)}\Big)^{t+\frac{\Delta t}{2}}_{i_L-\frac{1}{2}} \qquad (11.50)$$

Equation11.48 can be made consistent and correct as

$$\left(E_{y(tot)}\right)^{t+\Delta t}_{i_L} = \left(E_{y(tot)}\right)^{t}_{i_L} + \frac{c_0 \Delta t}{(\varepsilon)_i}\left[\frac{\left(\tilde{H}_{z(tot)}\right)^{t+\frac{\Delta t}{2}}_{i_L+\frac{1}{2}} - \left(\tilde{H}_{z(sca)}\right)^{t+\frac{\Delta t}{2}}_{i_L-\frac{1}{2}}}{\Delta x}\right] \qquad (11.51)$$

$$-\frac{c_0 \Delta t}{\varepsilon_0}\left(\tilde{H}_{z(inc)}\right)^{t+\frac{\Delta t}{2}}_{i_L-\frac{1}{2}}$$

This correction fixes the inconsistency at grid point i_L, arises because of TF/SF boundary. Similarly, the inconsistency arising at grid point $i_L - 1/2$ (Refer Figure 11.8), while calculating the magnetic field, can be avoided as

$$\left(\tilde{H}_{z(sca)}\right)^{t+\frac{\Delta t}{2}}_{i_L-\frac{1}{2}} = \left(\tilde{H}_{z(sca)}\right)^{t-\frac{\Delta t}{2}}_{i_L-\frac{1}{2}} + \frac{c_0 \Delta t}{(\mu)_i}\left[\frac{\left(E_{y(tot)}\right)^{t}_{i_L} - \left(E_{y(sca)}\right)^{t}_{i_L-1}}{\Delta x}\right] \qquad (11.52)$$

$$-\frac{c_0 \Delta t}{\varepsilon_0}\left(E_{y(inc)}\right)^{t}_{i_L}$$

Similarly, TF/SF sources can be implemented in 2D FDTD formulation also. Details of these FDTD update equations including the TF/SF correction can be found in [70].

11.5 NUMERICAL DISPERSION AND STABILITY

FDTD method basically provides an approximate solution for the fields behaviours for the real physical behaviours of the fields. In this method, derivative of a continuous function is approximated using finite difference scheme as already discussed. It was found there the solution accuracy depends on the step size Δx. The error associated with the non-zero step size Δx is known as numerical dispersion [71]. Though, this numerical dispersion, depends on the wavelength, time step (Δt) and the direction of propagation of the wave also. The effect of numerical dispersion is equivalent to filling the medium with a material whose dielectric constant is different from the dielectric constant of the actual material. As a result, even in a homogeneous free space, the numerical velocity of the wave differs from the free space velocity of the wave. In a source free region, Maxwell's curl equations for a *TEM* wave propagating in x direction can be written as

$$\varepsilon \frac{\partial E_y}{\partial t} = - \frac{\partial H_z}{\partial x} \tag{11.53}$$

$$\mu \frac{\partial H_z}{\partial t} = - \frac{\partial E_y}{\partial x} \tag{11.54}$$

These coupled first-order differential equations can be decoupled and eliminating H_z, second-order differential equation for E_y can be written as

$$\frac{\partial^2 E_y}{\partial t^2} = \frac{1}{\mu\varepsilon} \frac{\partial^2 E_y}{\partial x^2} \tag{11.55}$$

The solution of monochromatic sinusoidal traveling wave can be written as

$$E_y(x,t) = E_0 e^{j(\omega t - kx)} \tag{11.56}$$

Equations 11.55 and 11.56 can be discretized and it can be shown that the dispersion equation can be written as

$$\sin\left(\frac{\omega \Delta t}{2}\right) = \pm \left(\frac{c\Delta t}{\Delta x}\right) \left[2 \sin\left(\frac{\bar{k}\Delta x}{2}\right)\right] \tag{11.57}$$

Equation 11.57 is the dispersion equation because it relates ω and $\bar{k}$. It is to be noted here that as Δx and Δt tend to zero, the wave propagation tends to become dispersion less because then the equation gives the exact value of the wave vector in the limiting case. This time step $\Delta t = \Delta x/c$ is the magic time step and the phase velocity as $\left(\bar{v}_p\right)$ can be defined as

$$\frac{\bar{v}_p}{c} = \frac{\pi f \Delta x}{c} \frac{1}{\sin^{-1}\left[\left(\frac{\Delta x}{c\Delta t}\sin\left(\pi f \Delta t\right)\right)\right]} \tag{11.58}$$

Considering the wave in traveling in free space, Equation 11.58 can be written as

$$\frac{\bar{v}_p}{c_0} = \frac{\pi \Delta x}{\lambda_0} \frac{1}{\sin^{-1}\left[\left(\frac{1}{\alpha}\sin\left(\frac{\pi\alpha\Delta x}{\lambda_0}\right)\right)\right]} \tag{11.59}$$

where α (stability ratio) $= c_0\Delta t/\Delta x$ and λ_0 is the free space wavelength. It can be easily seen that if $\alpha = 1$, the time step becomes the magic time step and numerical dispersion can be avoided. Figure 11.9 shows the plot between the phase velocity and cell size $\Delta x/\lambda_0$. From the figure, it is clear that as the value of α deviates from one, the introduced error increases, i.e. numerical phase velocity becomes more and more different from the actual phase velocity. From the figure, it is also clear that for a particular value of α (other than one), the error can be reduced, if the grid cell size is reduced. For FDTD simulations, the problem space is discretized into cells and then material properties are introduced. Smaller cell size ensures uniform distribution

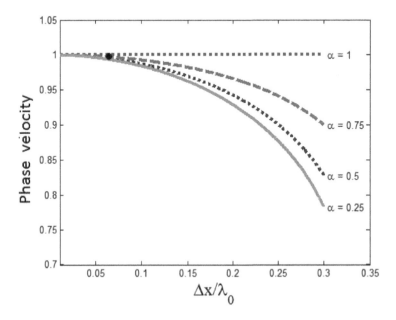

FIGURE 11.9 Phase velocity plotted against cell sizes for different values of α showing numerical dispersion.

material properties and thereby uniform distribution of field values. But, as the cell sizes are reduced, the cost of computation increases. As a compromise between the accuracy and computational cost, the rule of thumb is that in no case the cell size should exceed $\lambda/10$ at the highest frequency. It was found that for an accuracy of about 1%, the cell size should be $\lambda/20$ at the highest frequency. In FDTD simulation, cell size Δx and time step Δt are closely related. If v_{max} is the maximum phase velocity then the maximum time step (Δt) permitted is $\Delta x/v_{max}$. If $\Delta t > \Delta t_{max}$, it can be said that the distance travelled by the wave in time Δt is more than Δx. It means that the wave will miss the next node in simulation which leads to instability. So the condition

$$\Delta t < \frac{\Delta x}{v_{max}} \tag{11.60}$$

is required to be satisfied for stability of the simulation. This is the Courant–Friedrichs–Lewy (CFL) stability condition. Due to inhomogeneity of the medium, phase velocity may vary from cell to cell. Therefore Δt is chosen as

$$\Delta t < \left(\frac{1}{2}\right)\frac{\Delta x}{v_{max}} \tag{11.61}$$

It means that $\alpha = 1/2$ and the wave takes a time span of $2\Delta t$ to travel to the next node. From the following example the concept of numerical dispersion will be clear. Let us consider a wave (modulated Gaussian wave) of central frequency 600 THz, is traveling in free space. Then the free space wavelength (λ_0) will be 0.5 μm. Table 11.1 shows the information regarding the numerical error introduced for different values of stability factor (α) and cell size (Δx). Time step is calculated, maintaining the CFL condition. It can be seen from the table that for $\alpha = 1$, phase velocity remains unchanged even for different values of cell sizes ($\Delta x/\lambda$) and the leading edge of the wave travels a distance of 6μm in 20 fs. It is exactly equal to the distance what light should travel in 20 fs in free space. But as the cell sizes are reduced, time steps are also reduced and the results become more and more accurate. But the reduction in

---

TABLE 11.1

Numerical Dispersion of a Wave of Central Frequency 600 THz, Traveling in Free Space

	$\alpha = 1$			$\alpha = 0.5$		
$\Delta x/\lambda$	1/5	1/10	1/20	1/5	1/10	1/20
Δx	0.1 μm	0.05 μm	0.025 μm	0.1 μm	0.05 μm	0.025 μm
Problem space	50 μm	25 μm	12.5 μm	50 μm	25 μm	12.5 μm
Δt	0.33 fs	0.167 fs	0.083 fs	0.167 fs	0.083 fs	0.042 fs
v_p/c	1	1	1	0.9431	0.9873	0.9969
Change in Phase velocity	–	–	–	−5.69%	−1.27%	−0.31%
S (in 20 fs)	6 μm	6 μm	6 μm	5.659 μm	5.9238 μm	5.9814 μm

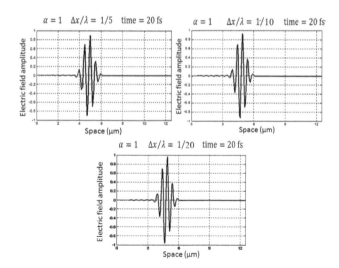

FIGURE 11.10   Traveling wave showing no numerical dispersion for $\alpha = 1$ even for different cell sizes.

cell size is associated with increased computational cost. Figure 11.10 shows the FDTD simulation results for the traveling wave discussed above for different values of $\Delta x/\lambda$ considering $\alpha = 1$. From the table, it is found that for $\Delta x/\lambda = 1/5$, the numerical phase velocity is 94.31 % of the actual phase velocity and the wave travels a distance of 5.659 μm instead of 6 μm. This error can be reduced if the cell sizes are reduced. It can also be seen from the table that when the ($\Delta x/\lambda$) value is changed to 1/20, the numerical phase velocity becomes 99.69 % of its actual value and the wave travels a distance of 5.9814 μm which is very close to the actual value. Figure 11.11 shows the numerical error introduced in FDTD simulation for $\alpha = 0.5$ for the same traveling wave. Results shown in Figures 11.10 and 11.11 are strictly in accordance with the results shown in Figure 11.9.

$$50\,\text{nodes} \qquad \lambda = 0.5\,\mu\text{m} \qquad f = 600\,\text{THz}$$

11.6   ABSORBING BOUNDARY CONDITIONS

In general, EM analyses of scattering structures are open region problems. But because of the limited memory of the computer, such unbounded problems can't be computed in the computer. Therefore, absorbing boundary conditions are needed for the truncation of the problem space domain. These boundary conditions absorb or suppress spurious back reflections and the spatial domain appears as infinite in extent.

11.6.1   MUR'S ABSORBING BOUNDARY CONDITIONS

Enquist and Majda[72] first worked on the analysis of this absorbing boundary condition which was later optimized by Mur [73]. A boundary is said to be reflection less

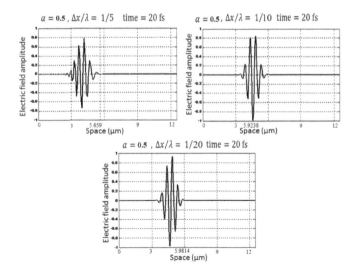

FIGURE 11.11   Traveling wave showing numerical dispersion for $\alpha = 0.5$ for different cell sizes.

when a wave incident on it and will continue to propagate in the forward direction only without any reflection. The TEM_x wave, considered in Equation.11.55 can be factorized and written as

$$\left(\frac{\partial E_y}{\partial t} + c\frac{\partial E_y}{\partial x}\right)\left(\frac{\partial E_y}{\partial t} - c\frac{\partial E_y}{\partial x}\right) = 0 \tag{11.62}$$

Applying central difference approximation, the left going wave can be written as

$$\frac{E(y)_{i+\frac{1}{2}}^{n+1} - E(y)_{i+\frac{1}{2}}^{n}}{\Delta t} = c\frac{E(y)_{i+1}^{n+\frac{1}{2}} - E(y)_{i}^{n+\frac{1}{2}}}{\Delta x} \tag{11.63}$$

If the field values at the half grid points and at half time steps are not available then they may be obtained by averaging as

$$E(y)_{i+\frac{1}{2}}^{n+1} = \frac{E(y)_{i+1}^{n+1} + E(y)_{i}^{n+1}}{2} \tag{11.64}$$

Similar type of expressions can be written for $E(y)_{i+\frac{1}{2}}^{n}$, $E(y)_{i+1}^{n+\frac{1}{2}}$, $E(y)_{i}^{n+\frac{1}{2}}$. It can be shown that Equation. 11.63 can be written as

$$\begin{aligned} E(y)_{i+1}^{n+1} + E(y)_{i}^{n+1} - E(y)_{i+1}^{n} - E(y)_{i}^{n} \\ = \frac{c\Delta t}{\Delta x}\left(E(y)_{i+1}^{n+1} + E(y)_{i+1}^{n} - E(y)_{i}^{n+1} - E(y)_{i}^{n}\right) \end{aligned} \tag{11.65}$$

As $\alpha = c\Delta t/\Delta x$, Equation. 11.65 can be written as

$$E(y)_i^{n+1} = E(y)_{i+1}^n + \frac{\alpha-1}{\alpha+1}\left(E(y)_{i+1}^{n+1} - E(y)_i^n\right) \tag{11.66}$$

Equation 11.66 is used to update the tangential field at boundary at node $x = 0$. Similarly, the tangential field at the boundary at $x = N$ (N = number of nodes in the FDTD problem space) can be written as

$$E(y)_N^{n+1} = E(y)_{N-1}^n + \frac{\alpha-1}{\alpha+1}\left(E(y)_{N-1}^{n+1} - E(y)_N^n\right) \tag{11.67}$$

Expression 11.66 and 11.67 are known as the first order analytical or first-order Mur's boundary conditions. It should be noted here that if $\alpha = 1$ Equation. 11.66 and Equation. 11.67 reduces to

$$E(y)_i^{n+1} = E(y)_{i+1}^n \tag{11.68}$$

$$E(y)_N^{n+1} = E(y)_{N-1}^n \tag{11.69}$$

These boundary conditions are stable for $\alpha = 1$ and for normal incidence only. For all other values of α, the residual reflection arises due to the phase velocity being different from c_0. Figure 11.12 shows the application of Mur's absorbing boundary condition to truncate the rightmost grid point so as to avoid any spurious reflection from the boundary. Here also, modulated Gaussian wave of central frequency 600 THz is used. Simulation is carried out keeping the CFL stability conditions in mind. Mur's second-order absorbing boundary conditions [71] for 2D cases can also be derived in a similar fashion.

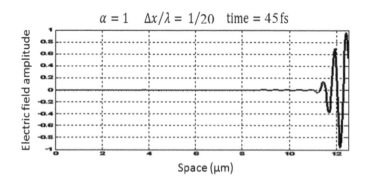

FIGURE 11.12 Application of Mur's absorbing boundary conditions (1D FDTD simulation).

11.6.2 PERFECTLY MATCHED LAYER

The concept of perfectly matched layer (PML) was first introduced in 1994 [74]. PML is an absorbing boundary condition (ABC) where the impedance of the absorbing layers is perfectly matched to the problem space. Different types of PML (split-field PML, uniaxial PML, and coordinate transformation PML) can be constructed for both electromagnetic wave[75–77] and acoustic wave problems [78,79]. A UPML (uniaxial PML) [80] can be achieved by incorporating a lossy medium at the boundaries of the problem space, i.e. an absorbing material which is artificial in nature is placed at the edges of the boundaries. After entering into the absorbing layer, the wave decays exponentially. Reflection from the lossy region is prevented because of impedance matching and as the outgoing waves are absorbed, reflection from the grid boundaries are also avoided. To prevent reflection at all angles at all polarization, diagonally anisotrpic absorbing material is placed at the boundary. This is main advantage of PML layer. Detail analytical discussion couldn't be done here but the readers may refer [65,66,70] for complete analysis on FDTD and PML. Here, in this chapter a simple example is given by developing 2D FDTD codes where a wave is shown which is propagating in a 2D free space (Figure 11.13). Dirichlet boundary condition is applied in X high and low boundary while Y high and low boundary is truncated by applying PML. The figure shows the absorption of light in the PML region.

11.7 PBGS WITH QUARTER-WAVELENGTH MATERIAL DUO

In the preceding section, it is discussed that the periodic nature of PBGS does not allow all the wavelengths to pass through it when a wave of certain range of wavelengths is incident on it. It can, then acts as a bandstop filter [3]. When a defect is introduced within the periodic structure, a passband, within the stopband is created and the PBGS acts as a transmission filter. The passband is because of the constructive interference of the trapped light within the defect area and due to this constructive interference, transmission spectra at the resonant wavelength is produced. This type of structures find various applications in optical communication [81]. Cavity

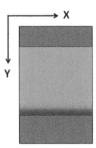

FIGURE 11.13   PML region showing the absorption of light (2D FDTD simulation).

resonator type PBGSs as narrowband transmission filter (NTF), with arbitrary material width have been proposed earlier [35,36] where the thin film material widths are mainly determined by trial and error method. In this section, optical filter based on Fabry–Perot resonator principle is shown which is basically a stack of multiple periodic layers of high (nH) and low (nL) refractive index (r.i.) materials. The widths of the film layers are the quarter wavelength at the operating wavelength. These types of resonating filters have wide applications in wireless and microwave regime [22,82,83]. As the layer widths of the material duo are predetermined, it is not required to choose layer widths arbitrarily.

11.7.1 CHARACTERISTICS OF THIN FILMS WITH SUBWAVELENGTH DIMENSIONS

To design the proposed PBGS as optical filter, quarter wavelength widths of two materials are used here. Therefore, before studying the characteristics of optical filters, designed with PBGS, the reflection, transmission characteristics of dielectric having sub-wavelength dimensions are studied. It is a well-known fact that a slab of quarter wavelength width at the working wavelength reflects maximum energy of the incident signal, whereas a slab of half-wavelength width at the working wavelength reflects minimum of the incident energy. The above statements are verified here with the developed FDTD codes.

To do so, a glass slab (r.i. 1.5) having quarter wavelength width, corresponding to the incident signal (modulated Gaussian wave) of central wavelength of 500 nm (frequency 600 THz) is chosen. The wave is made to incident normally on the slab and its reflection and transmission coefficient is noted. For this simulation, TF/SF source is used and Mur's first order boundary conditions are used to truncate the FDTD problem space. Figure 11.14 shows the simulation results of a quarter-wave glass

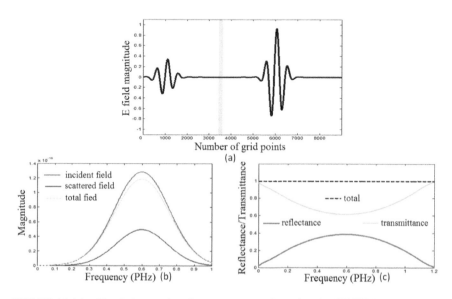

FIGURE 11.14   Simulation results of a quarter-wave glass plate (at 600 THz).

plate. From the figure, it is clear that this plate reflects maximum of the incident field energy at 600 THz. The slab and the medium are considered as lossless here. Therefore, the figure shows the conservation of the energy also (dotted black line).

(a) Shows the reflected and transmitted wave from a glass slab of quarter wavelength thickness at 500 nm (600 THz) (b) Fourier transform of the incident field, scatter filed and total field (c) reflectance (and transmittance) plot showing that the reflectance is maximum at 600 THz for the quarter-wave dielectric slab.

(a) Shows the reflected and transmitted wave from a glass slab of half-wavelength thickness at 500 nm (600 THz) (b) Fourier transform of the incident field, scatter filed and total field (c) reflectance (and transmittance) plot showing that the reflectance is minimum at 600 THz for the half-wave dielectric slab.

In the next step, a glass plate of thickness half wavelength (at 500 nm) is chosen and excited with the same source. The simulation results of this half-wave plate are shown in Figure 11.15. It can be seen from the figure that the reflection is minimum for a glass plate of half-wavelength thickness. But still the conservation is 100 % as the slab and medium is considered lossless. Therefore, the results shown in Figures 11.14 and 11.15 are in accordance with supporting theory and similar to the results shown in [84].

11.7.2 BIMATERIAL CAVITY RESONATOR WITH QUARTER-WAVELENGTH MATERIAL DUO

To design a bandstop a filter, a multilayer stack of two dielectric materials of suitable r.i. contrast, i.e. an N-fold periodic replication of high/low index layers of the types $(HL)^N$ or $(LH)^N$ can be considered, where H and L represent the high and low r.i layer, respectively, and N is the number of bilayers. The reflection bands arise because of

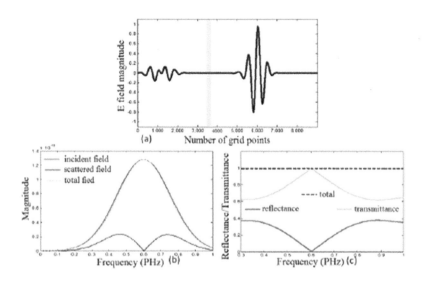

FIGURE 11.15 Simulation results of a half-wave glass plate (at 600 THz).

the periodic nature of the structure. Here, the layer widths are considered as the quarter wavelength widths at the working wavelength (1550 nm).

Now two types of cavity resonators namely $(LH)^N 2L(HL)^N$ and $(HL)^N L(HL)^N L$ can be considered to design the PBGS, which can act as narrowband filters. In the first type of structure, the 2L defect (cavity) is an absentee layer and so ignoring the half-wave absentee low index layer, the structure can be rewritten as $(LH)^{N-1} LHHL(HL)^{N-1}$. Again the structure can be considered as $(LH)^{N-2} LH LLHL(HL)^{N-2}$ by further ignoring the half-wave absentee high index layers. It can be seen here that the process of deletion of half-wave absentee layers (LL and HH) can be continued till it reaches a structure like LL. Therefore, it can be said that the structure $(LH)^N 2L(HL)^N$ basically behaves as half-wave low index layer at the working wavelength which is responsible for the reflection dip within the stopband as already stated.

If a similar technique is applied to a structure like $(HL)^N L(HL)^N$, it can be seen that the structure will finally behave like a single L layer at the working wavelength. Now, if an extra L layer is added at the end of same structure, the structure will be look like $(HL)^N L(HL)^N L$ and will finally act as LL which provides a transmission band within the stop-band. The above mentioned two types of structures are popularly known as Fabry–Perot resonators or 1D Bragg gratings.

11.7.3 Stopband Characteristics of the Structures $(HL)^N(HL)^N$ and $(LH)^N(LH)^N$

It is already mentioned that the reflection band arises because of the periodic nature of PBGS and two different structures like $((HL)^N(HL)^N$ and $(LH)^N(LH)^N)$ can be considered for creating the stopband. In Figure 11.16, those two types of PBGS along

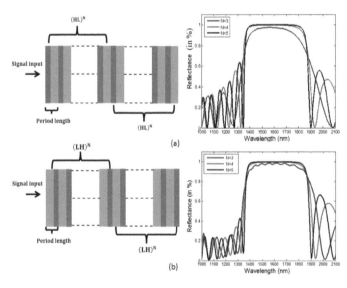

FIGURE 11.16 Schematic diagram for PBGSs like (a) $(HL)^N(HL)^N$ and (b) $(LH)^N(LH)^N$[8] along with their stopband spectra.

with their spectral characteristics are shown when the structures are excited with a Gaussian source of central wavelength 1550 nm. For this simulation, Lithium Niobate ($LiNbO_3$) and Magnesium Fluoride (MgF_2) are considered as high and low r.i. materials and their widths are the quarter wavelength width at the working wavelength, i.e. at 1550 nm. At 1550 nm, r.i of $LiNbO_3$ and MgF_2 are considered as 2.13 and 1.38 at 1550 nm [85,86], respectively. Here, in this work, extraordinary (n_e) r.i. of $LiNbO_3$ is assumed. From the figure, it can be seen that when the number of bilayers (N) are increased, the nature of the stopband improves. One more thing to be noted here that though, both the structures provide more or less similar type of reflection band, the stopbands are better for the structure $(HL)^N(HL)^N$ for same values of N.

11.8   TRANSMISSION FILTERS USING FABRY–PEROT CAVITIES

It is already mentioned that Fabry–Perot cavities can be formed utilizing structures like $(HL)^NL(HL)^N$ Lor $(LH)^N2L(HL)^N$. In that case, both the structures basically behaves as a half wavelength low r.i. layer at the working wavelength, provided both high and low index layer are considered to be of quarter wavelength width at the working wavelength. Therefore, a passband within the stopband can be obtained. This principle is used here to design the narrowband transmission filter. To do that, power reflectance (or transmittance) characteristics of both the structures are studied. Figure 11.17 shows the structures $(LH)^N2L(HL)^N$ and $(HL)^NL(HL)^NL$ and their power reflectance spectrum. Same source of light (Gaussian wave with central wavelength 1550 nm) is used here to excite the structures. Dotted black line and solid black line are the reflectance spectrum of the structure $(LH)^N2L(HL)^N$ and $(HL)^NL(HL)^NL$, respectively. Here, for the simulation, N is considered as three. There are some important inferences that can be obtained from these figures. Firstly, the peaks of the passband obtained within the stopband for the structures are not at the working

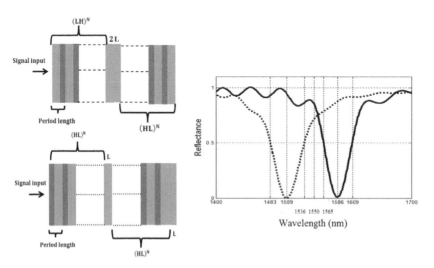

FIGURE 11.17   Comparison of the power reflectance spectra obtained from the Fabry–Perot cavities like $(LH)^N2H(HL)^N$ and $(HL)^NL(HL)^NL$.

wavelength. For the structure, $(LH)^N 2L(HL)^N$, the peak is at 1509 nm and whereas, the structure $(HL)^N L(HL)^N L$ provides a passband at 1586 nm. Therefore, the phase shift corresponding to peak wavelength change of +36 nm and -41 nm are, respectively, found for the structure $(HL)^N L(HL)^N L$ and $(LH)^N 2H(HL)^N$. It is also seen that the full width half maximum (FWHM) bandwidth of the transmittance spectrum, is 44 nm for the structure $(HL)^N L(HL)^N L$ and the same is 53 nm for the structure $(LH)^N 2L(HL)^N$. Therefore, it can be concluded that for designing the narrowband transmission filters (NTF), utilizing the quarter wavelength width, the structure $(HL)^N L(HL)^N L$ is always a better choice.

11.8.1 SHIFTING THE CENTRAL WAVELENGTH OF THE OUTPUT SPECTRA AT THE WORKING WAVELENGTH

It is clear that Fabry–Perot cavities like $(LH)^N 2L(HL)^N$ and $(HL)^N L(HL)^N L$ provide a phase shifted reflection dip within the reflection band and the phase shift is comparatively less for the structure $(HL)^N L(HL)^N L$. The phase shift can be adjusted by adjusting the width of the sandwiched L layer width at the middle of the structure $(HL)^N L(HL)^N L$. It is found here that if the width of the L layer is adjusted to $0.9347L$, the peak of the reflection dip shifts at the working wavelength and a transmittance spectrum centred at 1550 nm is obtained. Though the other structure $((LH)^N 2L(HL)^N)$ is not very viable to design the narrowband transmission filter with materials of quarter wavelength width, study reveals that the phase shift associated with that type of structure, can also be adjusted by adjusting the sandwiched $2L$ layer width. It was seen that if the $2L$ layer width of the structure $(LH)^N 2L(HL)^N$ is changed to $2.1L$, transmission spectrum centred at 1550 nm is obtained. From the above discussion, it can be concluded that by increasing and reducing the sandwiched low r.i. layer width, central peak of the reflection dip can be adjusted towards the higher or lower wavelength side, respectively. This is an useful information for the designer. Dotted black line and solid black line in the Figure 11.18 shows the power reflectance spectra of the structure $(LH)^N \mathbf{2.1L}(HL)^N$ and $(HL)^N \mathbf{0.9347L}(HL)^N L$.

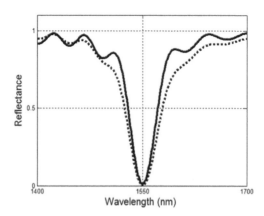

FIGURE 11.18 Comparison of the power reflectance spectra obtained from the Fabry–Perot cavities like $(LH)^N 2H(HL)^N$ and $(HL)^N L(HL)^N L$ after adjusting the defect layer width.

11.8.2 NARROWBAND OPTICAL FILTER AND ITS ANALYSIS

As already explained, a structure like $(HL)^N L(HL)^N L$ is a better option compared to a structure like $(LH)^N 2L(HL)^N$ to design an NTF. Therefore, a periodic structure $(HL)^N L(HL)^N L$, made up of $LiNbO_3$ (high refractive index layer) and MgF_2 (low refractive index layer) is considered where the layer thicknesses are the corresponding quarter wavelength widths at 1550 nm. Figure11.19(a and b) show the power reflectance spectrum of such a structure for different values of N. As it is clear from the figure that as the bilayer number of the PBGS is increased, FWHM bandwidths of the spectral responses is reduced which increases the Q value of the filtered output. It is clear in the figure that the FWHM bandwidth of the reflectance spectra of this structure is 43 nm for $N = 3$. But, the bandwidth of the spectrum can be reduced by increasing the bilayers (N). It is found that For N = 5, the FWHM is 8 nm. By further increasing the value of N, The FWHM bandwidth can be further reduced. Thus, the structure can be used as an NTF. It is also clear from the figure that for all values of N, peak of the transmitted band occurs at 1586 nm. However, when the sandwiched L layer width is considered as 0.9374L, peak of the transmission band is found at 1550 nm as already explained.

Simulations are carried out for other wavelengths also. It is found that when the defect width is not adjusted, the of phase shift of the spectral output varies with the input wavelength following a particular mathematical pattern. In Figure 11.20(a), it is seen that the position of the peak of the transmission bands are, respectively, at 1381 and 1791 nm when the central wavelengths of the input signals are 1350 and 1750 nm. So, the phase shifts of the peak of the transmission band for these two signals (with central wavelengths at 1350 and 1750 nm) are 31 and 41 nm and for a signal with central wavelength at 1550 nm, the phase shift is 36 nm. Therefore, it can be said that, when the central wavelength of the input signal is changed by ± 200 nm from 1550 nm, the amount of shift of the transmission peak is changed by ± 5 nm from 36 nm. Similar type of effects are observed for sources with other wavelengths also. Here also, it is observed that the peak wavelength of the transmission spectra can be made to occur at the working wavelength if the width of the defect layer is adjusted. Figure 11.21 shows a consolidated plot of reflectance spectra (N = 3) for five different input wavelengths (without adjusting the defect layer width) From the figure, it can be seen that as the central wavelength of the source changes from

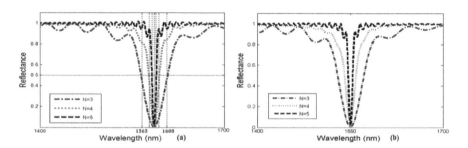

FIGURE 11.19 Reflectance spectra of the structure $(HL)^N L(HL)^N L$ for different number of bilayers (N) (a) when the defect layer is not adjusted (b) with adjusted defect layer width [8].

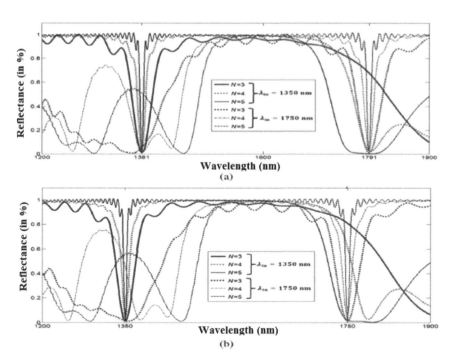

FIGURE 11.20 Reflectance spectra of the structure $((HL)^N L(HL)^N L)$ for input signal of different central wavelength(1350 and 1750 nm) (a) when the defect layer is not adjusted (b) with adjusted defect layer width [8].

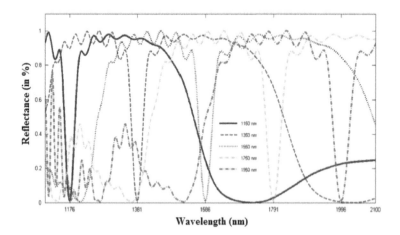

FIGURE 11.21 Phase shifted reflection spectra of an structure $((HL)^3 L(HL)^3 L)$ for input signal of different wavelengths showing that the shift of the wavelength peak is 36 ± 5 nm as the central wavelength of the source is varied by $\pm n200$ nm from 1550 nm for $n = 1,2,3,...$ [8].

$(1550 \pm n200)$ nm, the shift of the spectral output peak is $(36 \pm n5)$ nm, for $n = 1, 2,$ …. It is also observed from the figure that keeping the structure geometry same, if the central wavelength of the input signal is increased, the FWHM bandwidth of the transmission spectra is increased. However, from Figures 11.19 and 11.20, it is observed that as the number of bilayers (N) is increased, the bandwidth decreases. All these information are important to the designer. For the simulation, 1D FDTD update equations are used and the FDTD problem space is truncated using Mur's first order boundary condition. From the captured TF/SF, reflectance can be calculated as

$$R(f) = \left[\frac{\mathcal{F}(E_{scatter})}{\mathcal{F}(E_{total})} \right]^2 \tag{11.70}$$

Here, $\mathcal{F}$ stands for Fourier transform

11.9 CONCLUSION

In this chapter at first, a brief discussion on most of the important topics of FDTD method is presented. The discussion is started with the basics of Maxwell's equations and its approximated discretization. Finite difference equations for electric and magnetic fields based on Yee's algorithm are derived. A brief description is given on the sources used for simulation. For, simulation stability and accuracy, CFL criteria (to avoid numerical dispersion) is also discussed here. Absorbing boundary conditions (Mur's ABC and PML) for terminating the problem space are also presented.

After discussing the basics of FDTD, spectral analysis of a narrowband tunable transmission filter based on Fabry–Perot cavity resonator using quarter-wavelength materials ($LiNbO_3$ and MgF_2) is presented. It is shown here that inclusion of a defect of low r.i. material in a multilayer PBGS at suitable positions, narrow-band phase-shifted transmission peak can be obtained. However, the FWHM bandwidth of the spectral output depends on the bilayer numbers of the periodic structure. It is found that a particular mathematical pattern exists between the central wavelength of the source for this type of structure and the amount of phase shift of the output spectra. It is also shown that by suitable adjustment of the defect layer width, the phase shift of the spectral output can be avoided. Therefore, optical filters designed with these types of PBGS are very useful for WDM networks. With the available fabrication technologies, the device dimensions used in this analysis can be practically implemented.

REFERENCES

1. A. Taove and M. E. Brodwin. Numerical solution of steady-state electromagnetic scattering problems using the time-dependent maxwell's equations. *IEEE Trans. Microw. Theory Tech*, 23(8):623–630, 1975.
2. D.A.B. Miller. "Device requirements for optical interconnects to silicon". *Proc. IEEE*, 97(7):1166–1185, 2009.

3. J. D. Joannopoulos, S. G. Johnson, J. N. Winn, and R. D. Meade. *"Photonic crystals: molding the flow of light"*. Princeton university press, 2011.

4. E. Yablonovitch. "Inhibited spontaneous emission in solid-state physics and electronics". *Phys. Rev. Lett.*, 58(20):2059, 1987.

5. S. John. "Strong localization of photons in certain disordered dielectric superlattices". *Phys. Rev. Lett.*, 58(23):2486, 1987.

6. L. Rayleigh. "On the maintenance of vibrations by forces of double frequency, and on the propagation of waves through a medium endowed with a periodic structure". *The London, Edinburgh, and Dublin Philosophical Magazine and Journal of Science*, 24(147):145–159, 1887.

7. L. Rayleigh. "On the remarkable phenomenon of crystalline reflexion described by Prof. Stokes". *The London, Edinburgh, and Dublin Philosophical Magazine and Journal of Science*, 26(160):256–265, 1888.

8. R. Bandyopadhyay and R. Chakraborty. "Design of tunable transmission filter using onedimensional defective photonic crystal structure containing electro-optic material". *Opt. Eng.*, 54(11):117105, 2015.

9. H. Wang and K-Q Zhang. "Photonic crystal structures with tunable structure color as colori metric sensors". *Sensors*, 13(4):4192–4213, 2013.

10. X. Zhao, Y. Zhang, Q. Zhang, B. Zou, and U. Schwingenschlogl. "Transmission comb of a distributed bragg reflector with two surface dielectric gratings". *Sci. Rep.*, 6:21125, 2016.

11. R. Bandyopadhyay and R. Chakraborty. "Realization of mode independent multichannel transmission filter by controlling the photon localization in symmetric cavities". *Opt. Quant. Electron.*, 49(6):233, 2017.

12. B. Mason, J. Barton, G. A. Fish, L. A. Coldren, and S.P. DenBaars. "Design of sampled grating DBR lasers with integrated semiconductor optical amplifiers". *IEEE Photon. Technol. Lett.*, 12(7):762–764, 2000.

13. H. Deng, Dong, X., H. Gao, Yuan, X., Zheng, W., andZu, X. "Standing wave field distribution in graded-index antireflection coatings". *Appl. Sci.*, 8(1):65, 2018.

14. M.K. Hedayati and M. Elbahri. "Antireflective coatings: Conventional stacking layers and ultrathin plasmonic metasurfaces, a mini-review". *Materials*, 9(6):497, 2016.

15. K.J. Vahala. "Optical microcavities". *Nature*, 424(6950):839–846, 2003.

16. L. Florescu, K. Busch, and S. John. "Semiclassical theory of lasing in photonic crystals". *JOSA B*, 19(9):2215–2223, 2002.

17. F. Lemarquis, L. Abel-Tiberini, C. Koc, and M. Lequime. "400-1000 nm all-dielectric linear variable filters for ultra compact spectrometers". In *International Conference on Space Optics*, volume 4, page 8, 2010.

18. S. Nazarpour. *"Thin films and coatings in biology"*. pages 1–9. Springer, 2013.

19. M. Berggren, A. Dodabalapur, R.E. Slusher, Z. Bao, A. Timko, and O. Nalamasu. "Organic lasers based on lithographically defined photonic-bandgap resonators". *Electron. Lett.*, 34(1):90–91, 1998.

20. D. Tosi. "Review of chirped fiber bragg grating (CFBG) fiber-optic sensors and their applications". *Sensors*, 18(7):2147, 2018.

21. C.C. Ping, T. Anada, S. Greedy, T. M. Benson, and P. Sewell. "A novel photonic crystal band-pass filter using degenerate modes of a point-defect microcavity for terahertz communication". *Microw. Opt. Technol. Lett.*, 56(4):792–797, 2014.

22. S. Mandal, C. Bose, and M.K. Bose. "A generalized analytical approach for designing of one dimensional photonic crystal based optical filter with at-top response". *J. Lightwave Technol.*, 32(8):1519–1525, 2014.

23. M.R. Wu, C.J. Wu, and S.J. Chang. "Near-infrared photonic band structure in a semiconductor metamaterial photonic crystal". *Appl. Opt.*, 53(31):7285–7289, 2014.

24. C. Williams, G. Rughoobur, A.J. Flewitt, and T.D. Wilkinson. "Single-step fabrication of thin-film linear variable bandpass filters based on metal–insulator–metal geometry". *Appl. Opt.*, 55(32):9237–9241, 2016.

25. H.C. Hung, C.J. Wu, T.J. Yang, and S.J. Chang. "Analysis of tunable multiple-filtering property in a photonic crystal containing strongly extrinsic semiconductor". *J. Electromagnet. Wave.*, 25(14-15):2089–2099, 2011.

26. Y. Enami, B. Yuan, M. Tanaka, J. Luo, and A.K.Y. Jen. "Electro-optic polymer/TiO2 multilayer slot waveguide modulators". *Appl. Phys. Lett.*, 101(12):123509, 2012.

27. Y. Enami, J. Luo, and A. Jen. "Electro-optic polymer/ *TiO2* multilayer slot waveguide modulators for optical interconnections". In *Frontiers in Optics*, pages FTu4E–5. Optical Society of America, 2013.

28. S.J. Orfanidis. *Electromagnetic waves and antennas*. Rutgers UniversityNew Brunswick, NJ, 2002.

29. E. Rosenkrantz and S. Arnon. "Tunable electro-optic filter based on metal-ferroelectric nanocomposite for VLC". *Opt. Lett.*, 39(16):4954–4957, 2014.

30. L. Scholtz, D. Korcekl, L. Ladaanyi, and J. Muulleroa. "Tunable thin film filters for the next generation PON stage 2 (NG-PON2). In *ELEKTRO, 2014*, Rajecke Teplice, Slovakia, pages 98–102. IEEE, 2014.

31. H. Wang, J. Dai, H. Jia, S. Shao, X. Fu, L. Zhang, and L. Yang. "Polarization-independent tunable optical filter with variable bandwidth based on silicon-on-insulator waveguides". *Nanophotonics*, 7(8):1469–1477, 2018.

32. W.P. Risk, G.S. Kino, Y. Khuri, and Butrus T. "Tunable optical filter in fiber-optic form". *Opt. Lett.*, 11(9):578–580, 1986.

33. R.A. Soref, F. De Leonardis, and V.M.N. Passaro. "Tunable optical-microwave filters optimized for 100 mhz resolution". *Opt. Express*, 26(14):18399–18411, 2018.

34. A. Khattak, G. Tatel, and L. Wei. "Tunable and Switchable Erbium-Doped Fiber Laser Using a Multimode-Fiber Based Filter". *Appl. Sci.*, 8(7):1135, 2018.

35. R. Ghosh, K.K. Ghosh, and R. Chakraborty. "Narrow band filter using 1D periodic structure with defects for DWDM systems". *Opt. Commun.*, 289:75–80, 2013.

36. A. Banerjee. "Design of optical filters for coarse wavelength division multiplexing by using down binary number sequence multilayer structures". *JOSA B*, 26(3):537–540, 2009.

37. A.K. Chu, K.H. Huang, C.H. Chiang, W.C. Tien, and F.Z. Lee. "Strain-induced tunability of thin DWDM thin-film filters". *Electron. Lett.*, 48(18):1, 2012.

38. S. Golmohammadi, M. K. Moravvej-Farshi, A. Rostami, and A. Zarifkar. "Narrowband DWDM filters based on fibonacci-class quasi-periodic structures". *Opt. Express*, 15(17):10520–10532, 2007.

39. L. Domash, M. Wu, N. Nemchuk, and E. Ma. "Tunable and switchable multiple-cavity thin film filters". *J. Lightwave Technol.*, 22(1):126, 2004.

40. S.K. Awasthi and S.P. Ojha. "Design of a tunable optical filter by using a one-dimensional ternary photonic bandgap material". *Prog. Electromagn. Res.*, 4:117–132, 2008.

41. A. Banerjee, S.K. Awasthi, U. Malaviya, and S.P. Ojha. "Design of a nano-layered tunable optical filter". *J. Mod. Opt.*, 53(12):1739–1752, 2006.

42. R.M. Jopson, J. Stone, L.W. Stulz, and S.J. Licht. "Nonreciprocal transmission in a fiber Fabry-Perot resonator containing a magnetooptic material". *IEEE Photon. Technol. Lett.*, 2(10):702–704, 1990.

43. J.M. Jin. "*The finite element method in electromagnetics*". John Wiley & Sons, New Jersey, 2015.

44. U.S Dixit. "Finite element method: an introduction". Department of Mechanical Engineering, Indian Institute of Technology Guwahati-781, 39, 2007.

45. S. Shi, C. Chen, and D.W. Prather. "Plane-wave expansion method for calculating band structure of photonic crystal slabs with perfectly matched layers". *JOSA A*, 21(9):1769–1775, 2004.

46. A.J. Danner. "An introduction to the plane wave expansion method for calculating pho-
    tonic crystal band diagrams". University of Illinois, 2002.
47. R.A. Norton and R Scheichl. "Planewave expansion methods for photonic crystal
    fibres". *Appl. Numer. Math.*, 63:88–104, 2013.
48. I. O-de Julian, R. A. Mendez-Sanchez, B. Manzanares-Martinez, F. Ramos-Mendieta,
    and E. Baez-Juarez. "The plane wave expansion method applied to thin plates. *J. Acoust.
    Soc. Am.*, 130(4):2346–2346, 2011.
49. T. Zhan, X. Shi, Y. Dai, X. Liu, and J. Zi. "Transfer matrix method for optics in graph-
    eme layers". *J. Phys. Condens. Matter*, 25(21):215301, 2013.
50. J.B. Pendry and A. MacKinnon. "Calculation of photon dispersion relations". *Phys. Rev.
    Lett.*, 69(19):2772, 1992.
51. A.K. Ghatak and K. Thyagarajan. *Optical electronics*. Cambridge University Press,
    Cambridge, 1989.
52. C.T. Chan, Q.L. Yu, and K.M. Ho. "Order-N spectral method for electromagnetic
    waves". *Phys. Rev. B*, 51(23):16635, 1995.
53. A.J. Ward and J.B. Pendry. "Refraction and geometry in Maxwell's equations". *J. Mod.
    Opt.*, 43(4):773–793, 1996.
54. E.N.E.E. Lidorikis, M.M. Sigalas, E.N. Economou, and C.M. Soukoulis. "Tight-binding
    parametrization for photonic bandgap materials". *Phys. Rev. B*, 81(7):1405, 1998.
55. A. T. Paxton et al. "An introduction to the tight binding approximation–implementation
    by diagonalisation". *NIC Series*, 42:145–176, 2009.
56. K.M. Leung and Y. Qiu. "Multiple-scattering calculation of the two-dimensional pho-
    tonic band structure". *Phys. Rev. B*, 48(11):7767, 1993.
57. J. Xu, K. Hatada, D. Sebilleau, and L. Song. "An efficient multiple scattering method
    based on partitioning of scattering matrix by angular momentum and approximations of
    matrix elements". *arXiv preprint arXiv:1604.04846*, 2016.
58. R. Sainidou, N. Stefanou, I.E. Psarobas, and A. Modinos. "A layer-multiple-scattering
    method for phononic crystals and heterostructures of such. *Comput. Phys. Commun.*,
    166(3):197–240, 2005.
59. C-X. Zhang, Y. Yuan, T-J. Li, S-K. Dong, and H-P. Tan. "Analytical method to study
    multiple scattering characteristics in participating media". *Int. J. Heat Mass Transf.*,
    101:1053–1062, 2016.
60. N.A. Nicorovici, R.C. McPhedran, and L.C. Botten. "Photonic bandgaps for arrays of
    perfectly conducting cylinders". *Phys. Rev. E*, 52(1):1135, 1995.
61. S. He, M. Qiu, and C.R. Simovski. "An averaged-field approach for obtaining the band
    structure of a dielectric photonic crystal". *J. Phys. Condens. Matter*, 12(2):99, 2000.
62. C.R. Simovski, M. Qiu, and S. He. "Averaged field approach for obtaining the band
    structure of a photonic crystal with conducting inclusions". *J. Electromagnet. wave.*,
    14(4):449–468, 2000.
63. E. Yablonovitch, T.J. Gmitter, and K.M. Leung. "Photonic band structure: The face-cen-
    tered-cubic case employing nonspherical atoms". *Phys. Rev. Lett.*, 67(17):2295, 1991.
64. K. S. Yee. "Numerical solution of initial boundary value problems involving Maxwell's
    equations in isotropic media". *IEEE Trans. Antennas Propag.*, 14(3):302–307, 1966.
65. A.Z. Elsherbeni and V. Demir. *"The finite-difference time-domain method for electro-
    magnetic with MATLAB simulations"*. The Institution of Engineering and Technology,
    Boston, MA, 2016.
66. R.C. Rumph. *EE5320 computational electromagnetics*. http://emlab.utep.edu/
    ee5390cem.htm, 2014.
67. D.E. Merewether, R. Fisher, and F.W. Smith. "On implementing a numeric huygen's
    source scheme in a finite difference program to illuminate scattering bodies". *IEEE
    Trans. Nucl. Sci.*, 27(6):1829–1833, 1980.

68. K. Umashankar and A. Taove. "A novel method to analyze electromagnetic scattering of complex objects". *IEEE Trans. Electromagn. Compat.*, (4):397–405, 1982.

69. R. Holand and J. Williams. "Total-field versus scattered-field finite-difference". *IEEE Trans. Nucl. Sci.*, 30:4583–4588, 1983.

70. A. Taove and S.C. Hagness. *"Computational electrodynamics: the finite-difference time-domain method"*. Artech house, Boston, London, 2005.

71. R. Garg. *"Analytical and computational methods in electromagnetics"*. Artech house, Norwood, 2008.

72. B. Engquist and A. Majda. "Absorbing boundary conditions for numerical simulation of waves". *Proc. Natl. Acad. Sci.*, 74(5):1765–1766, 1977.

73. G. Mur. "Absorbing boundary conditions for the finite-difference approximation of the timedomain electromagnetic-field equations". *IEEE Trans. Electromagn. Compat.*, (4):377–382, 1981.

74. J.P. Berenger. "A perfectly matched layer for the absorption of electromagnetic waves". *J. Comput. Phys.*, 114(2):185–200, 1994.

75. E. Turkel and A. Yefet. "Absorbing PML boundary layers for wave-like equations. *J. Appl. Numer. Math.*, 27(4):533–557, 1998.

76. W. C. Chew and W. H. Weedon. "A 3D perfectly matched medium from modified Maxwell's equations with stretched coordinates". *Microw. Opt. Technol. Lett.*, 7(13):599–604, 1994.

77. S.D. Gedney. "An anisotropic perfectly matched layer-absorbing medium for the truncation of FDTD lattices". *IEEE Trans. Antennas Propag.*, 44(12):1630–1639, 1996.

78. Q.H. Liu and J. Tao. "The perfectly matched layer for acoustic waves in absorptive media". *J. Acoust. Soc. Am.*, 102(4):2072–2082, 1997.

79. S. Abarbanel, D. Gottlieb, and J.S. Hesthaven. "Well-posed perfectly matched layers for advective acoustics". *J. Comput. Phys.*, 154(2):266–283, 1999.

80. Z.S. Sacks, D. M. Kingsland, R. Lee, and J.F. Lee. "A perfectly matched anisotropic absorber for use as an absorbing boundary condition". *IEEE Trans. Antennas Propag.*, 43(12):1460–1463, 1995.

81. A.A.M. Saleh and J. Stone. "Two-stage Fabry-Perot filters as demultiplexers in optical FDMA LANs". *J. Lightwave Technol.*, 7(2):323–330, 1989.

82. J. Stone and L.W. Stulz. "High-performance fibre Fabry-Perot filters". *Electron. Lett.*, 27(24):2239–2240, 1991.

83. Q. Zhu and Y. Zhang. "Defect modes and wavelength tuning of one-dimensional photonic crystal with lithium niobate". *Optik*, 120(4):195–198, 2009.

84. H. Loui. "1D-FDTD using MATLAB". *ECEN-6006 Numerical Methods in Photonics Project*, 1, 2004.

85. M. Polyanskiy. Refractiveindex. info. *MediaWiki. http://refractiveindex. info, accessed*, 4, 2015.

86. D. E. Zelmon, D. L. Small, and D. Jundt. "Infrared corrected sellmeier coefficients for congruently grown lithium niobate and 5 mol.% magnesium oxide–doped lithium niobate". *JOSA B*, 14(12):3319–3322, 1997.

12 Super Achromatic Multi-Level Diffractive Lens: A New Era Flat Lenses

Ekata Mitra and Subhashish Dolai

CONTENTS

12.1 INTRODUCTION

The advent of diffractive optical elements plays a very important role in many modern systems, from biometrics security system, new-age mobile phones to spacecraft. Generally, optical diffraction is defined as when a bunch of rays experiences an obstacle and transmits in diverse directions at a different angle. The diffraction phenomenon can only be nurtured by the analysis of wave property (electromagnetic (EM) wave theory), whereas the corpuscular behaviour of light is responsible for reflection and refraction. As the conventional lenses are fundamental to the imaging systems, based on geometrical optics, a simple plano-convex lens requires larger bending angles or curvature which in turn requires of designing of larger thicknesses, it exploits the principle of refraction to focus light. But along with the trend towards the miniaturization (as seen in the semiconductor industry and other IOT devices), it is a new challenge to make a trade-off between the edge-thickness (i.e. structure and weight) of the optical element with the increment in numerical aperture (or resolution). Recently, by exploiting diffraction super achromatic 'flat lenses' are introduced and further improve its depth of focus by introducing multi-diffracting lenses, by space-based arranging 'zones' with an appropriate geometric phase (alsoknown as the Pancharatnam–Berry [1] phase) to scatter the transmitted waves such way that it can build constructive interference [2]. To start with the flat lenses, there must be a detailed analysis of two types of flat lenses, namely meta-lens and multi-diffractive lens (MDL) [3]. Another branch of the optical system is struggling with the challenge of focussing of all light at one single point and it was kind of impossible for engineers to achieve it using a flat lens for many years [4]. But in very recent days, after years of experiment, a flat, thin-shaped lens came into the picture which is capable of bending individual lights in a single plane and the ground-breaking incident in the optical industry and inter-disciplinary domains are known as super achromatic lens [5]. It is highly demanding and advantageous in terms of its ease of access and wide user base. Besides that, this flat device designed based on the principle of diffraction, fabricated in some transparent medium like glass or plastic, produces better optical performance rather than a conventional lens and intuitive, straightforward way of fabrication to tailor with the various devices.

12.2 HISTORY

The diffraction effect of light was first observed by Grimaldi in 1665 and later the effect was also identified by James Gregory which was a bit different from the common corpuscular nature of light. The phenomenon is explained by Christian Huygens in 1678 and he expressed that each wavefront of diffraction is resource of secondary wavefront and thus the ground-braking theory of light as wave was evolved. It was further exploited by double-slit experiment of Thomas Young and after that, Augustin Jean Fresnel was able to quantify amplitude and phase property of each wavefront and explained wave propagation and diffraction theory with mathematical accuracy[6].

The major outbreak of light was the EM behaviour of light, identified by Maxwell in1860. After invention of optical holography mainly by electron microscopy, it was further characterized by the growth of computer technology and heuristic optimizing algorithms. The proposal of super-achromatic optics and its design methodologies are exploited since 1959. Professor Herzberger came up with the idea of such a lens that has same back focal point for all wavelengths. He suggested a series of super-achromatic monoplets with non-zero air spaces that can able to focus all wavelength at single point with no aberration. Later Drucks proves that it is impossible to thin monoplet without air space. But In 1970, Goodman's experiment paved way a new era of optics. He first introduced the fabrication techniques to design diffractive optical element for holography and extends computational algorithms to optimize quantization of the phase and amplitude. Since 1980, many techniques are evolved both in fabrication and computational algorithms to quantify the image aberration free[6].

The aberration effect of simple curved glass or the conventional lens was used to make camera lenses as the structure can help the collimated light, coming from different angles, to bend such that they all sum up or converge to a common focal point on the optical axis. But the issue with such a thick lens is that it produces aberrations (defects like defocusing, distortions). For example, the lights when get captured at the edges of the curved surfaces do not align in-phase with the rest of the light, therefore creating a hazy aberrated focus at the edge of the optical system. To correct these aberrations, flat lens is introduced that could focus light with high optical efficiency within the visible spectrum.

The ground-breaking incident of optical science came with the hand of scientists at SEAS Lab [7], Harvard Institute of Technology. Primarily, the flat lens was designed as a series of small nanostructures (preferably named as nanoantennas), fabricated with a thin wafer of silicon, 60 nanometers thick coated with concentric rings of v-shaped gold nanoantennas and the radial distribution of phase discontinuities of the concentric rings used for controlling the focal length. These nanoantennas are systematically arranged such that the light hits these antennas on that thin silicon wafer gets refracted artificially by generating the spherical wavefronts so that it can end up forming an aberration-free image on the same focal plane.

Few years later, a more improved version of flat lens is fabricated which is thinner than the previous one and this lens is employed on a dielectric material rather than a metal so that it can able to focus on different wavelengths of light at the same point. This kind of lens used an achromatic meta-surface which yields a polychromatic image with improved optical efficiency and is able to compensate chromatic dispersion. [8]

Though the fabrication of metalens is not cost-effective in comparison to conventional diffractive lens, many applications including spectroscopy, polarization and imaging technology are very demanding for metalens and still research in this optical field is adaptable to good engineering practice. To the emerging field of flat optics, a recent explosion of MDLs over metalenses is primarily based on their sub-wavelength thicknesses and when designed properly exceeds optical function and performance. Moreover, fabrication cost of MDL is very low but as the thickness is little higher than the metalens, the journey of MDL is challenging in the field of imaging system.

12.3 PROBLEMS IN CONVENTIONAL DIFFRACTIVE LENS

To study the problems in classic traditional optics, starting with an ideal lens, the main objective is to analysehow the image is getting formed and the main challenges while forming the image. For an ideal case, when a collimated light is coming from any particular point, on an object, it would traverse through the lens and all ends up together at a single point in the optical axis on the image plane. However, a real spherical lens could not able to focus the white light exactly to a single point rather it is getting spread out over some region on the plane, even if they are perfectly made. These deviations in performance or dispersive image are known asaberrationsof the conventional lens. So, the challenges are now lined up with two main points, i.e. geometrical structure of the optical system and the white light which are of various wavelength. To discuss the challenge of geometrical optics, we have classically summed up its detrimental performance by the approximation of paraxial optics and thus the incident rays that are away from the centre of the spherical curvature miss the focal point after getting reflected [9].

Primarily aberrations can be classified into two ways: *monochromaticandchromatic*.

12.3.1 MONOCHROMATIC ABERRATION

The most common *monochromatic aberrations* are of different kind:

- Defocus
- Astigmatic aberration
- Spherical aberration
- Field curvature
- Comatic aberration
- Image distortion

12.3.2 DEFOCUS

In optics,the most common aberration in imaging system is that whether an image is out of focus, more specifically, if the image is spreading out off the focus along the optical axis away from the image plane, the phenomenon is optically known as defocus which reduce thequality (sharpness, contrast) of the image[9] (Figure 12.1).

12.3.3 SPHERICAL ABERRATION

Another type of image forming aberration which is mainly based on geometrical structure, known as the spherical aberration[9] (Figure 12.2).

As the edge of an ideal convex lens is having spherical curvature, the farther the light hits from the centre of the lens, the refraction of light becomes tighter in

Focused Image   De-focused Image

FIGURE 12.1   Defocus.

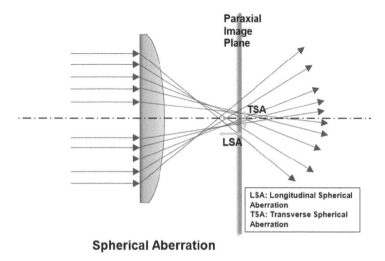

Spherical Aberration

LSA: Longitudinal Spherical Aberration
TSA: Transverse Spherical Aberration

FIGURE 12.2   Spherical aberration.

comparison to the refraction closer to the centre of the lens. This causes a deviation in the focal point and thus blurry image to be formed.

12.3.4   COMATIC ABERRATION

A 'Coma' (the name came from the tail (coma) of a 'Comet') is generally considered as the most problematic aberration due to the asymmetric blurry tail in the image that produced because of spreading in off-axis on image plane. This phenomenon is unlike to the spherical aberration, when the rays come at an angle to the lens, the rays miss the focus and it ends up meeting at lateral axis rather than optical axis. Actually, the degradation in image with comatic aberration is a consequence of difference in refraction as the ray passing through parabolic surface and the incident angle varies with the curve [9]. The problem of Coma is mostly dependent upon the shape and the structure of the lens. To eliminate this aberration, a strongly recommended solution is to use a combination of lenses that are arranged symmetrically around a central point so that light that comes at an angle, can be positioned before focal point (back focal point) as a parallel ray (parallel to optical axis) (Figure 12.3).

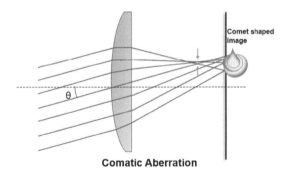

FIGURE 12.3 Comatic aberration.

As a result, when the light hits in the final lens, the peripheral rays are meet exactly at the focus point with zero coma.

12.3.5 ASTIGMATIC ABERRATION

Another type of distortion occurs when some bunch of oblique rays are striking on the lens rather than the optical axis and thus the lens becomes asymmetrical to the incident rays. As a matter of fact, the lens appears before the plane of incidence and the oblique rays formed an image in the sagittal plane (normal to the tangent plane) which is different than the image appeared in the tangential /meridional plane. This makes a deviation in an image, more precisely the image that formed in the sagittal plane is greater than the tangential plane as the incident rays traverse different spherical front and the planes do not coincide.

It is usually faced largely by the human eye as the outer lens (cornea) is having a spherical curvature and because of that light needs to bend or refract at an angel, uneven inside the eye, it produces blurry image [9] (Figure 12.4).

To correct astigmatism, a combination of glasses is required to nullify this effect.

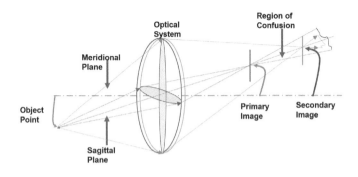

FIGURE 12.4 Astigmatic aberration.

12.3.6 FIELD CURVATURE

This type of distortion happens when the optical path between the object point and the lens increases with the object's height[10]. The planar object in front of an uncorrected lens will be mapped onto a curved surface or field curvature, known as Petzval surface. This very common optical problem makes a flat object to emerge sharp in some part of the frame and other remains blurred due to off-axis location. This aberration can be negating the effect by a combination of negative and positive lens (Figure 12.5).

12.3.7 IMAGE DISTORTION

Image distortion is an outcome of a phenomenon where deviation in rectilinear projection happen [9]. This optical distortion is due to different areas of lens or combination of different lens have different focal length and different magnification and this effect can be compensated by using a thin lens with zero distortion (Figure 12.6).

12.3.8 CHROMATIC ABERRATION

The aforementioned aberrations in image construction are an effect of monochromatic light. But in the case of white light, more precisely, a polychromatic light of

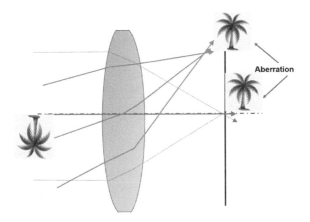

Petzval Field Curvature

FIGURE 12.5 Field curvature aberration.

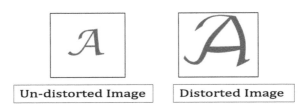

FIGURE 12.6 Distortion.

various wavelength causes chromatic aberration significantly while forming an image. As different rays will traverse the different optical paths and the focal length is directly related to refractive index which in turn is a function of wavelength, different light will meet different positions. As a result, dispersion will take place and image will be distorted.[9]

The common example of chromatic aberration can be understood very easily as the human eye has a substantial amount of dispute which causes chromatic aberration in some condition. Although this aberration is counterbalanced by the several psychophysical methodologies, it can be explained with the differentposition of our eyes and a purple dot. If the eyes are close enough to the object, here, the purple dot will appear as a blue dot is circled by a red colour ring but as the axial length between the eye and object are large, it seems different to our eyes, the same purple dot emerges as the core red dot is surrounded by the blue ring (Figure 12.7).

12.4   GRIN SYSTEM

The first attempt to combat the effect of aberration was to introduce a concept of gradually varying refractive index in a single lens. The idea came out from the simple human eye mechanism as our eye has the high refractive index in the central layer of the lens, approximately 1.406, and it varies up to 1.386 in the edges of lens. This controls the emergent ray of light at the entrance pupil and thus produce a good quality image with higher resolution and low aberration at both short and long distances[9] (Figure 12.8).

An ordinary lens always has homogeneous material which act as a medium, always reconstruct the wavefront in such a direction that converges at one point as the

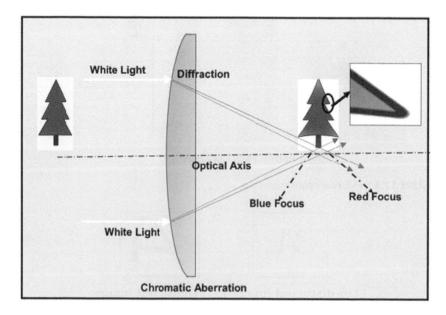

FIGURE 12.7   Chromatic aberration.

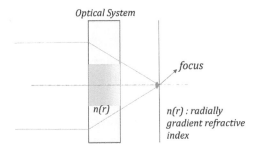

Optical System

focus

n(r)

*n(r) : radially
gradient refractive
index*

*The geometry of corresponding gradient lens
for bending of parallel rays*

FIGURE 12.8   Geometric GRIN system.

material has a different refractive index than surroundings. But as light transverse through inhomogeneous medium, the denser region slows down the speed of light and lighter medium helps the light to move faster than the other part. This principle of light in different medium allows engineer to customize the lens based on controlling a set of parameters, the lens is hence named as GRadient in the INdexed refraction. This optical system is very helpful in monochromatic application whereas useful in polychromatic environment if they are aspheric.

12.5   FLAT LENS

The objective of any optical system is to perform the one-to-one mapping between the image plane and the object so that a distortion-free image can be formed. Besides that, nowadays as optical elements are in trend, include themselves in terms of all essential devices, some of which are user-friendly IOT devices like mobile-camera, sensors for security purpose. Thus, the optical element, here, lens needs to be miniatured without compromising the main objective of proper focusing on the image plane and quality imaging mechanism. This trend came up with some challenges since conventional lens are bulky. One solution to deal with this drift of optical theory is Flat Optics.[3]

The flat optics started its journey by the advancement of conventional diffractive lens technology, established its new design as MDL, although a secondary approach has been explored later, which leverages the concept of using sub-wavelength theory. The recent approach in flat optics is the technology using metasurfaces, known as metalens, a binary (2D) structure which is very thin, approximately 100th times smaller than hair, able to polarize the light.[2]

12.6   METALENS

The term 'metalens', coined from 'metasurface', is designed carefully to deal with the challenges faced by the conventional 3D diffractive lens. Therefore, starting with a flat wafer and the nanostructured surface alter the phase of incident wavefront after reflection, creates a new wavefront and its phase profile allows to produce an image with no spherical aberration. However, diffractive lens is used to restrict the phase by continuously changing the optical element thickness, the metalens is used to shape the emerging wavefront by limiting the phase modulation to its minimum, 2π. In a metalens, the metasurfaceis fabricated to design nanostructures, known as, nanoantenna, and the phase is induced within the surface of the material via those nanostructures[2] (Figure 12.9).

Ideally, in metalens, a regular array of nano-antennas is required to be designed as nano-pillars in their subsequent location (subwavelength-spaced positions) for the desired phase alteration at each location. Moreover, the diameter of the nano-pillar is needs to be flexible for creating that phase with respect to the specified location. But as the phase is a function of position co-ordinates and ideally it is restricted to 0 to 2π, it is very elusive to demonstrate the ideal design approach in visible spectrum, i.e. the pillars are slightly different from others and aperiodic in nature. Besides that, the available material (TiO2) for nano-pillars is of much lower refractive index,[2] so that the pillars unable to confine the light, interacting more with the adjacent pillars. Another shortcoming of traditional technique is that the nano-structures may face angle-dependent effects. To achieve the target application of metalens in visible frequency, the metasurface needs to be more flexible and different parts of the metasurface alter in phase in the desired direction to meet the 100% efficiency.

However, high-NA visible-light lenses in the ideal design approach are paradoxical, using a simple optimization technique within a limited parameter, able to achieve a simulated result that enables a meta-lens with almost 80% efficiency. Therefore, in other application, near-infrared, broadband application, the nanoantennas are designed in such a way that it can control the incoming light more significantly by selecting the proper material and the size of the structure must be of subwavelength quasi-periodic structure opposite to a conventional lens which is based on a super-wavelength property. These metalenses are useful in the Internet of things, mobile, lidar automobile applications.

12.7 MULTI-DIFFRACTIVE LENS

To increase the optical efficiency, MDLs enable thinner, lighter, and simpler imaging systems over conventional lens. Although, the thickness of MDL is not as smaller as metalens, but the large-area and high-NA [3,6] flat lens are widely for its integration ability and compatibility with other Internet of Things. It overcomes the phase delay issues by optimizing the design algorithm by computational methodologies (Genetic

Schematic Diagram : cross-section of
nano-structured Meta Lens

FIGURE 12.9 Metalens in cross-sectional view.

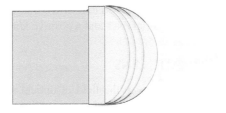

Schematic Diagram : cross-section of Multi-Diffractive Lens

FIGURE 12.10 MDL (*multi-diffractivelens*) in cross-sectional view.

Algorithm, Binary Search) and hence able to approximate the optimal continuous phase distribution (Figure 12.10).

Moreover, another challenge, i.e. the trade-off between high numerical apertures and decrement in focussing efficiency is also tackled by the heuristic approach of computational simulation as well as in experimental process. In addition to that, the multi-level diffraction lens offers a solution for broadband performance by modifying their structure using higher order diffraction.

12.8 ACHROMATIC LENS

The next challenge faced by the optical engineers to bring all visible colourspectrum together at one point to get a proper image on axis. As because they are different in wavelength and that are directly varies with the optical path length, it was impossible for engineers to breakthrough it without any aberration (chromatic aberration) (Figure 12.11).

The achromat is a lens [5] that formulated such a way to compensate the effects of chromatic and spherical aberration. They are made up of two lenses, specifically known as achromatic doublet, which are managed to focus two wavelengths (typically red (680 nm) and blue (470nm)) on the same plane[11]. In a design perspective,

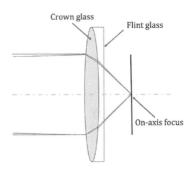

Achromatic Doublet

FIGURE 12.11 Schematic view of achromatic *l*ens.

the two individual diffractive lenses that used to form a single achromatic lens are the combination of concave lens of flint glasses and a positive lens of crown glass.

12.9 SUPER-ACHROMATIC LENS

The most emerging commercial lens in the field of optical system is super-achromatic lens. It is an advancement of achromatic lens; it allows four different rays to focus at a single point on a same plane [12]. Simultaneously, this ultra-thin lens is also capable of correcting the spherical and chromatic aberration in broadband application [3]. To experimentally characterize the lens, the concept on diffraction theory and quartic theory are realized in a manner that it could bend light even in a flat surface by inducing multi diffractive interface and with the microstructures in the lens are structured computationally by specially modified optimizing algorithms to measure the geometry of a flat surface so different colours can traverse through it and ends up meeting to a single point (on-axis-focus).[5] The resulting lens of any transparent material like as thin as glass or plastic, a breakthrough in imaging industry, known as 'super-achromatic lens', As it is ultra-thin and well-corrected in terms of aberration and distortion, it has its immense application in medical domain, smartphone industry and automobile industry. On top of it, this lighter lens has a power of capturing video which can be used for drones or satellites in essential security industry[5].

12.10 DESIGN AND METHODS

With the advent of optics, two types of sub-wavelength optical elements are vastly nurtured, namely MDL and metalens. With the hand of inverse design and fabrication, researchers are able to invent super-compact yet powerful and aberration controlled diffractive optical element of a few sub-wavelength edge thickness.

The relation between the inverse design and fabrication is so inter-dependent. The better design requires the good fabrication and that leads to a good optimized design result. The detailed steps of such diffractive optical element have to deal with three dimension and the problem has to be formulated in reverse direction, i.e. the DOE is to be designed as that would give a better performance and accurate result keeping the quantization level, fabrication process in account while selecting the proper algorithmic methods for optimizing the result (Figure 12.12).

The proposed computational method for designing MDLs based on optimization in complex amplitudes and phase of the beams generated by the lens for the desired wavelengths. The computational algorithm so far employed in this regard are categorized in the following way (Figure 12.13).

12.11 ITERATIVE DIRECT BINARY SEARCH

The iterative direct binary search (IDBS) is computationally intensive algorithm which is used to search all possible collection of discrete phase levels or complex amplitude in a pixel-by-pixel order. This algorithm starts with phase function and this process continues until the algorithm achieve optimum results. This traditional method produces better reconstruction phase at the cost of a huge number of iteration

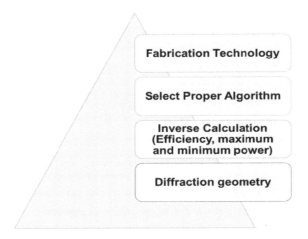

FIGURE 12.12 Schematic of *fabrication process*.

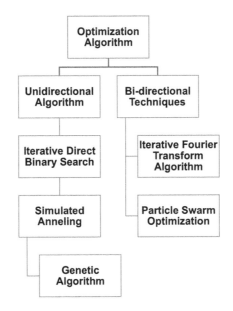

FIGURE 12.13 Classification of optimization algorithm.

steps and extensive computational calculations [13]. The major drawback of the IDBS algorithm is about scanning possible outcomes locally instead of being global best. So, if the sets are sorted then local optimum shows the best result.

12.12 SIMULATED ANNEALING

The idea of this process was formulated to overcome the demerits of the conventional binary search algorithm and expediates the process by reducing the number of steps

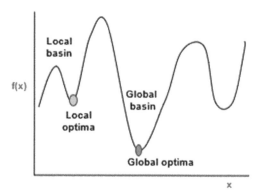

FIGURE 12.14 Graphical representation of computational algorithm.

[14]. In contrast with the previous algorithm IDBS, this algorithm includes randomness by employing probability function for each pixel. The principal of the algorithm employed in such a way that the sets are becoming smaller as the number of iterations increase. In this way, the algorithm converges global optimum solution by exploiting the all possible outcomes. Though the algorithm scans all the outcomes, the only drawback is the process converge slowly due to the randomness in subsequent parameters [14] (Figure 12.14).

12.13 GENETIC ALGORITHM

Genetic Algorithm is a metaheuristic computational process based on the idea of natural selection of Charles Darwin. The process starts with the random generation of population of chromosome in lieu with the genetics and evolution theory of Sir Darwin. Eachchromosome consists of genes or individual variables and a group of chromosomes are known as population. To generalize the function, it is better to organize the initial population in terms of a population matrix which can be programmed intuitively and then improvise through its subsequent process of mutation, crossover of mating pool, inversion and repeat until the outcome converge with the global optimum. The steps are as following:

12.13.1 INITIATE POPULATION

Ideally, the population should introduce every possible set to allow randomness and find out the best solution.

12.13.2 COST FUNCTION OR FITNESS FUNCTION

The chromosomes are required to be fitted into some cost function according to the problem (maximize or minimize function) for evaluation. Each chromosome then has an associated cost and eligible for comparing to the other from the same batch.

12.13.3 NATURAL SELECTION

Natural selection is an important step where the healthiest chromosomes are preferred to qualify (survive) for the next pool. Similar to biological process of evolution, at first, the chromosomes are prepared to invoke natural selection. The idea is to keep suitable chromosomes and reject the rest. The choice of suitable chromosome is based on cost and thereafter sorted to identify the corresponding fitness of the chromosomes.

12.13.4 SELECT POPULATION FOR MATING

From the natural selection, randomly chosen chromosomes are further evaluated to form as a parent. The criteria behind selection of parent reside with the lowest cost function and the fittest chromosome considered as parent.

12.13.5 GENERATE OFFSPRING

The offspring can be modulated from selected parents in various ways. For binary chromosomes, uniform crossover is the most general procedure. For simplification, a mask is considered to be a series of ones and zeros, generated for individual set of parents. The masking chromosome and the parent chromosome are composed of the same number of bits. If the bit in the mask is a one, then the corresponding bit from the mother is passed to first offspring and the respective bit from the father is transferred to second offspring. Similarly, when the bit in the mask is a zero, then the corresponding bit from the mother is transferred to second offspring and the corresponding bit from the father is passed to first offspring. To exemplify the concept of binary crossover, we have to consider two parents and masking bit in matrix format as an input bit:

mother = [1 0 1 1 1 0 1 1]
father = [1 1 0 0 1 1 0 0 1]
If the mask for single-point crossover is given by,
mask = [1 1 1 0 0 0 0 0]
then the offspring are:
first offspring = [1 0 1 0 1 1 0 0 1]
second offspring = [1 1 0 1 1 0 1 1 1]

12.13.6 MUTATE SELECTED MEMBERS FROM THE POPULATION

After selection of offspring, the mutation rate has to be induced and the portion of bits or values within a population, random variations are included in the population. To demonstrate this mutation effect, a binary mutation is taken for general consideration which converts a zero to a one or a one to a zero or to improve the situation.

12.13.7 TERMINATE WHEN OPTIMUM CONDITION IS REACHED

This process ultimately shows the global optimum result which is completely different from the initial chromosome to be produced in the next-generation population. Generally, the average objective function or cost function improved by the process of repetition and search algorithm and thus the best chromosomes are selected for breeding.

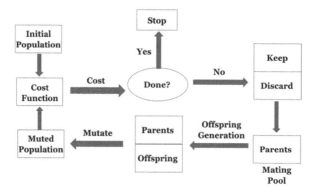

12.14   ITERATIVE FOURIER TRANSFORM ALGORITHM

Iterative Fourier transform algorithm (IFTA) is primarily concerned with transformation of phase distribution (U (x, y, z)) to design the diffractive optical element at desired focal point. The optical elements are hence characterized to perform in a desired phase distribution and the repetitive process of transformation and inverse transformation are required to achieve ultimate solution. [15].

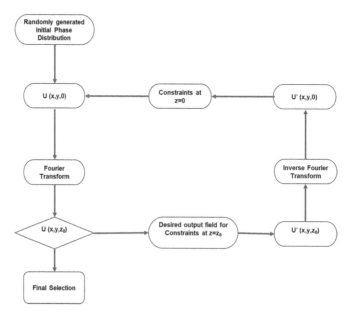

12.15   PARTICLE SWARM OPTIMIZATION

Evolutionary computational techniques are basically stochastic, robust in nature. This algorithm is adaptive to the dynamic nature like swarm movement, depends on the principle of intelligence converging to the feasible solution through experience and knowledge. The basic objective of optimization technique concerned about finding the best fitness function among the scarce resources.

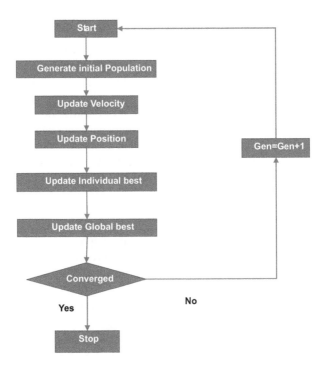

PSO is a population-based robust stochastic algorithm which is often hybridized with some non-differential objective function. It can handle large search space and specialized to formulate the random movement of bird flocking or fish schooling unlike classical optimization technique. As PSO is sensitive to certain parameters like C1, C2 and C3, these parameters are carefully chosen for better simulation.

$$V_i^{k+1} = w * V_i^k + C_1 . rand(0,1) * \left(LBest_i - X_i^k\right) + C_2 * rand(0,1)\left(GBest_i - X_i^k\right)(3.5)$$

$$X_i^{k+1} = X_i^k + V_i^k$$

Where w = the inertia factor, $0 < w < 1$,
   C_1 = local acceleration constant of a particle, C_2 = global acceleration constant of a particle,
   C_1, C_2: are usually selected in (0, 2)
   $LBest_i$ = local best position of a particle, $GBest_i$ = global best position of a particle

12.15.1 Fabrication Techniques

Irrespective of the algorithm are applied to design the optical element, fabrication parameters and their limitations in each process are needs to be taken into account. To demonstrate the entire design of optical element, the common fabrication parameters, technologies and constraints in modelling nanostructures are discussed below briefly.

12.15.2 Mask-Based Lithography

The traditional technique for fabrication in semiconductor industry is Mask-based lithography[6]. There are some basic steps which are required to be followed to produce an efficient diffractive element. The very first step is to deposit a photoresist layer on the substrate by using a 'controlled' level of spin coating. The reason behind the term 'controlled' is as featured as the thickness of the layer depends on the speed of the spinning. A binary mask is formed with pattern generator and then it is exposed to UV ray for chemical modification. Next step is to allow some photoresist to be etched away by specific solution and then by extracting the photoresist layer, the desired thickness of two-level structure is achieved.

12.15.3 Parallel Direct Writing

The advantage of using Parallel direct writing lithography[6] is very cost-effective and compatible to diffractive optical element. However, this cannot show any improvement on resolution limit over basic direct writing lithography, this method is still preferred for its other specification.

This method started with the idea of reconfigurable masking process which can be achieved by customized or programmable spatial light modulator (SLM). The modulator is basically a display device where the light is getting through a reduction lens and focused on the LCD. Using a reduction lens, it is very efficient to expose the structure to the parallel beam (write beam) at the same time and each pixel can be controlled by limiting the pixel size of the display. The next step is almost like any other fabrication technique, i.e. development and etching process. Finally, this structure can examine for its specific measurement under interferometric microscope and verified for meeting the estimated criteria of certain dimension.

12.15.4 Electron Beam Lithography

Electron Beam Lithography (EBL) [6] uses a focused electron beam, which provides higher pattern density unlike the optical lithography process. The 10-50 keV electron source causes the shorter wavelength, resulting in high-resolution patterning. The direct-write EBL is carried out by two techniques for scanning on the surface for pattern transfer. The raster scanning scans through the whole area sequentially unlike vector scanning, where the electron beam is only directed to the requested pattern and hops from feature to feature. The smallest dot that can be achieved with an e-beam lithography is ~ 100 Å. The mask is software defined as against photoresist in case of optical lithography. However, electron beam resist is available which is poly-methyl methacrylate (PMMA). Thin layer of gold on e-beam resist is used to prevent charging up. After the pattern is written, the sacrificial gold layer is then removed by chemical etching. The only drawback of this technique is the slow exposure speed and since it involves using electron requires vacuum system, the overall complexity of the system increases.

12.16   COMPARATIVE ANALYSIS AND DISCUSSION

As the design methodology evolves with the semiconductor industry, the constant demand of optical engineering is migrated from its traditional lens to flat lens to meet the current trend of IOT devices and its uses. The application of camera lenses is travelled a lot from mobile industry to medical industry, automobile to information and security devices. The main criteria are to achieve a significant good image with less aberration and comparatively thin lens than the traditional glass.[2] With the help of diffraction theory, MDL has produced a huge improvement on various application and the use of metamaterial, the metalens are also developed good image with reduced width, typically 100nm (Table 12.1).

12.17   CONCLUSION

In summary, we demonstrate the idea of flat lens and its importance over the traditional lens. Additionally, we have also discussed the detailed classification of aberration to understand the necessity of correction of default lens so that the desired specification can be achieved based on the knowledge. The comparative study analysis shows that the MDL is comparatively better in case of CMOS compatibility but

TABLE 12.1
Comparative Analysis

Properties	Diffractive Lens	Multi-Diffractive Lens	Metalens
Lens edge thickness	In the range of few mm	In the range of few μm	Range of few nm
Fabrication complexity	Simple as Glass	Complexity is less as it can be made up of low-index polymer	Two-level structure; made of high index materials
CMOS compatibility	Not compatible	Compatible	Not compatible
High numerical aperture (NA) and efficiency	For achromatic rays, efficiency is low, aberrated image production	Higher than Metalens both in Narrowband and Broadband	Better focussing efficiency than diffractive lens
Polarization sensitivity	No degree of freedom	Polarization selective, ideal focus without spherical aberration	Degrade performance for unpolarized light, Works for left or right circularly polarized light

as Metalens is much thinner that MDL, it is often preferred more in IOT or Automobile industry. Using this concept of flat lens, the idea of a super-achromatic lens is proposed. The idea of focussing at least four wavelengths on a single point is challenging for Optical Engineers. Recently, the newly introduced evolutionary computational techniques are used to estimate the performance (field of depth, focus length, Depth of view, optical efficiency, High NA), so it is better to optimize the relevant parameters accordingly to construct the image with accuracy.

REFERENCES

1. *"Polarization-controlled optical holography using flat optics | Light: Science & Applications."* https://www.nature.com/articles/s41377-020-00373-w (accessed Jul. 30, 2020).
2. J. Engelberg and U. Levy, "The advantages of metalenses over diffractive lenses," *Nat. Commun.*, vol. 11, no.1, 2020, doi: 10.1038/s41467-020-15972-9.
3. S. Banerji, M. Meem, A. Majumder, F. G. Vasquez, B. Sensale-Rodriguez, and R. Menon, "Imaging with flat optics: metalenses or diffractive lenses?," *Optica*, vol. 6, no. 6, p. 805, 2019, doi: 10.1364/optica.6.000805.
4. *"Flat Lens for Focus-Free Imaging | Optics & Photonics News."* https://www.osa-opn.org/home/newsroom/2020/march/flat_lens_for_focus-free_imaging/ (accessed Jun. 15, 2020).
5. *"New lens ready for its close-up | UNews."* https://unews.utah.edu/new-lens-ready-for-its-close-up/ (accessed Jun. 15, 2020).
6. G. N. Nguyen, *"Modeling , design and fabrication of diffractive optical elements based on nanostructures operating beyond the scalar paraxial domain Giang Nam Nguyen To cite this version:,"* 2015. Accessed: Jun. 15, 2020. [Online]. Available: https://hal.archives-ouvertes.fr/tel-01187568.
7. *"An economy of algorithms | Harvard John A. Paulson School of Engineering and Applied Sciences."* https://www.seas.harvard.edu/news/2017/01/economy-of-algorithms (accessed Jun. 15, 2020).
8. S. Banerji, M. Meem, A. Majumder, B. Sensale-Rodriguez, and R. Menon, "Extreme-depth-of-focus imaging with a flat lens," *Optica*, vol. 7, no.3, p. 214, 2020, doi: 10.1364/optica.384164.
9. E. Hecht, *Optics (4th Edition).* Addison-Wesley, 2001.
10. *"Petzval field curvature - Wikipedia."* https://en.wikipedia.org/wiki/Petzval_field_curvature (accessed Jun. 15, 2020).
11. "Achromatic Lens," *SpringerReference*, 2011. https://en.wikipedia.org/wiki/Achromatic_lens (accessed Jun. 15, 2020).
12. "Superachromat - Wikipedia." https://en.wikipedia.org/wiki/Superachromat (accessed Jun. 15, 2020).
13. B. K. Jennison, D. W. Sweeney, and J. P. Allebach, "Efficient design of direct-binary-search computer-generated holograms," *J. Opt. Soc. Am. A*, vol. 8, no.4, p. 652, Apr. 1991, doi: 10.1364/josaa.8.000652.
14. *"Diffractive Optics: Design, Fabrication, and Test | (2003) | O'Shea | Publications | Spie."* https://spie.org/Publications/Book/527861?SSO=1 (accessed Aug. 02, 2020).
15. F. Wyrowski and O. Bryngdahl, "Iterative Fourier-transform algorithm applied to computer holography," *J. Opt. Soc. Am. A*, vol. 5, no.7, p. 1058, Jul. 1988, doi: 10.1364/josaa.5.001058.

13 Adaptive Repetitive Control of Peristaltic Pump Flow Rate with an Optical Flow Sensing System

Naiwrita Dey, Ujjwal Mondal
and Anindita Sengupta

CONTENTS

13.1   INTRODUCTION

Peristaltic pump is a well-known type of positive displacement pump and can be operated in reversible mode. Working method of this pump involves the occlusion of a flexible tube in a periodic interval which leads to the transfer of the fluid enclosed within the tube. It is commonly known as roller pump. It is independent of the pressure head or viscosity. The stroke to volume ratio remains constant as the peristaltic pump motor holds a linear response to the driving frequency of it. Human intestines work as such a pumping mechanism. Unlike regular pumps, liquid flowing via the tube section of roller pump remains uncontaminated as it does not make any contact point with the metallic part of the pump [1,2]. Therefore it is widely used in various biomedical applications and chemical process industry. Mathematical model of the pump include time-varying parameters and an unstructured uncertainty along with some periodic disturbances. Static pressure drop and unmodelled dynamics is taken into account in the overall system model as a form of periodic disturbance. Any precise fluid dispensing application or biomedical application demands precise control of peristaltic pump flow rate along with periodic disturbance elimination thus this present work is considered under the framework of repetitive control.

Repetitive controller is a special controller which deals with periodic signals. It has been implemented to several applications for disturbance rejection which is periodic in nature [1,2].Several works have been carried out for minimizing tracking error and rejection of periodic disturbances. Discrete wavelet transform-based repetitive controller is implemented to minimize the memory utilized in the repetitive controller, hence minimizing the tracking error [3]. To control the rotating axis of peristaltic pump, repetitive controller was applied by Hillerstorm [4].Use of stepper motor has been reported to regulate the peristaltic pump system at distinct speed level for different flow rates[5].Parameter variation is also considered in disturbance rejection. A feedback controller along with an observer has been applied for the same [6]. rpm of the pump head is taken into account as an indirect feedback of the flow rate. A robust repetitive control law is obtained to reduce the oscillations of the liquid flow rate[7]. Application of disturbance observer has been further studied for peristaltic pump to reduce the effect of periodic load disturbance on overall control performance[8].

Different control methods for disturbance elimination are discussed in the literature previously. Conventional repetitive controller is modified with an augmentation of adaptive control in this present work to control flow rate and desired roller position in peristaltic pump head. The adaptive nature of repetitive control algorithm can efficiently handle unknown system parameter drifts along with periodic disturbance rejection. The adaptive control law is to be designed to track the varying total peripheral resistance and update the controller parameters accordingly. Design procedure is elaborated for the proposed method and simulation study is shown.

This chapter is structured in the following manner. Sections II and III describe different parameters of the pump in terms of mathematical formulation and mathematical modelling of the peristaltic pump. Section IV shows different application-based case studies followed by Section V optical flow sensing system for peristaltic pump. Section VI gives insights to the proposed controller design for the pump to maintain flow rate at a desired level. The corresponding simulation study and results are described in Section VII. To sum up conclusions are stated and further improvements are suggested.

13.2   ROTARY PERISTALTIC PUMP

It contains a length of silicon tubing or any elastic tubing placed interior a curved raceway which is located at the travel perimeter of the rollers. A rotor is attached to the rollers. They are mounted on the rotating arms ends arranged in such a way that one roller compressed the flexible tubing at all times. Compression and relaxation of a trapped fluid volume in that tubing section forces it to be pushed ahead and move forward. Continual rotations of rollers produce continuous fluid flow. Separation between the rollers and raceway is known as occlusion. For optimum function occlusion adjustment is required.

The roller action is highly periodical in nature. Rotary peristaltic pump output is obtained by two parameters that are the pump motor rpm and the volume amount dispensed in motor's each revolution. Tubing dimension and size, total track length are also the factors causing the variance of fluid volume. A 3 roller rotary peristaltic pump has been considered for the present study. Figure 13.1 shows the three roller unit. Figure 13.2a and 13.2b show the inner arrangement of the roller pump head

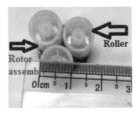

FIGURE 13.1   Three roller unit.

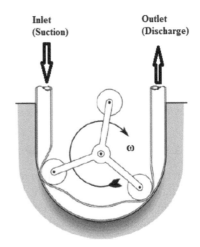

FIGURE 13.2   (a) Pump head of 3 roller peristaltic pump(within casing) (b) Pump head with different parameters.

model .. Occlusion can be adjusted by balancing the compression within the flexible tubing caused by the pump rollers.

13.2.1 DIFFERENT PUMP PARAMETERS

The transported liquid volume through the pump may vary and get affected by different attributes peristaltic pump. Flow determinates are precisely listed as pump head and tubing diameter, roller RPM and length of tubing in contact with rollers. It is relatively independent of circuit resistance and hydrostatic pressure. Flow rate varies proportionally with pump motor rotational speed and tube's inner diameter. Change in inner diameter of the tube and roller speed makes a significant change in the flow rate. Velocity of fluid through the tubing is determined by the velocity of movement of the bearings along the tubing. The linear velocity V (mm/s) of the bearings along the tubing can be calculated by the following equation:

$$V = \omega * r \tag{13.1}$$

Peristaltic pump flow rate (Q)is expressed by the following equations:

$$Q = A * V \tag{13.2}$$
$$Q = \pi t^2 * L * \omega * r \tag{13.3}$$

Flow rate considering periodic disturbance is given bywhere

Ω = is the angular velocity of the rotor (radians/s) onto which the bearing is mounted,

R = is the radius from the centre of the shaft to the external diameter of the bearing,

A = the cross-sectional area of the inner tubing,

T = the tubing inner radius and

L = length of the tubing clamped in one rotation.

More accurate relations include the number of rollers and different tubing deformation. Graphical representations of different pump parameters variation with respect to each other have been reported by many researchers, as shown in Figures 13.3–13.4[9–11].

The relationship between flow rate(mL/min) and motor rotation speed (rpm) is shown in Figure 13.5. It can be seen from the figure that the total region can be divided into three zone: linear zone, inflection point zone and nonlinear zone.

13.3 PERISTALTIC PUMP MODEL

An offline model identification method is used in here to obtain the pump dynamics. A three-roller 5000 rpm, 12V rated rotary peristaltic pump is studied here. Input output data set is taken for a certain applied voltage. Pump motor rpm and flow rate are the input output variables, respectively. Using MATLAB system identification toolbox a second-order plant model having relative degree two is chosen here with

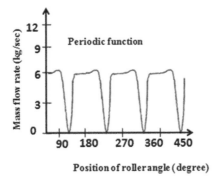

FIGURE 13.3 Mass flow rate of peristaltic pump is periodic with position of roller angle [9].

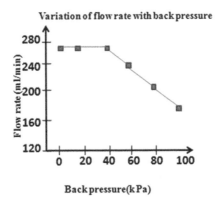

FIGURE 13.4 Relation between pump flow rate and backpressure [10].

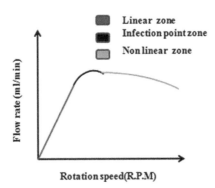

FIGURE 13.5 Graphical representation of flow rate and rotational speed relationship [11].

91% fitness percentage. Equivalent state space representation of the pump model is given by

$$\dot{X} = A_P x\left(t\right) + B_P u\left(t\right)$$
$$Y = C_P x\left(t\right) + D_P u\left(t\right)$$

with the following state matrices, given by

$$A_P = \begin{bmatrix} -285.3000 & -464.8000 \\ 1.0000 & 0 \end{bmatrix}, B_P = \begin{bmatrix} 1 \\ 0 \end{bmatrix}, C_P = \begin{bmatrix} 0 & 30.68 \end{bmatrix} D_P = \begin{bmatrix} 0 \end{bmatrix} \quad (4)$$

13.3.1 DISTURBANCES AND UNCERTAINTIES

Positioning of the pump at the calculated distance from the inlet(suction tank) and the outlet (delivery pipeline) has a prime contribution to facilitate the discharge of the pumped fluid. As the rotating rollers compress the flexible tubing section alternatively, this introduces a considerable amount of resistance on the pump motor which significantly aggregates as pressure disturbance in the fluid flow. This may cause an uneven flow rate in peristaltic pump [12]. This is named as load disturbance which follows a periodic nature. Therefore attenuation of this disturbance on peristaltic pump motor is needed along with a precise controller design for pump flow rate regulation . Frequency of this periodic load disturbance is proportionally related to the pump motor rotational speed. It is also associated with the angle of the roller position.

G(s) is a nominal plant model. Figure13.6 is showing plant along with input disturbance and uncertainty in the form of uncertain model dynamics which is the subject for simulation study.

Flow rate considering periodic disturbance is given by

$$Q = \pi t^2 * L * \omega * r - F\left(\omega\right) \quad (13.5)$$

In literature, it is reported that driving frequency for peristaltic pump can be considered from 1 to 4 Hz[12,13].Therefore, periodic disturbance induced in the system is generally low-frequency signal.

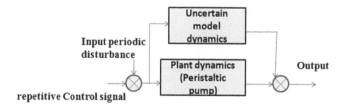

FIGURE 13.6   Plant model along with input disturbance and uncertain model dynamics.

13.4   OPTICAL SENSOR-BASED APPLICATION INCLUDING PERISTALTIC PUMP

Pump parameter monitoring has always been an important thing for maintaining desired performance level. Peristaltic pumps are used in many medical applications and chemical industries as the liquid flowing through this remains unpolluted. Optical sensor-based non-invasive measurement techniques have been experimented and reported by many researchers. Few such application based cased studies of rotary peristaltic pump has been discussed here.

13.4.1   Case I. Continuous Monitoring of Intrapulse Measurement of Blood Flow

A cardiac circulation set-up is simulated with an optical flow sensor-based measuring arrangement to measure the pulsatile blood flow rate. It comprises multimode fibres, fibre Bragg grating(FBG)sensor and a light-emitting diode for illumination, as can be seen in Figure13.7. FBG is used here to have a track on temperature change. Rotating peristaltic pump flow rate has a linear relation with rotational frequency of the pump. As flow rate increases, the pulse frequency also increases. This has been validated experimentally in this work[14] .

13.4.2   Case II. Optical Fibre-Based Spectrophotometer

Periodically pulsed peristaltic pump plays a vital role in liquid sample spectrometry as it generates samples in absorption spectrometry and the sample is only in contact with the tubing section and can be transferred over longer distance in the ICP spectrometer[15]. Peristaltic pump can overcome the matrix effect to a great extent. Rotating roller head causes the transportation of the entrapped liquid volume continuously towards the direction of nebulizer. An optical fibre-based sensor-based experimental set-up has been reported for spectroscopy measurements. Remote-sensing set-up with optical fibre is experimented and validated in the flow channel[16].The

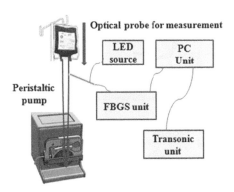

FIGURE 13.7   Flow rate measurement of peristaltic pump using optical probe for blood flow monitoring in human body.

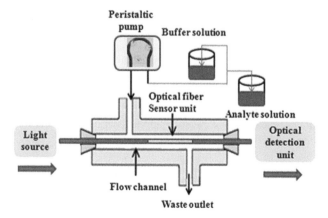

FIGURE 13.8   Experimental set-up of optical fibre-based spectrophotometer.

experimental set-up has been shown in Figure13.8. Peristaltic pump interference noise harmonics are caused due to pulsation which is inductively coupled with plasma mass spectrometry[17].

13.4.3   CASE III: ULTRASONIC VASCULAR VECTOR FLOW MAPPING FOR 2D FLOW ESTIMATION

2D flow estimation is carried out both acoustically and optically. Peristaltic pump is used to generate a periodic and pulsatile flow. Particle image velocimetry (PIV) system is compared with the vascular vector flow mapping(VFM) method. Figure13.9 depicts the experimental set-up[18].

13.4.4   CASE IV: OPTICAL FIBRE SENSOR-BASED COLORIMETRIC DETERMINATION

An optical sensor-based upon evanescent wave interaction is used in the application, reported here for chlorine concentration detection in water. A miniaturized and

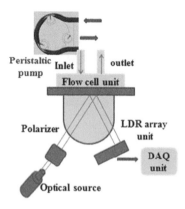

FIGURE 13.9   2D flow estimation using VFM method.

portable system with detection platform is developed which is based upon microfluidic channel. Two peristaltic pumps were involved in the reagents and water carrier flow channel. The experimental set-up is shown in Figure13.10[19].

13.4.5 CASE V: FLUIDIC SYSTEM USED TO TEST THE pH SENSOR

Peristaltic pump is used in pH-measuring system as fluid passing through remains uncontaminated by any means[20]. Fluorescence sensor-based pH-measuring system is designed and experimented with an integrated optoelectronic transduction system enabled with online measurement feature in this reported work. Sensor dynamic response has been observed for the same[21]. A rotary peristaltic pump is used in the set-up shown in Figure13.11.

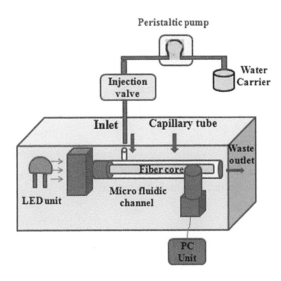

FIGURE 13.10   Experimental set-up of optical fibre-based calorimetric determination.

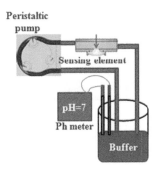

FIGURE 13.11   Experimental set-up of pH sensing of fluidic system.

13.5   PROPOSED METHOD OF OPTICAL FLOW SENSING SYSTEM (OFSS) FOR PERISTALTIC PUMP

An optical sensing arrangement for flow rate measurement of peristaltic pump is proposed here. This system will be placed at the outlet of the roller pump. Pair of infrared transmitter and photodetector forms the optical sensing unit and they need to be placed facing each other with a specific distance. The implementation of the overall system two such pair of optical sensing unit: one capillary tube and a timing calculator digital circuitry. The working principle of the measuring unit as follows: the two optical sensing units will be placed at start and end position of the capillary tube and they will act as start and stop gate. Start gate signal is activated when the interface of liquid and air come across it through the capillary tube and the timing circuit will be turned on accordingly. On the other end, the stop gate signal is turned on When the interface of liquid and air come across it at the end of the tube and the timing circuit is turned off. As the intensity of the infrared varies, an equivalent voltage signal is transmitted to the timing circuit unit. The time counted by the circuit is the travel time of the liquid volume from start gate point to stop gate point. Diagram of proposed method is shown in Figure 13.12. Thus the pump flow rate can be equated as follows:

$$Q' = \frac{\pi R_c^2 h}{\theta t} \tag{13.6}$$

where
  Q' = Flow rate,
  Rc = Radius of the capillary tube,
  h = Distance between start gate and stop gate and
  θt = Time counted.

13.6   CONTROLLER DESIGN FOR PERISTALTIC PUMP

Peristaltic pump motor driver controls the speed .Precisely the flow rate of the roller pump is controlled by regulating the position of rollers.

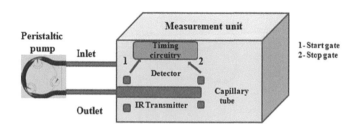

FIGURE 13.12   Proposed method of OFSS for peristaltic pump.

13.6.1 REPETITIVE CONTROL LOOP DESIGN

Repetitive control is a special class of learning control which is based on internal model principle. This learning control methodology is designed to obtain a desired tracking performance when reference signal is periodic in nature. This controller also aims to enhance the control performance by reducing the effect of periodic disturbance signal on the closed-loop system[1,22]. Inclusion of the low pass filter Q(s) modifies the conventional RC loop into a stable finite dimensional RC loop (FDRC) [23].

The transfer function of the given RC loop in above Figure13.13a can be derived as

$$u(s) = \frac{1}{1 - e^{-sN}} e(s) \tag{13.7}$$

Transfer function of the FDRC loop is given by

$$u(s) = \frac{1}{1 - Q(s)e^{-sN}} e(s) \tag{13.8}$$

The plugged in low pass filter Q(s) within the conventional RC loop shown in Figure13.13b must hold the following condition:

$$Q(s) = \frac{\omega_c}{s + \omega_c} \tag{13.9}$$

$$\left| f(j\omega) \right| \approx 1, \omega \le \omega_c$$

where ω_c is the low pass filter cut off frequency.

Internal model principle says that for accurate tracking of periodic reference command or attenuation of periodic disturbance the sampling time period of the signal should be as same as the delay block of the FDRC loop. This fine for fixed frequency disturbance rejection but does not hold a desired performance for unknown or time-varying frequency disturbance sinusoids. Present work focused on the control problem formulation considers attenuation of time-varying and constant-frequency

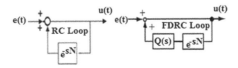

FIGURE 13.13 (a) Conventional repetitive control system. (b) Finite dimensional repetitive control system.

disturbance signal to regulate the peristaltic pump flow rate at a specified value. The proposed method augments an adaptive control law along with the existing repetitive control law. Control loops are shown in Figures13.14a and b.

Error dynamics of the plant model is given by

$$e_1(t) = x(t) - x_r(t)$$
$$e_2(t) = \dot{e}_1(t) + ke_1(t) \tag{13.10}$$

K is a feedback gain. The repetitive control law is given by

$$u_{rc}(t) = u_{rc}(t - N) - k_{rc}e_2(t) \tag{13.11}$$

Periodic sinusoidal disturbance model varying with time can be written as

$$V(t) = A_1\cos(\omega * x) + A_2\sin(\omega * x) \tag{13.12}$$

A_1 and A_2 are the parameters to be estimated by the adaptive law. The control law can be written by augmenting the adaptive law as

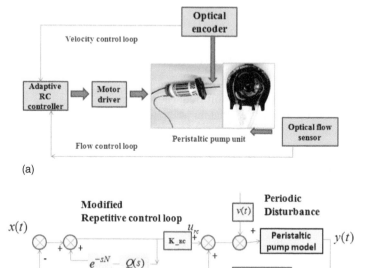

(a)

(b)

FIGURE 13.14　(a) Flow and velocity control loop of the peristaltic pump (b) Closed-loop feedback control of peristaltic pump with adaptive repetitive controller.

$$u(t) = u_{rc} - k_{rc}e_2(t) + u_{ad}$$

$$u_{ad} = \hat{A}_1 \cos(\omega * x) + \hat{A}_2 \sin(\omega * x) \tag{13.13}$$

$$\hat{A}_1 = -k_1 e_2 \cos(\omega * x), \hat{A}_2 = -k_2 e_2 c \sin(\omega * x)$$

where u_{ad} is the adaptive term.

13.7 SIMULATION RESULTS

Simulation has been carried out using MATLAB. The following simulation parameters have been considered.

A simple sinusoidal disturbance model is given below

$$v_1(t) = 2\sin(2 * \pi * t).$$

and a time-varying sinusoid disturbance model given below is considered for the present work.

$$v_2(t) = 2\cos(2 * \pi * \sin(0.2 * pi * t))$$

where $\omega_c = 10$ rad/sec, $N = 1$ sec.

The reference signal is given to the system is a 4V step signal to maintain a constant flow rate shown in Figure 13.15. Both the disturbance signals are shown in Figure 13.16a and b. Error has been shown in Figure 13.17a and b with the implementation of FDRC considering both the cases. It is observed that due to the intervention of $V_2(t)$ the peristaltic pump control loop is exhibiting higher value of tracking error. The effectiveness of the controller response is specified by obtaining few performance indices under disturbance signals. Integral of the square error and absolute

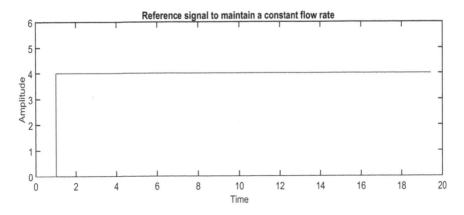

FIGURE 13.15 Reference signal applied to the system to maintain a constant flow rate.

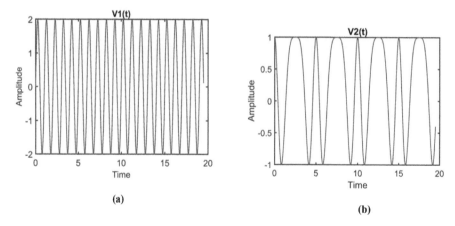

FIGURE 13.16 (a) Sinusoidal disturbance signal $V_1(t)$. (b) Varying frequency sinusoidal disturbance signal $V_2(t)$.

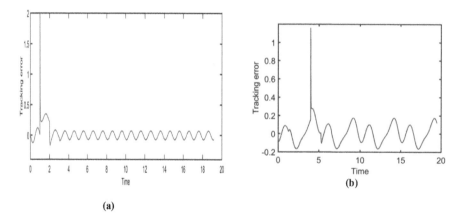

FIGURE 13.17 (a) Tracking error considering $V_1(t)$ using FDRC (b) Tracking error considering $V_2(t)$ using FDRC.

TABLE 13.1

Performance Indices of the Overall System Using FDRC Considering Different RC When $V_2(t)$ Is Present

	Performance Indices	
RC Gain	**Integral Absolute Error (IAE)**	**Integral Square Error (ISE)**
$K_{RC} = 2$	6.845	2.691
$K_{RC} = 5$	3.765	0.833
$K_{RC} = 8$	2.6	0.4006
$K_{RC} = 10$	2.157	0.2767

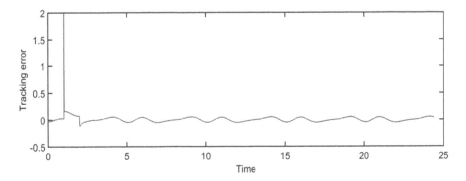

FIGURE 13.18 Tracking error considering $V_2(t)$ using adaptive FDRC(AFDRC).

error is calculated separately. They are named as ISE and IAE. Table 13.1 shows the performance indices obtained for the proposed controller. Figure 13.18 depicts the minimization of tracking error with implementation of adaptive finite dimensional repetitive controller.

13.8 CONCLUSION

Different peristaltic pump-based fluidic application with optical sensing and measuring arrangement reported in this present study. The simulation model of a three-roller pump based on offline system identification method has been carried out. A modified repetitive controller is proposed here with an augmented adaptive control law and it has been implemented for disturbance attenuation of varied frequency and constant frequency periodic signals. Controller performance has been measured in terms of performance index. It can be concluded that proposed controller works efficiently to maintain a desired flow rate with the proposed OFSS for peristaltic pump. Further, this work can be extended considering a delay-based model of pump.

REFERENCES

1. Francis, B.A. Wonham, W.M. The internal model principle for linear multivariable regulators. *Appl. Math. Optim.* 1975, 2, 170–194.
2. Tomizuka, M. Zero phase error tracking algorithm for digital control. *ASME J. Dyn. Syst. Meas. Control*, 1987, 65, 65–65, 68.
3. Mondal, U.; Sengupta, A.; Pathak, R.R. Servomechanism for periodic reference input: Discrete wavelet transform-based repetitive controller. *Trans. Inst. Meas. Control.*2016, 38, 14–22.
4. Hillerstrom, G. and Jan Sternby, "Application of repetitive control to a peristaltic pump", Transactions of the ASME, Vol. 116, DECEMBER1994.
5. Aruna Kommu, Raghavendra Rao Kanchi and Naveen Kumar Uttarkar (2014), *"Design and Development of Microcontroller based Peristaltic Pump for Automatic Potentiometric Titration"*, International Conference on Communication and Signal Processing, April 3–5, 2014, India
6. L. de Maré, S. Velut, P. Hagander, M. Åkesson (2001), *"Feedback Control of Flow Rate From a Peristaltic Pump Using Balance Measurements"*, European Control Conference, Portugal.

7. Dongdong Fei, Chen Deng, Qinshun Zhang, Zhuliang Hunag, "Repetitive control to the flow rate of peristaltic blood pumps", *Int. Adv. Res. J. Sci, Eng. Technol.*, Vol. 2, Issue 8, August 2015

8. Dey, N.; Mondal, U. and Sengupta, A, *Modified Repetitive Controller for Disturbance Rejection of Peristaltic Pump, International Conference on Opto-Electronics and Applied Optics (Optronix)*, 2019, 10.1109/OPTRONIX.2019.8862338.

9. Christos Manopoulos, Giannis Savva, Achilleas Tsoukalis, Georgios Vasileiou, Nikolaos Rogkas, Vasilios Spitas and Sokrates Tsangaris, "Optimal design in roller pump system applications for linear infusion", *Comput. Des.*2020, 8, 35; doi:10.3390/computation8020035.

10. K. B. Vinayakumar, Girish Nadiger, Vikas R. Shetty, N. S. Dinesh, M. M. Nayak, and K. Rajanna, Packaged peristaltic micropump for controlled drug delivery application, *Rev. Sci. Instrum.*88, 015102 (2017); doi:10.1063/1.4973513.

11. www.leadfluid.com; https://leadfluid.com/how-to-calculate-the-flow-rate-of-leadfluid-peristaltic-pump/

12. Sangkwon Na, Shane Ridgeway and Li Gaol, *"Theoretical and Experimental Study of Fluid Behavior of a Peristaltic Micropump", Proceedings of the 15th Biennial University/Government/ Industry Microelectronics Symposium*, ISBN: 0-7803-7972-1, USA, July 2003.

13. Tomostar M. (2016) ,"A Novel Blood Pump Design and Characterization", *J Bioeng. Biome. Sci.* 6: 199. doi:10.4172/2155-9538.1000199

14. Albert Ruiz-Vargas, Scott A. Morris, Richard H. Hartley and John W. Arkwright, " Optical flow sensor for continuous invasive measurement of blood flow velocity", *J. Biophotonics* ·May 2019.

15. López-García, I., Sánchez-Merlos, M., Viñas, P., and Hernández-Córdoba, M. (1996). Peristaltic pumps-Fourier transforms: a coupling of interest in continuous flow flame atomic absorption spectrometry. *Spectrochim. Acta B At. Spectrosc.*, 51(14), 1761–1768. doi:10.1016/s0584-8547(96)01557-1.

16. Kanso, M., Cuenot, S. and Louarn, G., Sensitivity of optical fibre sensor based on surface plasmon resonance: Modeling and experiments. *Plasmonics*3, 49–57 (2008). doi:10.1007/s11468-008-9055-1

17. Erik Björn, Tobias Jonssona and Daniel Goitoma, "The origin of peristaltic pump interference noise harmonics in inductively coupled plasma mass spectrometry", *J. Anaty Atomic Spectrim*,10, 2002 DOI:10.1039/B204771C

18. Kashyap, R., Chakraborty, S., Zeng, S., Swarnakar, S., Kaur, S., Doley, R., and Mondal, B. (2019). Enhanced biosensing activity of bimetallic surface plasmon resonance sensor. *Photonics*, 6(4), 108

19. Xiong, Y., Tan, J., Wang, C., Wu, J., Wang, Q., Chen, J., Duan, M. (2017). A miniaturized evanescent-wave free chlorine sensor based on colorimetric determination by integrating on optical fibre surface. *Sensors Actuators B Chem.*, 245, 674–682. doi:10.1016/j.snb.2017.01.173

20. Milanovic, J.Z., Milanovic, P., Kragic, R., Kostic, M. (2018) "Do-It-yourself" reliable pH-stat device by using open-source software, inexpensive hardware and available laboratory equipment. *PLoS One*13(3): e0193744. doi:10.1371/journal.pone.0193744

21. Luca Ferrari, Luigi Rovati, Paola Fabbri, Francesco Pilati (December 2012) "Disposable fluorescence optical pH sensor for near neutral solutions, *Sensors* 13(1): 484-499 DOI:10.3390/s130100484

22. Inoue T., Nakano M, and Iwai S. *High accuracy control of servomechanism for repeated contouring*. In *Proceedings of the 10th Annual Symposium: Incremental Motion Control Systems and Devices*. 1981; 285–291.

23. Tomizuka M., Tsao T.-C., and Chew K.-K.. Analysis and synthesis of discrete-time repetitive controllers. *ASME Journal of Dynamic Systems, Measurement, and Control*. 1989; 111: 353–358.

14 Conclusion

Pampa Debnath, Arpan Deyasi and Siddhartha Bhattacharyya

CONTENT

The present decade is the exploration of photonics and plasmonics devices for the search of robust telecommunication network design for accurate and efficient information transfer [1–3]. Owing to the present development of nanophotonic devices and their experimental realization, researchers are now thinking in terms of achieving the mission of quantum information processing by resorting to photonic neuromorphic computing. Moreover, the present research trends may be considered as an inspiring force for attaining solutions to big data problems. This volume entails several such initiatives to help achieve these daunting propositions.

A bandgap engineering of sol–gel spin-coated TiO_2 thin film on glass substrate results in an increase in the optical conductivity due to annealing [3]. In the UV region, optical conductivity increases sharply as absorption becomes high. Thus, TiO_2 thin film with tune-able optical properties may pave the way for use in a wide variety of optical applications. Metamaterials and metasurfaces[4–7] viz., absorber, polarization converter and antennas have been found to be conducive for preparing ultra-thin structures which in turn boosts the preparation of compact miniature structures. This is extremely useful in modern-day 5G and higher data-rate communication system where a number of devices are embedded in a single chip to constitute the complete system.

Joint modulation of a single light beam is occurred by using two different sawtooth pulses at a time. Applying two biasing voltages jointly rather than application of a single biasing signal to the KDP based Pockels cell parallel to the Y- and Z-directions, the phase part of the light signal can be significantly controlled. One can get zero phase difference after the operation. One can also organize a good digital phase modulation using the scheme [8]. When an intense light signal is passing through a Kerr type of non-linear medium multiple times, changes occur at the intensity of the central frequency of the light signal. The light intensity for the central frequency decreases more sharply in the case of second passing the light in contrast to the first passing. The intensity of the harmonics in the case of second passing increases at the cost of decreases of intensity of central frequency. This proposed scheme may be useful for amplification of harmonic powers of light. All-optical X-OR gate is developed by using phase-encoded mechanism with optical tree architecture. The scheme can easily and successfully be extended and implemented for any higher number of input digits by proper use of electro-optic modulator and phase encoding technique. A phase-encoded algebraic operation can be conducted with the proposed system.

Slotted photonic crystal waveguides [9] have been an effective platform for efficient light–matter interactions owing to the ultra-high optical confinement as attained after merging the strong spatial and temporal confinement of the slotted and the photonic crystal waveguides, respectively. Light–matter interactions, especially the nonlinear optical phenomena typically require high operating power and/or large footprint of the underlying waveguides which are the fundamental constraints limiting their applicability in integrated photonics. Stimulated Raman Scattering has been such a potential nonlinear distributed light–matter interaction; that conventionally takes kilometers of length of the optical fibre along with the pump power lying in the order of few hundreds of Watt for observable gain. Despite the untiring research being carried out over the past several years in miniaturizing the effective interaction length and the threshold power of Raman amplifiers, the miniaturization is yet to be competent enough to PICs for exploring on-chip active optical functionalities. Use of SRS in SiNC/SiO2 embedded SPCW may circumvent the limitations. The ultra-high SRS gain in SiNC/SiO2 material is fortified further by the intense spatio-temporal confinement of SPCWs, which, thereby, is able to provide unprecedented miniaturization of the effective length and threshold power.

Simulated findings reveal that communication system designed at 1310 nm has potential advantages than the conventional 1550 nm when both the systems have equal input power and embedded with RAMAN amplifier [10]. When both the systems are compared at 5 Gbps data rate for 80 km fibre length, Q-factor for the system operated at 1310 nm is almost 3 times higher the counterpart. However, eye height is approximately 10% greater for the former, which speaks for less sensitivity. For both the systems, gain is reduced with increasing bit rate. This novel design can be effectively used in real communication system to provide lower data loss.

The optical comb filters have been used for long time as multichannel devices and have shown impact in the areas of WDM/DWDM system [11], multi-wavelength lasers and optical signal processing. To have better performance in those specific applications, these comb filters are required to use in ultra-narrow band operations. Different techniques for achieving comb spectrum are in place. A specific design of optical comb filter using Gaussian-sampled periodically-chirped fibre Bragg grating can be realized for use as a multichannel device.

Several systems have been proposed where a location estimation theory with a wireless remote monitoring system is used for non-destructive damage assessment (NDDA) of impact response on aerospace structures in real-time. To identify the coordinates of an applied impact, the estimation theory accounts for the sensor information and physical properties of the surface [12]. An inexpensive and computationally efficient method can be visualized which includes a simplified interval analysis to account for the presence of noise in the system response and a heuristic algorithm for optimization. Any deviation found because of noise can be estimated using least square regression and linear approximation techniques. A wireless remote monitoring system which enables to retrieve, process, and transmit information from an array of Fibre Bragg Grating optical sensors is reported along with a remote system capable of receiving, processing, and display of the transmitted data in real-time.

Gap solitons[13] exist in different types of photorefractive optical lattices, i.e.non-centrosymmetricphotorefractives, centrosymmetricphotorefractives and pyroelectricphotorefractives. A theoretical foundation using the Helmholtz equation can function as a general framework for photorefractive crystals having different nonlinearities and configurations. It is observed that the gap solitons are either single humped and symmetric or double humped or even multi humped and asymmetric. Generally, the first finite bandgap supports entirely positive single-humped solitons and the second finite bandgap supports asymmetric and double-humped or multi-humped solitons. The stability of gap solitons is examined in all three configurations by linear stability analysis. The growth rate of perturbation can be formulated by a general eigenvalue problem which can be modified to suit each configuration of photorefractive crystal. The eigenvalue problem can be solved numerically by first discretizing the Sturm Liouville system by finite differences and then finding out the eigenvalues of the resultant pentadiagonal matrix. While the pyroelectric gap solitons are found to be stable across the two finite bandgaps, gap solitons in other configurations are found to be both stable and unstable for different values of frequency within the two finite bandgaps.

It is observed from experimental studies that inclusion of adefect of low refractive index material in a multilayer PBGS at suitable positions, narrow-band phase-shifted transmission peak can be obtained. However, its FWHM bandwidth of the spectral output depends on the number of bilayers of the periodic structure [14]. It is found that a particular mathematical relationship exists between the central wavelength of the source input and the amount of phase shift of the output spectra. It is also seen that by suitable adjustment of the defect layer width, the phase shift of the transmission peak can be avoided. Therefore, opticalfilters designed with these types of PBGS are very useful for WDM networks. With the available fabrication technologies, the device dimensions used in this analysis can be practically implemented.

Flat lenses are now becoming more popular than the traditional lenses. Although MDL is comparatively better in case of CMOS compatibility but as Metalens[15] is much thinner that MDL, it is often preferred more in IOT or Automobile industry. Super-achromatic lenses have come up based on the concept of flat lenses. However, the idea of focussing at least four wavelengths on a single point using the super-achromatic lenses is a challenging proposition for the optical engineers. Recently, the newly introduced evolutionary computational techniques can be used to optimize the parameters like field of depth, focus length, depth of view, optical efficiency, high numerical aperture, etc., in order to facilitate construction of images with more accuracy.

Different peristaltic pump-based fluidic applications rest on precise optical sensing and measuring arrangements. A modified repetitive controller [16] comprising an augmented adaptive control law can be implemented fordisturbance attenuationof varied frequencyand constant frequency periodic signals. It is found that such a controller works efficiently to maintain a desired flow rate with an OFSS for peristaltic pumps as well.

REFERENCES

1. Smit, M. K., Williams, K. A. (2020) Indium phosphide photonic integrated circuits, *Optical Fibre Communication Conference*, W3F.4
2. Loudon, R. (1970) The propagation of electromagnetic energy through an absorbing Dielectric, *Journal of Physics A*, 3, 233-245
3. Yablonovitch, E. (1987) Inhibited spontaneous emission in solid-state physics and electronics, *Physical Review Letters*, 58, 2059-2061
4. Faruque, M. R. I., Islam, M. T., Misran, N. (2012) Design analysis of new metamaterial for EM absorption reduction, *Progress in Electromagnetics Research*, 124, 119-135
5. Elsheakh, D. M. N., Elsadek, H. A., Abdullah, E. A. (2012) Antenna Designs with Electromagnetic Bandgap Structures, *Metamaterial*, Ed. Jiang, X. Y., InTech, Rijeka, Croatia
6. Faruque, M. R. I., Islam, M. T., Misran, N. (2010) Evaluation of em absorption in human head with metamaterial attachment, *Applied Computational Electromagnetics Society Journal*, 25(12), 1097-1107
7. Yang, F., Rahmat-Samii, Y. (2009) *Electromagnetic Bandgap Structures in Antenna Engineering*, Cambridge University Press, Cambridge, UK
8. A. Yariv, P. Yeh, *Electro-optic modulation in LASER beams. Photonics – optical electronics in modern communications*. New York: Oxford University Press; 2007.
9. Armenise, M. N., Campanella, C. E., Ciminelli, C., Dell'Olio, F., Passaro, V. M. N.Phononic and photonic bandgap structures: modeling and applications, *Physics Procedia*, vol. 3(1), pp. 357-364, 2010
10. Miller, S. E., Kaminow, I. P. (1988) *Overview and summary of progress', Optical Fibre Telecommunications II*, 1-27, Academic Press, Boston
11. J. E. Rothenberg, H. Li, Y. Li, J. Popelek, Y. Sheng, Y. Wang, R. B. Wilcox, and J. Zweiback, "Damman fiber Bragg gratings and phase-only sampling for high-channel counts," *IEEE Photon. Technol. Lett.*, vol. 14, pp. 1309–1311, Sept. 2002.
12. Li, H., Sheng, Y., Li, Y., Rothenberg, J.E., "Phased-only sampled fibre Bragg gratings for high channel counts chromatic dispersion compensation," *J. Lightw. Technol.*, vol. 21, 2074-2083, 2003
13. S. Trillo and W. E. Torruellas, Eds., *Spatial Solitons, Springer Series in Optical Sciences*, vol. 31. Springer, Berlin, 2001
14. W.P. Risk, G.S. Kino, Y. Khuri, and ButrusT. "Tunable optical filter in fibre-optic form".*Opt. Lett.*, 11(9):578–580, 1986
15. J. Engelberg and U. Levy, "The advantages of metalenses over diffractive lenses," *Nat. Commun.*, vol. 11, no.1, 2020, doi: 10.1038/s41467-020-15972-9
16. Hillerstrom, G. and Jan Sternby, "Application of repetitive control to a peristaltic pump", *Transactions of the ASME*, 116:786–789, December 1994

Index